Biotechnological Applications of the Yeast

Biotechnological Applications of the Yeast

Editor: Harvey Marshall

Callisto Reference
www.callistoreference.com

Callisto Reference,
118-35 Queens Blvd., Suite 400,
Forest Hills, NY 11375, USA

Visit us on the World Wide Web at:
www.callistoreference.com

ISBN: 978-1-64116-801-4 (Hardback)

Cataloging-in-Publication Data

Biotechnological applications of the yeast / edited by Harvey Marshall.
p. cm.
Includes bibliographical references and index.
ISBN 978-1-64116-801-4
1. Yeast fungi. 2. Yeast fungi--Biotechnology. 3. Biotechnology. I. Marshall, Harvey.
QK617.5 .B56 2023
579.562--dc23

Table of Contents

Preface

Yeasts are a type of eukaryotic unicellular microfungi that are found abundantly in nature. They do not belong to any single phylogenetic or taxonomic group. Yeasts are used primarily in the production of alcoholic beverages and in baking. Saccharomyces cerevisiae is a species of yeast, which is widely utilized in the food industry. It is an extensively studied eukaryotic microorganism and serves as a key model organism in contemporary cell biology research. It has the capability to reproduce sexually by fusing with the cells of opposite mating types and asexually by budding. It is also known as baker's yeast, and has facilitated the invention of effective manufacturing techniques for biopharmaceuticals, ethanol, antibiotics, biofuels, and technical enzymes. This book contains some path-breaking studies on yeast. It explores all the important biotechnological applications of yeast in the present day scenario. This book will serve as a reference to a broad spectrum of readers.

The researches compiled throughout the book are authentic and of high quality, combining several disciplines and from very diverse regions from around the world. Drawing on the contributions of many researchers from diverse countries, the book's objective is to provide the readers with the latest achievements in the area of research. This book will surely be a source of knowledge to all interested and researching the field.

In the end, I would like to express my deep sense of gratitude to all the authors for meeting the set deadlines in completing and submitting their research chapters. I would also like to thank the publisher for the support offered to us throughout the course of the book. Finally, I extend my sincere thanks to my family for being a constant source of inspiration and encouragement.

Editor

1

Brewer's Spent Yeast (BSY), An Underutilized Brewing By-Product

Alice Jaeger [1], Elke K. Arendt [1,2], Emanuele Zannini [1] and Aylin W. Sahin [1,*]

1 School of Food and Nutritional Science, University College Cork, T12K8AF Cork, Ireland; 116458826@umail.ucc.ie (A.J.); e.arendt@ucc.ie (E.K.A.); e.zannini@ucc.ie (E.Z.)
2 APC Microbiome Institute, University College Cork, T12YT20 Cork, Ireland
* Correspondence: aylin.sahin@ucc.ie

Abstract: The repurposing of by-products and the reduction of waste from food processing streams is an ever-increasing area of interest. Brewer's spent yeast (BSY) is a prevalent by-product of the brewing industry. The spent yeast cells are removed at the end of the bulk fermentation. A small amount of it is used to start the next batch of fermentation; however, the majority of the spent yeast is discarded. This discarded yeast is high in nutrients, in particular proteins, vitamins and minerals, as well as containing functional and biologically active compounds such as polyphenols, antioxidants, β-glucans and mannoproteins. At present, BSY is mainly used in animal feed as a cheap and readily available source of protein. This review explores alternative, value-added applications for brewer's spent yeast including nutritional ingredients, functional food additives as well as non-food applications. A major challenge in the utilization of BSY in food for human consumption is the high level of RNA. An excess of RNA in the diet can lead to an increase in uric acid in the bloodstream, potentially causing painful health conditions like gout. This issue can be overcome by RNA degradation and removal via additional treatment, namely heat treatment and enzymatic treatment. There is potential for the use of BSY ingredients in various food applications, including meat substitutes, bakery products and savory snacks.

Keywords: fermentation; brewer's yeast; saccharomyces cerevisiae; yeast β-glucans; food applications

---

1. Introduction

Yeast is a single-celled organism used in beer manufacture for the transformation of sugars from grain to alcohol. Brewer's spent yeast (also known as residual yeast or surplus yeast) is a prevalent by-product of the brewing industry, created when the yeast used in fermentations is no longer useful and must be disposed of. It is estimated that 15 to 18 tons of surplus yeast are produced per 10,000 hL of finished beer [1]. Other major by-products include brewer's spent grains and hot trub. Brewer's spent yeast (BSY) is a rich source of B vitamins, proteins (45–60%) and other compounds that could be applicable to both food and non-food products [2]. However, the high level of nucleic acids in BSY limits its use as a protein supplement in humans due to the negative health effects caused by an excess of uric acid [2,3]. Therefore, to date, BSY is mainly utilized in animal feed formulations as a low-cost source of protein.

Yeast extracts formed by the breakdown of the yeast cell membrane can be used as additives in food products as a source of free amino acids, peptides or for their antioxidant properties [2]. Compounds such as β-glucans and mannoproteins can also be derived from BSY. These have many functional uses in food products such as textural stabilization, as well as potential health benefits [4]. Due to these properties, there is a huge range of possible applications for BSY as a functional food additive.

During the production of beer, yeast ferments the simple sugars present in the cooled wort to alcohol. In brewing, yeasts are generally classified into two broad categories: Top-fermenting 'ale' yeasts which consists of *Saccharomyces cerevisiae* strains and bottom-fermenting 'lager' yeasts of *Saccharomyces pastorianus* and *Saccharomyces carlsbergensis* strains [5]. The yeast strain used can have a significant impact on the overall quality of the beer. Along with ethanol and carbon dioxide, the yeast produces other compounds that can play a key role in the sensory properties of the beer, including a wide variety of esters, ketones, aldehydes and higher alcohols [6]. Generally, *Saccharomyces* species are used in brewing, however, in recent years there has been increasing interest in the use of non-*Saccharomyces* species in the production of low-alcohol or alcohol-free beers [7]. Yeast is separated from the bulk at the end of fermentation by a process known as flocculation which occurs towards the end of the fermentation process. During this process, thousands of yeast cells form 'flocs' which then either rise to the surface (top-fermenting) or sediment on the bottom of the vessel (bottom fermenting) [8]. Good flocculation properties are essential in an industrial strain as yeast can be reused multiple times during the brewing process [9]. A small amount of yeast from the previous fermentation is used to start the next fermentation in a process known as re-pitching. Due to only a small percentage being re-used in the manufacturing process, there remains a large excess of spent yeast. Due to BSY having a high chemical oxygen demand (COD) value of 0.53 kg/hL, it cannot be disposed of into wastewater streams without prior treatment as it would have a severe negative effect on the environment [1].

Essentially, brewer's spent yeast is an underutilized by-product of the brewing industry that has many potential uses. Food and beverage manufacturers are increasingly interested in waste reduction and reducing their environmental impact and so, due to its low cost and high nutrient density, brewer's spent yeast is a very valuable raw material. BSY could be utilized by a wide variety of industries in the production of nutritional supplements, functional food ingredients and other value-added products.

2. Use of Yeast in Brewing and Fermentation—An Overview

In order to understand the value of BSY and its potential applications, it is important to comprehend how spent yeast is formed, the role yeast plays in the brewing process and the processes involved in creating the by-product.

2.1. *Yeast*

Yeast is a eukaryotic, single-celled organism classified as a member of the Fungi kingdom [1]. There are thousands of naturally occurring wild yeast strains. Brewing originated by harnessing these wild strains in conjunction with other microorganisms to produce a fermented beer product. Nowadays, the strains used for brewing are much more strictly controlled, widely studied, and carefully maintained. Yeast cells are approximately 10 μm in diameter and are complex organisms, consisting of several organelles, each with a defined purpose. *Saccharomyces cerevisiae*, the main strain used in brewing, contains 16 chromosomes within the nucleus of the cell [10]. Yeasts can be haploid, diploid or aneuploid, meaning they contain one, two, or three copies of each gene, respectively. Brewer's yeasts are aneuploid, therefore making them more stable, less prone to mutations and consistent over several generations which is beneficial in brewing to produce a consistent product [10]. The yeast cell is surrounded by a cell wall and a cell membrane which gives structure to the cell but also plays a key role in certain processes, namely flocculation. Disruption of this membrane is a key step in processing of BSY to produce yeast extract. Yeast can grow aerobically or anaerobically but only the anaerobic 'fermentative' mode of growth is applicable in brewing. During fermentation, yeast cells grow and multiply producing valuable by-products, namely alcohol, carbon dioxide and flavor compounds [11].

2.2. *A Brief Overview of Fermentation*

Fermentation is the process by which sugars in the wort are transformed into alcohol, carbon dioxide and other by-products by the yeast cells. The first stage of fermentation is known as the 'primary fermentation'. During this process, the yeast cells use nutrients supplied by the wort to grow and multiply. The by-products of this process are alcohol, carbon dioxide and numerous flavor compounds. Yeast metabolizes sugar according to the following:

$$C_6H_{12}O_6 \rightarrow 2C_2H_5OH + 2CO_2 + \text{Energy} \quad (1)$$

However, this is a simplified equation as many flavor compounds are also produced as by-products. The product of this primary fermentation is called 'green beer'. This is an immature beer containing undesirable substances such as diacetyl, aldehydes and sulfur compounds which give the beer an unclean and unbalanced taste [1].

In order to reduce the levels of these undesirable flavor compounds, the beer must go through what is known as 'secondary fermentation', also referred to as maturation, conditioning or lagering. While these are traditionally thought of as two independent processes, in reality, there is a significant overlap. While the yeast has completed its primary goal of alcohol production during the first fermentation, during the maturation process it continues fermentation and removes the undesirable flavor compounds produced in that first stage. The main compound that is targeted in this process is diacetyl, a molecule with an intense buttery off-flavor. Diacetyl is metabolized to 2,3-butandiol, a compound with a much higher flavor threshold than diacetyl, therefore having less impact on the overall flavor of the final beer [12]. The extent and rate of the removal of these compounds can be controlled by manipulating yeast concentration, temperature and levels of agitation. Generally, optimal yeast health, an increased temperature and an increased exposure of yeast to diacetyl (by, i.e., agitation) lead to an increased rate of diacetyl removal [12].

2.3. *Pitching*

The inoculation of wort with yeast cells is called 'pitching'. In a study by Verbelen et al. (2009), the degree to which yeast performance and final beer flavor were affected by pitching rate was investigated. It was found that while a higher initial cell count leads to a faster fermentation, there were limited physiological and metabolic differences observed in the yeast cells in a higher density. The main difference came in the enhanced diacetyl levels in the beers with a higher pitching rate [13]. In contrast to this, a study by Kucharczyk and Tuszyński (2015), showed a similar increase in fermentation rate but no significant difference in the levels of diacetyl produced [14]. Serial re-pitching of the yeast can also have an effect on the final beer composition. It has been shown that the polypeptide profile of the beer is directly influenced by the generation of yeast used in inoculation [15].

2.4. *Yeast Viability*

Beer quality is greatly influenced by the biochemical performance of the yeast and therefore it is important for brewer's to be able to accurately predict yeast viability (also known as yeast vitality). In order to optimize fermentations, it is vital to determine yeast viability in order to adjust the pitching rate accordingly [16]. Most commonly, methylene blue staining is used to determine the health of the yeast cells. Yeast cells that are viable contain an enzyme that decolorizes the methylene blue stain [17]. Therefore, it is easy to distinguish between living and dead cells under the microscope; live cells are colorless whereas dead cells remain stained. While methylene blue staining is the most commonly accepted method for determining yeast viability, a study by Smart et al. (1999) investigated and proved the use of citrate methylene violet staining as an alternative method of determining yeast viability [18]. Similarly, in a study by Bouix and Leveau, (2001), a flow cytometric method was used to monitor efflux percentage of yeast cells during fermentation. The results showed that the efflux percentage is a reliable indicator of cell energy and vitality [16].

3. Formation and Processing of Brewer's Spent Yeast (BSY)

In order to utilize to their full extent the nutrients and functional ingredients within the yeast cells, the BSY must be separated effectively from the bulk, followed by processing into yeast extract and other yeast products.

3.1. *Flocculation*

Flocculation is defined as the tendency of yeast cells to associate with one another [10]. This allows for easier removal of the yeast from the 'green' beer at the end of fermentation from the top or the bottom of the vessel, depending on yeast type. It is crucial that yeast flocculation occurs at the correct time, towards the end of the bulk fermentation in the brewing process. Premature flocculation or incomplete flocculation may negatively impact the fermentation process leading to a slower, 'sluggish' fermentation [19]. Yeast flocculation is a highly complex phenomenon, influenced by the expression of yeast flocculation genes, namely, *FLO1*, *FLO5*, *FLO8* and *FLO11* [8]. However, these genes are highly variable and differ from one yeast strain to the next. Flocculation is also impacted by factors which affect the composition and morphology of the cell wall, including nutrient availability, dissolved O_2 levels, fermentation temperature and yeast handling and storage conditions. Conditions that increase cell collisions within the medium (i.e., agitation) can also have an effect on flocculation [8].

3.2. *Production of Yeast Extract*

In order to access the yeast components with the most potential for food and nutraceutical applications, surplus yeast may be processed into yeast extract. Yeast extract is defined as the soluble content of a yeast cell that remains once the cell wall has been destroyed and removed [20]. The method chosen to disrupt the yeast cell wall can have an impact on the future composition of the yeast extract. There are mechanical methods including using the cell mill or sonotrode, as well as non-mechanical methods such as autolysis. On an industrial scale, non-mechanical methods are most frequently used.

The methods of manufacturing yeast extracts can be divided into autolysis and hydrolysis. Yeast hydrolysates are yeast extracts prepared with the addition of hydrochloric acids or proteolytic enzymes. Yeast autolysis is the self-digestion of the cell by its endogenous enzymes and occurs naturally at the end of the cell's life cycle. During autolysis, proteins, glycogen, nucleic acids and other cell components are degraded [9]. These endogenous enzymes consist of cell proteases, nucleases, glucanases and phospholipases [21]. Autolysis of the yeast cell is triggered when the cell growth cycle is completed and the death or decline phase begins. These intracellular enzymes lead to partial degradation of the yeast cell wall, allowing extraction of the valuable materials including proteins, vitamins and carbohydrates without damaging their native structure [22]. The temperature at which autolysis occurs is highly important, with 50 °C for 24 h being deemed desirable in terms of content and yield of protein and α-amino nitrogen [23]. Plasmolysis refers to a modified autolysis process, with the addition of 'accelerators' to aid cell breakdown. These accelerators include inorganic salts and organic solvents. There are some disadvantages to yeast autolysis including low extract yield, difficult phase separation due to the high amount of residue in the autolysate, an increased risk of microbial contamination and poorer sensory characteristics [24].

In a study by Jacob et al. (2019), various methods of yeast cell wall disruption and their effects on yeast extract composition were investigated [25]. A cell mill was used as the mechanical method of cell wall disruption and autolysis was used as the non-mechanical method. Autolysis was induced by heating the material to 50 °C and holding for 24 h, following the addition of sodium chloride and ethyl acetate at the beginning of the process [25]. Cell disruption via cavitation was also carried out using an ultrasonic sonotrode. After disruption, the cell walls were then separated from the internal elements of the yeast cell via centrifugation. It was observed that for many amino acids that play a key role on the aroma properties of a beer, release is most effective as a result of autolysis. These amino acids include leucine, isoleucine, valine, histidine, proline, cysteine and glutamine. This study showed that the method of preparation of yeast extract can have an impact on the amount of free amino acids as well as the protein fraction after processing. However, due to the diversity of the starting material, comparison between differing BSY sources is difficult [25].

Other novel processing methods have been investigated, including a combination of autolysis, enzymatic hydrolysis and selective membrane filtration. The production of brewer's spent yeast fractions by ultrafiltration was investigated by Amorim et al. (2016). One hundred liters of brewer's spent yeast autolysate was filtered and dehydrated by lyophilization to give 1 kg of dry matter. Four different fractions were obtained, each of different sizes and molecular weights, as well as differing nutritional compositions [26]. While very little literature exists exploring this method of yeast processing, it could potentially be used to create targeted fractions, each with a differing composition and therefore differing potential uses.

3.3. Modification of Yeast Extract: Debittering

A challenge that stands in the way of yeast extract production from brewer's spent yeast is the strong, bitter flavor associated with the brewery waste. Bitter substances arising from hops, namely, iso-α-acids (such as humulones and iso-humulones), tannins and resins, adsorb onto the yeast cell walls during fermentation [27,28]. Generally, these compounds are removed by washing with alkaline or organic solvents prior to the extraction process. However, these methods of debittering are not economical and require the cells to be cleaned with water, creating a large amount of wastewater. A study by Shotipruk et al. (2005) investigated the potential for using alternate methods of debittering brewer's spent yeast [28]. It was discovered that, since the bitter substances adsorb solely onto the cell wall, debittering of spent brewer's yeast may be combined with cell debris separation. The effect of homogenization of the spent yeast prior to separation was also investigated. While the protein transmission and debittering efficiency were higher for the homogenized autolysate, there was still a strong bitter flavor when compared to cells washed with an alkaline solution. This is theorized to be due to the release of bitter substances from ruptured cell walls, as a result of the homogenization [28]. A high pH (pH 10) in combination with a heat treatment, over 50 °C, maximized the solubilization of the bitter constituents as debittering is more profound at higher temperatures [29]. In terms of residual yeast composition, the debittering process does not appear to cause any significant compositional difference between raw spent brewer's yeast and debittered brewer's yeast [29].

4. BSY Composition and Properties

Brewer's spent yeast is most commonly used in the food industry as either a flavor enhancer or as a nutritional supplement. Due to its high nutritive value, it has many potential uses in the functional food industry.

4.1. Protein

Brewer's spent yeast is a high-protein by-product. It contains 45–60% protein and is generally regarded as safe (GRAS). In a study by Vieira et al., 2016, BSY was found to contain 64.1% protein (dry weight) with a high percentage of essential amino acids as seen in Tables 1 and 2. It was determined that 40% of the total amino acid count was comprised of essential amino acids [20]. The levels of essential amino acids are in line with the requirements set out by FAO/WHO (1990) for adult humans [30]. Generally, the sulfur (S–) containing amino acids (methionine and cysteine) are the limiting factors in the amino acids composition but in this study it was found that S-amino acids were above the target set by the FAO/WHO [20]. A high level (34%) of flavor-enhancing amino acids was also determined, namely, glutamic acid, aspartic acid, glycine and alanine. A high level of these particular amino acids presents the potential of BSY or BSY extracts as flavor-enhancing ingredients.

Table 1. Chemical Composition of Brewer's Spent Yeast.

Chemical Composition	Vieira et al. (2016) (g/100 g dw)	Cabellero-Cordoba and Sgarbieri (2000) (%)	Saksinchai et al. (2001) (%*w*/*w*)	Marson et al. (2020) (% dw)	Mathias et al. (2015) (% dw)	Jacob et al. (2019) (g/100 g)
Protein	64.1+/−0.2	47.19	n.d.	40.8	45.6	74.3+/−0.5
α-amino nitrogen	12.9+/−0.1	n.d.	4.5+/−0.2	n.d	4.09+/−0.04	5.81+/−0.10
Ash	14.0+/−0.2	8.55	13.3+/−0.7	7.0+/−0.1	5.9+/−0.05	13.5+/−2.3
Fat	1.32+/−0.04	3.53	n.d.	0.21	n.d.	0.67+/−0.01
Moisture	7.70+/−0.12	n.d.	n.d.	n.d	n.d.	6.8
Carbohydrate	12.9+/−0.1	21.52	26.8+/−0.3	n.d	n.d.	14.7
RNA	4.00+/−0.16	7.04	21.3+/−0.4 (by orcinol) 23.2+/−0.6 (by UV_{260nm})	1.9+/−0.1	n.d.	5.518+/−0.10
Soluble Fiber	n.d.	9.65	n.d.	6.6+/−0.1	n.d.	n.d.
Insoluble Fiber	n.d.	2.57	n.d.		n.d.	n.d.

n.d. stands for 'not determined'; dw represents 'dry weight'.

Table 2. Mineral Composition of Brewer's Spent Yeast Extract [20,31,32].

Macrominerals	Vieira et al. (2016) mg/100 g dw	Jacob et al. (2019) mg/100 g dw	RDA (FSAI)
Sodium (Na)	1228+/−22	88.1+/−0.001	1600 mg
Potassium (K)	9148+/−69	6248.7+/−0.001	2000 mg
Calcium (Ca)	27.1+/−0.40	16.4+/−0.001	800 mg
Magnesium (Mg)	273+/−2.31	210.2+/−0.001	375 mg
Trace Elements			
Chromium	0.019+/−0.00	n.a.	40 μg
Iron	1.76+/−0.03	3.672+/−0.01	14 mg
Manganese	0.564+/−0.013	0.15+/−0.01	2 mg
Cobalt	0.030+/−0.001	0.252+/−0.01	n.a.
Molybdenum	0.003+/−0.00	n.a.	50 μg
Zinc	11.9+/−0.29	9.963+/−0.01	10 mg
Copper	0.364+/−0.001	0.221+/−0.011	1 mg
Selenium	0.030+/−0.00	n.a.	55 μg

n.a.: not available; dw: based on dry weight; RDA: recommended dietary allowance.

In an earlier study by Caballero-Cordoba and Sgarbieri (2000), a protein content of 47.19% was determined. The amino acid profile of the *Saccharomyces* strain used in this study generally corresponds with the results obtained by Vieira et al. (2016) above. However, it was found that the limiting essential amino acid in this trial was lysine with only 5.69 g/100 g available when the FAO/WHO standard is 5.8 g/100 g. Similar results were reported by Podpora et al. (2016), whereby the protein content in two yeast extracts from spent yeast was found to be 62.5–63.8% [2]. The amino acid profiles of these two extracts agree with other values found in literature. When compared to *Saccharomyces cerevisiae* and rootlets, another by-product of the brewing industry, BSY has a high level of essential amino acids, as seen in Figure 1.

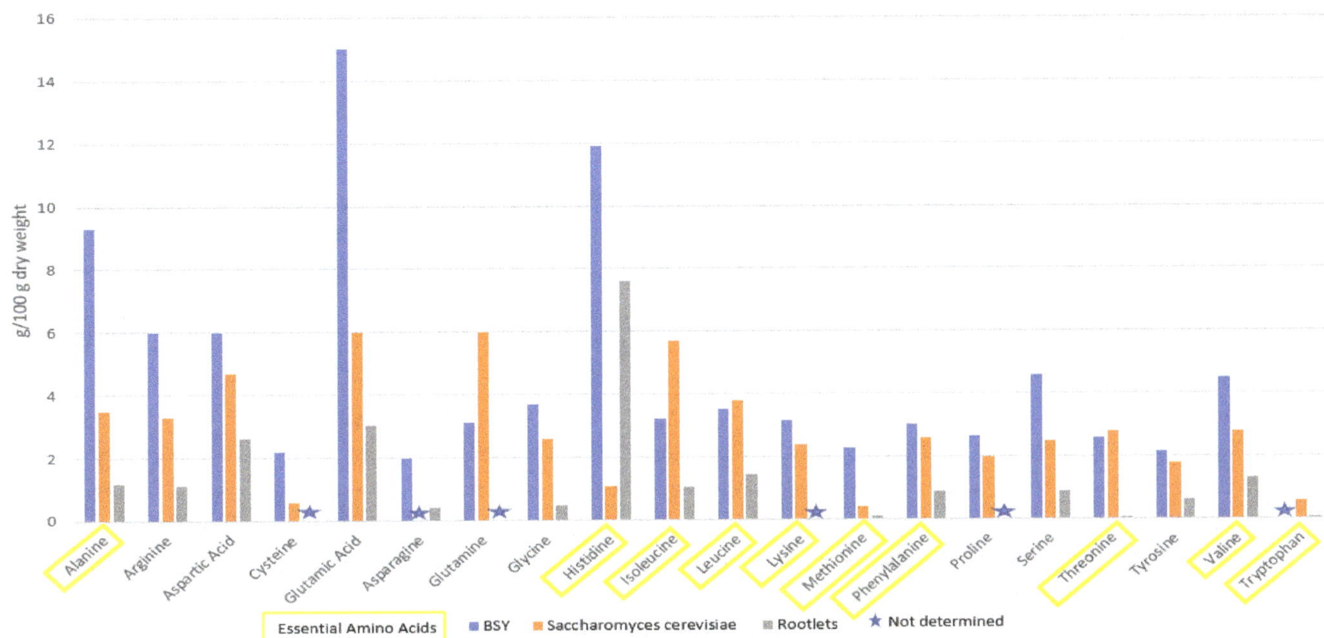

Figure 1. Amino Acid Profiles of Brewer's Spent Yeast (BSY) in comparison to *Saccharomyces cerevisiae* and rootlets, with the essential amino acids highlighted in yellow [20,33,34].

4.2. Carbohydrates

A high percentage of BSY polysaccharides are insoluble, approximately 83% of total polysaccharide content. Even following harsh alkaline treatment, they can only be partially solubilized, leaving an insoluble fraction of 24% of total polysaccharides [5]. BSY polysaccharides are mainly comprised of β-glucans, mannoproteins and glycogen. Chitin is also present in the cell walls in a small concentration but greatly contributes to the insolubility of the polysaccharides. In terms of *Saccharomyces cerevisiae*, the cell wall makes up about 25–50% of the volume and 15–30% of the dry weight of the cell [35]. According to Pinto et al. (2015), the cell wall consists of 50–60% β-glucan, 35–40% mannoproteins, 1–3% chitin and 1–23% glycogen [5].

The β-glucan component of the cell wall is comprised of mainly a β-(1,3) glucan backbone (approx. 85% of yeast glucan component) and a minor β-(1,6) glucan (approx. 3% of glucan component) [36]. β-glucans have been linked with many health benefits including pre-biotic, anti-inflammatory, anti-diabetic, anti-cancer and immune-modulating effects [5,37]. Functionally, β-glucans are also useful in altering the texture, mouthfeel and rheological properties of food products, as well as being effective stabilizers, emulsifiers and potential fat replacers [4,38]. Interestingly, when compared with commercially available β-glucans from baker's yeast, the BSY β-glucans exhibited a higher apparent viscosity, water holding capacity and increased emulsion stabilizing capacities [39]. Alkaline treatment is most commonly used as the method for β-glucan extraction and in a study by Thammakiti et al. (2004), it was discovered that homogenization can increase the effectiveness of β-glucan extraction as well as increasing the apparent viscosity [39]. This is theorized to be due to the fragmentation of the yeast cell walls. Glycogen, an α-glucan, is present in two forms in the yeast cell wall: a soluble and an insoluble form. The latter due to a covalent linkage between glycogen and cell wall β-(1,3) glucan via a β-(1,6) linkage [40]. Mannoproteins, one type of glycoprotein, also present potential uses as emulsifiers in the food industry [41].

4.3. Mineral and Vitamin Content

In a study by Vieira et al. (2016), the mineral contents of brewer's spent yeast extract was determined by atomic absorption spectrometry (AAS) and inductively coupled plasma mass spectrometry (ICP-MS). This same method was also used by Jacob et al. (2019) to determine the mineral levels of an industrially produced spent yeast to use as a control in the examination of the effect of differing preparation methods on key extract characteristics [31]. A comparison of these studies is outlined in Table 2.

The sodium content was 1228 mg/100 g based on dry weight (dw) and the potassium content was 91.48 mg/100 g dw. These minerals play a key role in ribosomal protein synthesis as well as being essential in acid-base balance and water retention in the cell. A study by Jacob et al. (2019) showed a significantly lower sodium content, however, the levels of the other minerals in question were very similar to those determined by Vieira et al. (2016) [20]. Other elements are vital for bone development, muscle function, blood coagulation and neurological function. These include calcium and magnesium which are present at a levels of 27.11 mg/100 g dw and 273.6 mg/100 g dw, respectively. According to the FSAI (2020), as described in the 2008 Council Directive [32], the recommended daily intake of calcium for adequate health is approx. 800 mg per day, so at present BSY does not seem to be a significant source of calcium for the human diet, though it could contribute. However, the daily recommended intake of magnesium is 350–420 mg per day, so BSY could potentially be a very good source of magnesium, depending on the bioavailability of the mineral. This is also true of potassium with a recommended daily intake of 2000 mg. Magnesium plays an important role in the uptake and utilization of vitamins B and E as well as maintaining electrolyte and fluid balances [20]. Regarding trace elements, yeast extract could be a substantial source of these key nutrients, with mean iron levels of 1.76 mg/100 g dw, zinc levels of 11.9 mg/100 g dw, manganese levels of 0.564 mg/100 g dw and copper levels of 0.364 mg/100 g dw, among others. These trace elements, while only required in very small concentrations, are essential in maintaining life.

Vieira et al. (2016) used HPLC to determine the levels of B vitamins in brewer's spent yeast extract. The results are outlined in Table 3. BSY is known as a good source of B vitamins, with high mean content of vitamin B_3 (72 mg/100 g), B_6 (55.1 mg/100 g) and B_9 (3.01 mg/100 g). A similar study by Jacob et al. (2019) determined relatively similar values of an autolyzed industrial BSY product via the same method. However, the value for vitamin B_6 was significantly different to that of Vieira et al. (2016). A study by Pinto et al. (2013) also examined the vitamin contents of residual brewer's yeast following alkaline treatment and lyophilization [42]. However, many of the results obtained by Pinto et al. (2013) are significantly lower than those determined by Vieira et al. 2016. In particular, the levels of nicotinic acid (B_3) (77.2 mg/100 g vs 0.79 mg/100 g) and pyridoxine (B_6) (55.1 mg/100 g vs 9.99 mg/100 g). Vitamin B_2 was also present at a level of 2.34 mg/100 g which contradicts the results from Vieira et al. where the level of B_2 was non-quantifiable. This could be due to the natural variation in BSY composition due to cell health or age, differing preparation methods or alternative analytical methods. The level of cyanocobalamin (vitamin B_{12}) was non-quantifiable in both studies. Significant differences are also present between the figures detailed and the results obtained by Jacob et al. (2019). While the levels of vitamin B_3, B_9 and B_{12} were quite similar, the levels of pyridoxine and riboflavin were significantly different in comparison.

Table 3. Vitamin Composition of BSY extract.

Vitamins	Vieira et al. (2016) (mg/100 g)	Jacob et al. (2019) mg/100 g dw	Pinto et al. (2013) (mg/100 g)	RDA (FSAI)
Thiamine	NA	7.46+/−0.5	NA	1.1 mg
Nicotinic acid (B3)	77.2+/−1.1	78.6+/−2.0	0.79+/−0.06	16 mg
Pyridoxine (B6)	55.1+/−2.5	5.90+/−0.5	9.99+/−0.06	1.4 mg
Folic Acid (B9)	3.01+/−0.02	5.29+/−0.1	0.25+/−0.03	200 μg
Riboflavin (B2)	NQ (0.329)	10.6+/−0.5	2.34+/−0.03	1.4 mg
Cyanocobalamin (B12)	NQ (0.256)	0.16+/−0.1	NQ	2.5 μg

dw: based on dry weight; NQ: not quantified; RDA: recommended dietary allowance.

4.4. *Phenolic Compounds*

Phenols are compounds which contain a hydroxyl (−OH) group bound directly to an aromatic ring. Polyphenols are phenols which contain more than one hydroxyl group. Polyphenols are of significant interest to the food and nutraceutical industries due to their antioxidant properties. Vieira et al. (2016) separated thirteen phenolic compounds from BSY via HPLC. Two compounds (gallic acid and (±)-catechin) were quantified in the yeast-free fraction whereas in the bound fraction, six compounds were quantified. (+)-catechin was the most abundant with 24.6 mg/100 g dw, followed by protocatechuic acid (13.1mg/100 g dw), p-coumaric acid (10.3 mg/100 g dw), ferulic acid (9.22 mg/100 g dw), gallic acid (2.11 mg/100 g dw) and cinnamic acid (1.24 mg/100 g dw) [20]. The remaining four compounds were not detected. The levels of these polyphenols can be affected by temperature, the nature of extraction and the method of separation. A lyophilization pre-treatment, prior to the extraction of polyphenols, leads to a reduced yield of gallic acid, *p*-coumaric acid and naringin due to losses or transformations during vacuum drying. Similarly, the use of filtration also caused significant polyphenol losses and is therefore unsuitable for use as a separation method [43].

4.5. *Glutathione*

Glutathione, also known as GSH, is a compound found in plants, animals and fungi with significant antioxidant and other metabolic properties. It is an endogenous peptide, comprised of glutamate, cysteine and an additional amino acid with a low molecular weight, possibly serine or glycine [44,45]. It is known that certain yeast species can produce significant amounts of glutathione during fermentation. Glutathione in yeasts is formed by metabolization of glucose to pyruvate, followed by pyruvate and Acetyl-CoA entering the citric acid cycle (TCA). Glutamate is formed during this process and then combines with two other amino acids, one of which is cysteine, to form glutathione [45]. In a study by Liu et al. (1999), the optimal conditions for glutathione production by *Saccharomyces cerevisiae* were studied. It was discovered that glucose was the best carbon source for GSH production while fructose and sucrose lead to increased yeast growth. A combination of peptone and yeast extract was found to be the most efficient nitrogen source for GSH production by *S. cerevisiae* [46].

5. BSY Applications

BSY, in its raw and processed forms, has had very limited applications to date. While many potentially exciting technologies have been investigated, few have been implemented on an industrial scale, with animal feed being the main outlet for BSY currently. Figure 2 provides an overview of the current and potential applications of BSY.

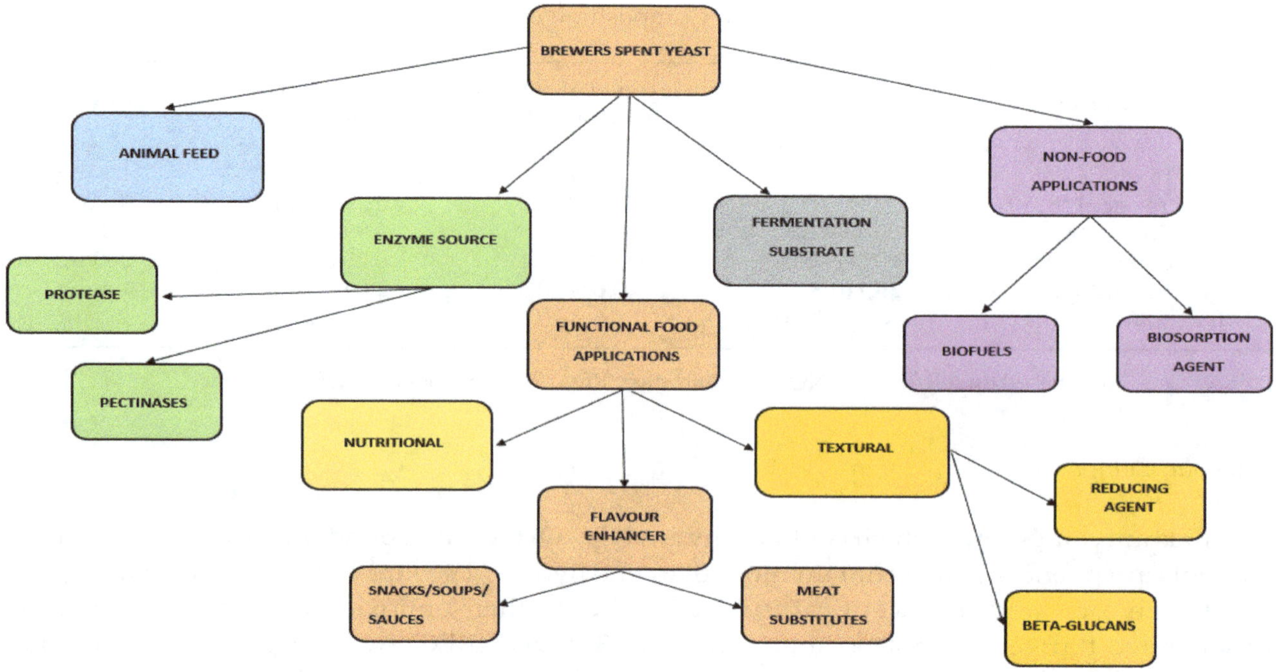

Figure 2. Overview of Current and Potential BSY Applications.

5.1. Animal Feed

Currently, the most well known and commonly used outlet for brewer's spent yeast is as a constituent of animal feeds, mainly as a cheap source of protein as well as a source of minerals and B vitamins. The spent yeast may be incorporated as a wet slurry or can be dried before feeding. The addition of yeast to animal feeds has been shown to have positive effects on feed quality and feed utilization.

In ruminant animals, in addition to a nutritional benefit, it has been shown that the addition of yeast to the bovine diet can have a positive effect, including increased milk production and improved digestion. In a study by Wohlt et al. (1998), it was determined that the supplementation of bovine feed with 10–20 g per day of yeast can improve milk yield significantly. It is theorized that this is due to the effects of yeast metabolites produced within the rumen and their subsequent interaction with other rumen microbes [47]. It has also been shown that the addition of yeasts to ruminant diets as a probiotic can improve the gut microflora of the animal, especially in situations where sub-therapeutic antibiotic treatment is frequently used. This improvement in gut health can lead to increased fiber digestibility, increased weight gain and reduced methane production [48]. A study by Harlow et al. (2016) also investigated the effects of incorporating BSY into ruminant caprine feed. The primary focus of this study was to determine whether the increased hops levels in modern craft beer could have a beneficial effect on the animal through the spent yeast. Due to the increased level of hops, a high level of α- and β-acids are present in the yeast and these compounds may act as antimicrobials, particularly inhibiting hyper ammonia producing bacteria (HAB). These bacteria cause major amino acid degradation, leading to a loss of valuable proteins. This study showed that the hops acids in the BSY protected proteins from degradation by HAB [49]. However, this was only shown in vitro so it is unclear if the same effect would be noted in live animals. Regardless, it does show potential for further study into rumen-protected proteins. It has also been shown that cost reduction in the sheep diet can be achieved by substitution of dried BSY for up to 100% of cornmeal in the feed formulation with no observable adverse effects on digestibility or feeding behavior [50].

There have also been several studies into the effect of yeast addition in poultry diets. In a study by Line et al. (1998), the effect of adding a yeast (*Saccharomyces boulardii*) to the animals diet was investigated. The key focus of this study was to reduce the frequency of gut colonization with pathogenic bacteria, namely *Salmonella* and *Campylobacter*. *Salmonella* is responsible for a high level of human foodborne infection and poultry is known to act as a reservoir for these types of pathogens. The broiler chickens were fed varying amounts of the yeast and were challenged by 'seeder' birds, animals known to carry pathogens. It was observed that there was a decreased frequency of *Salmonella* colonization in the animals that had been fed the yeast-supplanted diet while the frequency of colonization of *Campylobacter* in all animals was unaffected. While it is unclear exactly how this occurs, it is theorized to be due to the presence of the sugar mannose in the yeast cell wall. Certain pathogens, including *Salmonella*, have an affinity to bind to mannose and therefore the mannose in the yeast cell could act as a bacterial inhibitor, preventing the pathogen from adhering to the intestinal cell wall. There was also no observed statistically significant change in animal weight [51]. However, in other studies, an increase in weight in addition to overall health was noted [48].

The addition of yeast into the diet of an animal for meat production can have a positive effect on meat quality. In a study by Milewski et al. (2013), the effect of yeast addition to the diet in lambs was examined. An increase in positive meat performance indicators was observed, including body weight, daily gains, growth rate, muscle dimension and fat thickness. There was also a marked increase in humoral immunity indicators including, lysozyme, ceruloplasmin and gamma globulin [52].

Brewer's spent yeast as animal feed has also been considered in aquaculture. Aquafeeds are generally heavily reliant on marine ingredients such as fish meal and oils [53]. In a study by San Martin et al. (2020), the potential of BSY as a replacement for fishmeal is explored. Due to its plentiful availability and its low cost, as well as its nutrient density, on paper, BSY presents itself as a viable option in formulating aquafeeds. The spent yeast used in this study underwent a pre-treatment to increase digestibility, including a hydrolysis step and an enhanced drying step, as it has a lower intestinal absorption rate than the typically used fishmeal. The results showed that the BSY had good digestibility and could potentially be used as a fishmeal alternative in aquafeeds. In a similar study, BSY was explored as an option to replace up to 60% of the diet of the giant freshwater prawn in different environments. There was no observable significant change in growth or survival from 0% substitution to 40% in either environment (clear water vs. biofloc system). However, an increase in fish growth was observed in the fish with 60% BSY substitution being raised in the biofloc system [54]. The effect of BSY feeding on fish digestive enzymes was investigated by Castro et al. (2013). In this study the effects of feeding 1–2% BSY to two fish species (White sea bream and meagre) were investigated. The results showed that BSY feeding led to an increased level of amylase and protease activity in the pyloric caeca of the sea bream while in the meagre only amylase activity was increased. While the mean weight of the fishes was unaffected, the feed efficiency with the BSY supplementation was significantly increased for the meagre [55].

5.2. *Brewer's Spent Yeast as an Enzyme Source*

Brewer's spent yeast can also be used as a medium for the extraction of enzymes for use in other processes. A series of studies have looked into the use of BSY proteases in the production of Sardine Protein Hydrolysates from low-value sardine by-products. Vieira and Ferreira (2017) found that hydrolysates produced from sardine sarcoplasmic proteins via BSY proteases exhibited antioxidant and ACE inhibitory properties [56]. In addition to this, a separate study determined that treatment with BSY proteinases gave the sardine protein hydrolysates increased emulsion, foaming and oil-binding capacities [57]. It was also observed in a follow-up study that Sardine Protein Hydrolysates (SPH) produced using residual yeast proteinases can have an anti-inflammatory effect in an intestinal-endothelial co-culture cell model [58]. Although it is unclear exactly how this effect occurs, the results suggest that it is a combination of the bio-active compounds present in the hydrolysates as well as the influence of the metabolites produced by the Caco-2 cell peptidases. This exciting discovery could potentially be used in the treatment and prevention of inflammatory diseases; however, more research is required. The use of BSY proteases to produce brewer's spent grain (BSG) hydrolysates also resulted in protection of the target cells from free-radical cytotoxicity [59].

Residual brewer's yeast has been shown to be effective as a growth medium for lactic acid bacteria in the production of proteolytic enzymes. In a study by Mathias et al. (2017), three brewery wastes were investigated as alternative mediums for the cultivation of lactic acid bacteria to compare their potential for protease production. Residual yeast showed the most potential for the production of extracellular proteolytic enzymes, with a final purified proteolytic extract of 145.5 U/g of precipitated protein. The high level of proteins and the absence of carbon led the microorganism to release extracellular proteases easily, making BSY a good potential medium for protease production by lactic acid bacteria [60].

Saccharomyces cerevisiae can also be a source of pectinases for use in the extraction of fruit juices. Fruits high in pectin benefit from the addition of pectinases for a higher juice yield. Crude pectinases extracted from yeasts have been shown to be effective in significantly increasing the juice yields of pineapple and pawpaw [61,62]. Brewer's spent yeast could potentially be a low-cost and abundant source of these pectinases.

5.3. Functional Food Applications

5.3.1. β-Glucans and Mannoproteins as Functional Food Additives

Brewer's spent yeast β-glucans have demonstrated abilities in thickening and emulsifying as well as exhibiting a good water holding capacity and potential fat replacing capabilities. β-glucans are also nutritionally non-functional, meaning they are very low in calories. This is a very desirable characteristic for a functional food additive [38]. In a study by Thammakiti et al. (2004), the potential uses and extraction methods for BSY β-glucans as a functional food ingredient were investigated. The β-glucans were extracted from a BSY slurry by autolysis and homogenization followed by alkaline extraction. The characteristics of these ingredients were then characterized and compared to commercially available products. The β-glucans from BSY exhibited an increased apparent viscosity, increased water holding capacity and better emulsion stabilizing properties than the commercial products tested. However, the oil binding capacity was unaffected [39].

Spent yeast β-glucans have also been implemented in bakery products, such as bread. A β-glucan-rich flour (2.02 g β-glucans/100 g flour) was used to create loaves with a higher specific volume and a more uniform pore structure than the control loaf [63]. This could be due to the stabilizing effect of the β-glucans on the gas cells in the dough, similar to the effect of other cereal β-glucans on bread dough. The β-glucan-rich loaf also had an increased fiber content (+39%) which adds certain nutritional benefits related to gut health and motility. In the same study, the effect of proteins/proteolytic enzymes as bread additives was also explored, but with less successful outcomes. The proteolytic enzymes degraded the gluten structure leading to a decreased specific volume and little significant functional or nutritional gain from the additional protein. There is very limited literature regarding the application of brewer's spent yeast and in cereal products.

Studies by Silva Araújo et al. (2014) and Worrasinchai et al. (2006) investigated the use of spent yeast β-glucans and mannoproteins as a fat-replacing and emulsion stabilizing ingredient in mayonnaise. In the first study, mannoproteins were used to replace xanthan gum as an emulsifying and stabilizing agent. No negative effect on the sensory properties of the mayonnaise were detected by the sensory panel, even after prolonged storage [4]. This study showed that BSY mannoproteins have potential in the food industry as a synthetic emulsifying and stabilizing agent. The potential uses of yeast mannoproteins as bio-emulsifies was also explored by [41]. Worrasinchai et al. (2006) used yeast β-glucans to replace oil in a reduced fat mayonnaise at levels of 25%, 50% and 75% substitution. As fat is a key textual component of many foods, particularly high-fat foods such as mayonnaise, a low-calorie alternative to oil would be extremely useful in low-fat food product development. In contradiction to the previous study, sensory characteristics of the reduced fat mayonnaise were slightly reduced. However, levels of up to 50% substitution of oil for β-glucan were deemed acceptable by a sensory panel [38].

A study by Pancrazio et al. (2016) investigated the use of BSY extract in cooked hams as a potential textural agent and flavor compound. The results showed that 1% BSY extract addition significantly increased hardness and chewiness, as well as protein and free amino acid levels in comparison to the control hams. While more research is required, these results show the potential of BSY extract as a gel stabilizer in hams, potentially improving sliceability and water holding capacity [64].

5.3.2. Reducing Agents

As described earlier in this review, yeast cells are a good source of glutathione, a tri-peptide antioxidant compound [44]. Glutathione, as well as cysteine, can be used as reducing agents in the production of bread dough. Reducing agents are used in dough formulations to reduce mixing time, increase dough elasticity, and improve extensibility. In a bread dough, the structure and 'strength' of the system is determined by the formation of di-sulfide (−SS−) bonds. Reducing agents work by disruption these bonds, therefore weakening the dough structure [65]. In a study by Yano (2012), reduced glutathione improved the gas retaining properties of a gluten free rice batter dough in a yeast leavening process [66]. Verheyen et al. (2015) explored the destabilising effect of reducing agents on wheat doughs, including cysteine and reduced glutathione. The results showed a reduced dough development time and an increase in softness of the dough as a result of free thiol (−SH) groups [67]. Due to the presence of significant amounts of glutathione and cysteine in BSY, there is potential for the application of BSY in dough production as dough improvers.

5.3.3. Health and Nutritional Applications

B-Glucans

In addition to use as a functional ingredient, yeast β-glucans can be used as an ingredient to deliver a health benefit. β-glucans contain a specific combination of β-(1 → 4) and β-(1 → 3) linkages, creating long chain polysaccharides with a high molecular weight. These generate viscosity in the gut and this is believed to be the main reason behind the health benefits of β-glucans [68]. The perceived health benefits of β-glucans include stimulating and modulating the immune system, improving blood cholesterol levels and reducing the postprandial glucose response, improving satiety [68,69]. The effect of BSY and extracted β-glucans on lipid metabolism was examined in rats. Two ingredients were fed to rats in differing doses, with one group receiving 50 mg/kg dried BSY daily and the other receiving 10/100 mg/kg β-glucans which had been extracted from the residual yeast. At the end of the study, both groups showed a significant reduction in the concentration of total cholesterol as well as LDL-cholesterol and triacylglycerols in the blood when compared to the control [70]. In a study conducted by Bell et al. (1999), the effects of β-glucans from yeast and oats on serum lipids was investigated. It was determined that both oat-derived, and yeast-derived β-glucans reduce serum cholesterol levels in the blood. The yeast β-glucans may be more efficient due to the higher concentration of β-glucan in the derivative (86%) when compared to the oat β-glucan extract (15.9%). Yeast β-glucans and oat β-glucans also differ significantly in other ways, including β-linkages. Oat β-glucan is a polysaccharide comprised of β-(1 → 4) linked glucose units separated by β-(1 → 3) linked glucose every one to three units, whereas yeast β-glucans comprise of a β-(1 → 6) branched, β-(1 → 3) polysaccharide [71]. It is important to note that both the β-glucan source and extraction method can have a significant impact on key β-glucan characteristics, namely structure and molecular weight [72]. Molecular weight has an impact on the gelling capacity of the β-glucan, where a low molecular weight exhibits a better gelling capacity and a larger molecular weight, poorer gelling capacity [73]. A study by Salgado et al. (2017) compared the capabilities of both yeast β-glucan and barley β-glucan in producing aerogels for drug delivery. It was determined that yeast β-glucans had increased stability and elasticity, a stronger resistance against compression stress and a high water absorption capacity [73].

Encapsulating Agent

BSY has also shown potential as a starting material to be used as an encapsulating agent. In a study by Marson et al. (2020), Maillard-treated BSY cell debris was used as an encapsulating agent for ascorbic acid, a commonly used but highly unstable ingredient used widely in the food industry. The yeast material was prepared by enzymatically hydrolyzing *Saccharomyces* species from Lager Pilsen beer production. The enzymes used were Alcalase, Protamex (from Novozymes) and Brauzyn (from Prozyn). A Maillard reaction was induced by heating the hydrolyzed cells in wet medium at 75 °C for 12 hrs., followed by cooling in an ice bath. The Maillard reaction is a reaction between reducing sugars and amino acids. While it causes browning and the production of aroma compounds, it can also improve the functional properties of the protein, such as solubility [74]. Following this treatment, the mixture was spray dried in a mini spray drier, with a double-fluid-type atomizer nozzle. The microstructure and shape of the particles were examined using Scanning Electron Microscopy and the sample showed very similar properties to the control (particles produced without the addition of ascorbic acid) giving particles with a high encapsulation yield, a low water activity (Aw) and good hygroscopicity [74]. Fourier transform infrared technology was used to analyze the produced particles as well as the dried yeast cell wall debris, ascorbic acid, and maltodextrin (a commonly used encapsulation agent). This confirmed the presence of yeast cell debris on the particle surface, meaning that encapsulation was successful. This study shows the exciting potential of hydrolyzed BSY debris as an encapsulating agent but further research into particle viability and storage stability, as well as nutritional value, is needed.

Cryoprotectant for Probiotic Bacteria

Da Silva Guedes et al. (2019) investigated the possibility of using BSY β-glucans as a protectant for probiotic lactobacilli during freeze-drying, storage and in the gastrointestinal tract. In order for probiotics to have a positive health benefit, a high number of cells must reach the gut alive and viable, after having gone through the processing, storage and digestion processes. Yeast β-glucans were compared with fructooligosaccharides (FOS), an established cryoprotectant in their ability to protect the probiotic strains from degradation. The β-glucans used were extracted from brewery slurry using enzymatic treatments and sonication. It was found that the cryoprotectant properties of BSY β-glucans were very similar to those of FOS, making it an effective cryoprotectant for probiotic bacteria. However, after 90 days of storage, FOS showed superior protective abilities than the yeast β-glucan [75].

Antioxidant Properties

Yeast hydrolysates have potential as food or nutraceutical ingredients due to their antioxidant properties [76]. This antioxidant activity is due to the presence of a high level of polyphenolic compounds. Autolysis of the BSY can enhance total phenolic content (TPC), ferric reducing antioxidant power (FRAP) and ACE-inhibitory activity [77]. A study by Marson et al. (2019) investigated the effect of sequential enzymatic hydrolysis on the antioxidant properties of spent yeast. The results showed that, after enzymatic hydrolysis, internal cell components, including nutrients and enzymes, were more effectively released. This led to a 63% increase in antioxidant properties, as determined by FRAP and 1,1-diphenyl-1-picrylhydrazyl (DPPH) assays. Crude protein and total protein recovery were also improved by 50% and 83%, respectively. The influence of re-pitching on the hydrolysis was also investigated and it was determined that it took the repatched yeast up to 3.5 times longer to reach the same level of hydrolysis as the non re-pitched yeast [78]. In comparison to liquid beer waste, BSY extracts have been shown to contain higher levels of polyphenols, antioxidants, and flavonoids, indicating BSY as a potential industrial source of polyphenols for the food and nutraceutical industries [43].

5.3.4. Flavor-Enhancing Ingredient

Yeast, in particular brewer's spent yeast, is known as good source of nucleotides due to its high RNA content (approx. 8–11% of dry mass). Certain nucleotides, in particular 5′-monophosphates, are well-known flavor enhancers [79]. During autolysis, endogenous enzymes break down nucleic acids into nucleotides, creating 5′-nucleotides. They can also be partially hydrolyzed by the addition of enzymes, mainly 5′-phosphodiesterase [79]. Guanosine 5′-monophosphate is an active flavor enhancer and adenosine 5′-monophosphate is a precursor to inosine 5′-monophosphate, a flavor enhancer with 100 times the flavor-enhancing power than monosodium glutamate [80]. Differing processing methods can create a wide variety of flavor compounds in the spent yeast, as a result of interactions between amino acids, nucleotides, carbohydrates and peptides [81]. The flavors created in preparing yeast extracts are widely applicable in the food industry, in particular for meaty, 'umami' flavors. These flavors are produced by the synergistic action of 5′-nucleotides, glutamic acid and cysteine [81,82]. The flavor-improving capacity of glutamate can be improved by up to 10–15 times due to the addition of 5′-nucleotides [81]. Yeast extracts enriched in these compounds can be used to add savory meat and mushroom type flavors to soups, sauces, cheeses and other prepared foods. Thermal treatment can have a significant impact on the umami proteolytic compounds in a yeast extract, as explored by [83]. Thermal treatment can accelerate the breakdown of the yeast cells (autolysis) and can enhance the production of new aroma compounds through reactions such as the Maillard reaction. It was determined that a temperature of 110 °C is the optimum temperature for producing umami peptides [83]. The heat-induced reaction between sulphureous amino acids (cysteine and methionine) and sugar is also known to create characteristically 'meaty' flavors [82]. As brewer's spent yeast is classified as a GRAS product, yeast extract can be used in food products safely and without an acceptable daily intake (ADI), making BSY a significant source of flavor-enhancing compounds for use in a wide variety of food products. Certain yeast extract fractions are in high demand from the food industry to be used in the manufacture of these flavor-enhancing compounds [26]. In order to produce these flavor-enhancing properties, the yeast cells must undergo cell degradation via autolysis or hydrolysis.

5.3.5. The RNA Issue—Purines

While these nucleotides can provide flavor-boosting properties, they also present some challenges. If the intake of yeast is relatively high, for example in the case of protein supplementation, the increased purine content can result in an increased uric acid levels in the blood, potentially leading to hyperuricemia and gout [3]. Generally, only a small amount of the purines in RNA are degraded into uric acid daily. An increase in uric acid in the body can be caused by on overabundance of purines in the diet but can also occur where there is an inherited genetic deficiency of purine salvage enzymes, such as hypoxanthine-guanine phosphoribosyl transferase (HPRT) and adenine phosphoribosyl transferase (APRT) as well as catabolic enzymes such as xanthine dehydrogenase (XDH) [84]. Degradation of RNA and the removal of the resulting mononucleotides needs investigation regarding the use of BSY as a significant ingredient in food products.

Several methods exist to degrade and/or remove nucleotides from yeast cells, though very limited literature exists on the application of these technologies to BSY. Some chemical methods for reducing the purine content of yeast cells includes undergoing alkaline hydrolysis or heat treatment in the presence of NaCl, as seen in Table 4 [85]. A reduction of up to 53.1% of purines has been observed as a result of this method. Canepa et al. (1972) describe a method to significantly reduce the nucleic acid content of yeast involving yeast suspension in a 50 mM phosphate solution, followed by a heat shock treatment, and dialysis/washing with phosphate solution. It was found that as a result of this treatment, 50–60 g of the treated yeast would only contain 1 g of nucleic acids, of which purines only comprised approx. 0.3 g, well within an acceptable range [86]. It would be extremely interesting to see if a similar method would be so successful if used on brewer's spent yeast. The use of a heat shock step in the degradation of RNA is also described by Wiebe (2002). In this case, the yeast cells are heated to

64–65 °C for 20–30 min. This causes the RNA to degrade into monomers that can then diffuse out of the cell and be separated by centrifugation or similar technologies [87]. While this leads to the loss of other elements of the fungal biomass, it could still potentially be applied to BSY. Several autolytic methods for RNA reduction in yeast have also been investigated. These revolve around the use of the cells intracellular enzymes to degrade RNA. This process can be controlled by temperature and is more favorable in many situations than a chemical treatment [88]. Hydrolysis of the yeast cells has been shown to reduce RNA content by up to 33% when compared to a non-hydrolyzed sample [74].

Table 4. Reduction of Purines in Yeast Cells by Applying Heat Treatment in the Presence of NaCl; adapted from Trevelyan et al. (1976) [85].

Sample	Suspension Medium	Total Solids Recovery (%)	Guanine (μmol/g)	Adenine (μmol/g)	Total Purines as RNA (%)
Pressed yeast cells after autoclaving	Water	73.8	29	30	5.14
	1.0 M NaCl	69	5.9	6.6	1.14
	Water	69.4	24.4	25.4	4.62
	0.4 M NaCl	66.3	12.1	12.6	2.39
	0.8 M NaCl	66.1	4.5	4.8	0.91
	1.2 M NaCl	66.8	3.2	4.2	0.71
	1.6 M NaCl	67.2	2.6	3.3	0.57
	2.0 M NaCl	69.1	3.9	3.4	0.7

5.3.6. Meat Substitute Applications

In addition to providing flavorings in savory products such as soups and sauces, BSY also holds potential as a significant ingredient in the formulation of meat substitutes. A study by Gibson and Dwivedi (1970) examined the use of debittered and dried BSY, prepared by a modified alkaline wash process, as a base ingredient in the production of meat substitutes. Isolated soy protein was also added to enhance textural properties and increase protein content. The meat analogue production process was based on the unfolding of vegetable proteins via a heat treatment and the increased formation of cation bridges between these polypeptides via pH manipulation [89]. A follow-up study looking into the nutritional profile of BSY-based meat substitutes examined the protein quality of the product by completing growth studies in mice. The BSY-based meat analogue was substituted as the protein source in the diets at varying levels (0%, 25%, 50%, 75%, 100%). The average weight gain and feed intake of the mice showed that the BSY-based meat substitute had very good protein digestibility and a good biological value. It was determined that the meat analogue, if substituted with methionine, would be of a similar protein quality to whole-egg protein or other high-quality animal proteins [90].

5.4. Fermentation Substrate

Due to its high nutritional value and low cost, BSY has shown potential for use as a fermentation substrate, as a means of providing nitrogen for microorganism growth. BSY also has a low carbon-to-nitrogen ratio which is desirable for an additive to fermentation media [91]. A study by Mathias et al. (2017) investigated the use of brewing by-products, including BSY, as growth media for the growth of lactic acid bacteria. Success was characterized by an increased acidity, as a result of lactic acid production by the lactic acid bacteria. BSY showed the most potential in the trial when compared to the other by-products tested. The addition of BSY extract on its own or in combination with other brewery by-products has a significant positive effect on lactic acid bacteria fermentation. Results suggest that BSY, in combination with other brewery by-products, can be a cost-effective fermentation medium for batch and fed-batch lactic acid fermentation [92]. Sawisit et al. (2012) produced succinate via *Actinobacillus succinogenes* and suggested that BSY extract could potentially be used as a nitrogen source for this process [93].

5.5. *Biofuels—Methane Production*

BSY is a brewing by-product containing high amounts of biodegradable organic matter [94]. Several studies have been conducted investigating the production of biofuel in the form of methane from the anaerobic fermentation of BSY. A study by Sosa-Hernández et al. (2016) compared the methane production of three different BSY samples (SY1, SY2 and SY3) and one sample (SY1) was examined in differing dilutions (25%, 50% and 75%) after an anaerobic digestion treatment. The results showed clear differences among the samples. While samples SY2 and SY3 produced a higher volume of gas compared to SY1 (665 mL, 360 mL and 80-117 mL, respectively), SY1 produced gas of a higher quality. As sample SY1 came from a source producing a 'hoppy' beer, the suspected inhibitory effect of hops could aid in explaining the difference in gas volume and composition between the three samples. The methane producing micro-organisms convert the readily available substrate from the sample while fermentation was inhibited by the hops, reducing the production of other gases (CO_2, H_2 etc.) [95]. A good-quality biogas should contain 50–80% methane so in this case, although the overall volume of gas produced by SY1 was smaller, the quality of the gas was more desirable. Similar existing studies reflect similar results to this, showing that BSY has potential as a biofuel-producing medium and could be used by breweries to offset on-renewable energy costs [96,97]. The inclusion of the brewery wastewater stream in the BSY fermentation has also been explored and does not seem to cause any negative effect on the biogas production. However, more research at a commercial scale is required [94,97].

5.6. *Biosorption Agent*

Brewer's yeast (*Saccharomyces cerevisiae*) has been shown to be an effective bio-sorbent. Yeast cells that have been specially modified using water-based magnetic fluid stabilized with perchloric acid have been shown to effectively aid in the biosorption and biodegradation of water-soluble dyes from wastewater streams [98]. The textile industry produces a large amount of wastewater containing dyes which would have a detrimental environmental impact if not removed. Since dead yeast cells appear to have increased biosorption efficacy, BSY could present a suitable source for these materials.

Yeast cell biomass has also been shown to be efficient in absorbing and removing heavy metals, such as chromium, lead and uranium, from wastewater streams [99,100]. The removal of toxic contaminants from industrial waste is incredibly important to minimize environmental damage, as the release of these compounds into the environment would have a significant negative impact on marine wildlife and eco-systems. The natural capacity of the yeast cells to flocculate gives a fast, easy and cheap method of separation, without the need for an energy input [101]. While little research exists on the application of these methods for industrial wastewater treatment, the use of a cheap, abundant material such as BSY to sequester heavy metals from wastewater streams is a very promising technology.

5.7. *Supplementation of Ethanol Fermentation*

A limited number of studies exist exploring the effect of BSY supplementation on ethanol fermentation. Brewer's spent yeast contains high levels of nitrogen and essential minerals, aiding the growth of fermenting yeast. In comparison to the use of yeast extracts and peptone, more commonly used nutritional supplements for fermentations, the results were very similar [102]. Supplementation of high-gravity fermentations with spent yeast (wet and/or dry) leads to a higher ethanol concentration and a significant acceleration in fermentation time, with the final stages of fermentation being reached after only 48 h as opposed to approx. 70 h in the un-supplemented fermentation. However, there were some manufacturing issues as a result of BSY addition. The increase of hydrocolloids because of the BSY addition caused a significant increase in the viscosity of the mash. However, this could be controlled by altering BSY addition or further treatment [103].

6. Conclusions

In recent years, there has been increasing interest in sustainability, in particular the reuse and valorization of food processing by-products. Brewer's spent yeast is of particular interest in this regard as it is produced in huge volumes as a by-product of the brewing industry. Spent yeast is currently extremely underutilized and has limited outlets, mainly being used in animal feed formulations. Due to its nutritive value and physical properties, there is evidence that residual yeast has untapped potential in a variety of applications, including functional food ingredients. There is also potential for use as a nutritional supplement in human and microbial nutrition as well as a raw material for enzyme production, biofuel production and as a biosorption agent. The incorporation of brewer's spent yeast in such applications is a very cost-effective prospect, while also reducing the volume of waste associated with the brewing industry.

Brewer's spent yeast is high in protein as well as minerals and vitamins, making it an attractive ingredient for food products to enhance nutritional value. Due to its high level of β-glucans, it also presents the possibility for use as a textural agent, a stabilizer or a gelling agent. However, the limiting factor preventing the use of a significant proportion of BSY in food products is the high level of RNA nucleotide bases, namely purines, present in the raw material. Excessive consumption of purines may lead to increased uric acid in the blood stream which has detrimental effects on human health. While this is a significant hurdle, it is possible to remove purines via additional processing such as heat, enzyme or alkaline hydrolysis treatments but further study must be carried out on the application of these technologies to brewer's spent yeast.

In essence, brewer's spent yeast is an underutilized brewing by-product with a variety of exciting potential applications in the food industry and beyond. Increased awareness of this potential may instigate further research into these new applications and create new opportunities for their exploitation.

Author Contributions: Conceptualization, A.W.S. and E.K.A.; writing—original draft preparation, A.J.; review and editing, A.W.S.; project administration, E.Z. All authors have read and agreed to the published version of the manuscript.

Acknowledgments: The authors would like to acknowledge Patrick O'Riordan. Gregory Belt, Steffen Muench and Thomas Monin for their input and scientific discussions.

References

1. Kunze, W. *Technology Brewing and Malting*, 2nd ed.; VLB Berlin: Berlin, Germany, 1999.
2. Podpora, B.; Świderski, F.; Sadowska, A.; Rakowska, R.; Wasiak-Zys, G. Spent brewer's yeast extracts as a new component of functional food. *Czech J. Food Sci.* **2016**, *34*, 554–563. [CrossRef]
3. Edozien, J.C.; Udo, U.U.; Young, V.R.; Scrimshaw, N.S. Effects of High Levels of Yeast Feeding on Uric Acid Metabolism of Young Men. *Nat. Cell Biol.* **1970**, *228*, 180. [CrossRef] [PubMed]
4. Araújo, V.B.D.S.; De Melo, A.N.F.; Costa, A.G.; Castro-Gomez, R.H.; Madruga, M.S.; De Souza, E.L.; Magnani, M. Followed extraction of β-glucan and mannoprotein from spent brewer's yeast (*Saccharomyces uvarum*) and application of the obtained mannoprotein as a stabilizer in mayonnaise. *Innov. Food Sci. Emerg. Technol.* **2014**, *23*, 164–170. [CrossRef]
5. Pinto, M.; Coelho, E.; Nunes, A.; Brandão, T.; Coimbra, M.A. Valuation of brewers spent yeast polysaccharides: A structural characterization approach. *Carbohydr. Polym.* **2015**, *116*, 215–222. [CrossRef] [PubMed]
6. Pinho, O.; Ferreira, I.; Santos, L.H. Method optimization by solid-phase microextraction in combination with gas chromatography with mass spectrometry for analysis of beer volatile fraction. *J. Chromatogr. A* **2006**, *1121*, 145–153. [CrossRef] [PubMed]
7. Bellut, K.; Arendt, E.K. Chance and Challenge: Non-*Saccharomyces* Yeasts in Nonalcoholic and Low Alcohol Beer Brewing—A Review. *J. Am. Soc. Brew. Chem.* **2019**, *77*, 77–91. [CrossRef]

8. Verstrepen, K.J.; Derdelinckx, G.; Verachtert, H.; Delvaux, F.R. Yeast flocculation: What brewers should know. *Appl. Microbiol. Biotechnol.* **2003**, *61*, 197–205. [CrossRef]
9. Ferreira, I.; Pinho, O.; Vieira, E.; Tavarela, J. Brewer's *Saccharomyces* yeast biomass: Characteristics and potential applications. *Trends Food Sci. Technol.* **2010**, *21*, 77–84. [CrossRef]
10. Bamforth, C. *Beer: Tap into the Art and Science of Brewing*, 1st ed.; Insight Books: New York, NY, USA, 1998.
11. Lewis, M.J.J.; Young, T.W. *Brewing*, 1st ed.; Chapman & Hall: London, UK, 1995.
12. Wainwright, T. Diacetyl-a review: Part i-analytical and biochemical considerations: Part ii-brewing experience. *J. Inst. Brew.* **1973**, *79*, 451–470. [CrossRef]
13. Verbelen, P.J.; Dekoninck, T.M.L.; Saerens, S.M.G.; Van Mulders, S.E.; Thevelein, J.M.; Delvaux, F.R. Impact of pitching rate on yeast fermentation performance and beer flavour. *Appl. Microbiol. Biotechnol.* **2009**, *82*, 155–167. [CrossRef]
14. Kucharczyk, K.; Tuszyński, T. The effect of pitching rate on fermentation, maturation and flavour compounds of beer produced on an industrial scale. *J. Inst. Brew.* **2015**, *121*, 349–355. [CrossRef]
15. Vieira, E.; Moura, C.; Almeida, T.; Meireles, S.; Brandão, T.; Pinho, O.; Ferreira, I. Influence of Serial Repitching on Beer Polypeptide Profiles. *J. Am. Soc. Brew. Chem.* **2012**, *70*, 275–279. [CrossRef]
16. Bouix, M.; Leveau, J.-Y. Rapid Assessment of Yeast Viability and Yeast Vitality During Alcoholic Fermentation. *J. Inst. Brew.* **2001**, *107*, 217–225. [CrossRef]
17. Painting, K.; Kirsop, B. A quick method for estimating the percentage of viable cells in a yeast population, using methylene blue staining. *World J. Microbiol. Biotechnol.* **1990**, *6*, 346–347. [CrossRef] [PubMed]
18. Smart, K.A.; Chambers, K.M.; Lambert, I.; Jenkins, C.; Smart, C.A. Use of Methylene Violet Staining Procedures to Determine Yeast Viability and Vitality. *J. Am. Soc. Brew. Chem.* **1999**, *57*, 18–23. [CrossRef]
19. Straver, M.H.; Aar, P.C.V.D.; Smit, G.; Kijne, J.W. Determinants of flocculence of brewer's yeast during fermentation in wort. *Yeast* **1993**, *9*, 527–532. [CrossRef]
20. Vieira, E.F.; Carvalho, J.; Pinto, E.; Cunha, S.; Almeida, A.A.; Ferreira, I. Nutritive value, antioxidant activity and phenolic compounds profile of brewer's spent yeast extract. *J. Food Compos. Anal.* **2016**, *52*, 44–51. [CrossRef]
21. Babayan, T.L.; Bezrukov, M.G. Autolysis in yeasts. *Acta Biotechnol.* **1985**, *5*, 129–136. [CrossRef]
22. Podpora, B.; Swiderski, F.; Sadowska, A.; Piotrowska, A.; Rakowska, R. Spent Brewer's Yeast Autolysates as a New and Valuable Component of Functional Food and Dietary Supplements. *J. Food Process. Technol.* **2015**, *6*, 6. [CrossRef]
23. Tanguler, H.; Erten, H. Utilisation of spent brewer's yeast for yeast extract production by autolysis: The effect of temperature. *Food Bioprod. Process.* **2008**, *86*, 317–321. [CrossRef]
24. Bayarjargal, M.; Munkhbat, E.; Ariunsaikhan, T.; Odonchimeg, M.; Uurzaikh, T.; Gan-Erdene, T.; Regdel, D. Utilization of spent brewer's yeast *Saccharomyces cerevisiae* for the production of yeast enzymatic hydrolysate. *Mong. J. Chem.* **2014**, *12*, 88–91. [CrossRef]
25. Jacob, F.F.; Hutzler, M.; Methner, F.-J. Comparison of various industrially applicable disruption methods to produce yeast extract using spent yeast from top-fermenting beer production: Influence on amino acid and protein content. *Eur. Food Res. Technol.* **2019**, *245*, 95–109. [CrossRef]
26. Amorim, M.; Pereira, J.O.; Gomes, D.; Pereira, C.D.; Pinheiro, H.; Pintado, M.M.E. Nutritional ingredients from spent brewer's yeast obtained by hydrolysis and selective membrane filtration integrated in a pilot process. *J. Food Eng.* **2016**, *185*, 42–47. [CrossRef]
27. Dwivedi, B.; Gibson, D. Processing of Spent Brewers' Yeast for Food Use. *Can. Inst. Food Technol. J.* **1970**, *3*, 110–112. [CrossRef]
28. Shotipruk, A.; Kittianong, P.; Suphantharika, M.; Muangnapoh, C. Application of rotary microfiltration in debittering process of spent brewer's yeast. *Bioresour. Technol.* **2005**, *96*, 1851–1859. [CrossRef]
29. Nand, K. Debittering of spent brewer's yeast for food purposes. *Food/Nahrung* **1987**, *31*, 127–131. [CrossRef]
30. FAO/WHO. Report of a joint FAO/WHO/UNU Expert Consultation technical report FAO/WHO and United Nations University. In *Energy and Protein Requirements*; World Health Organization: Geneva, Switzerland, 1985.
31. Jacob, F.F.; Striegel, L.; Rychlik, M.; Hutzler, M.; Methner, F.-J. Yeast extract production using spent yeast from beer manufacture: Influence of industrially applicable disruption methods on selected substance groups with biotechnological relevance. *Eur. Food Res. Technol.* **2019**, *245*, 1169–1182. [CrossRef]

32. Fsai Commission Directive. 2008/100/ECof 28 October 2008 amending Council Directive 90/496/EEC on nutrition labelling for foodstuffs as regards recommended daily allowances, energy conversion factors and definitions. *Off. J. Eur. Union* **2008**, *51*, 1–4.
33. Waters, D.M.; Kingston, W.; Jacob, F.; Titze, J.; Arendt, E.K.; Zannini, E. Wheat bread biofortification with rootlets, a malting by-product. *J. Sci. Food Agric.* **2013**, *93*, 2372–2383. [CrossRef]
34. Martini, A.E.V.; Miller, M.W.; Martini, A. Amino acid composition of whole cells of different yeasts. *J. Agric. Food Chem.* **1979**, *27*, 982–984. [CrossRef]
35. Lipke, P.N.; Ovalle, R. Cell Wall Architecture in Yeast: New Structure and New Challenges. *J. Bacteriol.* **1998**, *180*, 3735–3740. [CrossRef] [PubMed]
36. Manners, D.J.; Masson, A.J.; Patterson, J.C. The structure of a β-(1 to 3)-d-glucan from yeast cell walls. *Biochem. J.* **1973**, *135*, 19–30. [CrossRef] [PubMed]
37. Jayachandran, M.; Chen, J.; Chung, S.S.M.; Xu, B. A critical review on the impacts of β-glucans on gut microbiota and human health. *J. Nutr. Biochem.* **2018**, *61*, 101–110. [CrossRef] [PubMed]
38. Worrasinchai, S.; Suphantharika, M.; Pinjai, S.; Jamnong, P. β-Glucan prepared from spent brewer's yeast as a fat replacer in mayonnaise. *Food Hydrocoll.* **2006**, *20*, 68–78. [CrossRef]
39. Thammakiti, S.; Suphantharika, M.; Phaesuwan, T.; Verduyn, C. Preparation of spent brewer's yeast beta-glucans for potential applications in the food industry. *Int. J. Food Sci. Technol.* **2003**, *39*, 21–29. [CrossRef]
40. Arvindekar, A.U.; Patil, N.B. Glycogen—A covalently linked component of the cell wall in *Saccharomyces cerevisiae*. *Yeast* **2002**, *19*, 131–139. [CrossRef]
41. Cameron, D.R.; Cooper, D.G.; Neufeld, R.J. The mannoprotein of *Saccharomyces cerevisiae* is an effective bioemulsifier. *Appl. Environ. Microbiol.* **1988**, *54*, 1420–1425. [CrossRef]
42. Pinto, L.; Lopes, M.; Carvalho Filho, C.; Alves, L.; Benevides, C. Determinação do valor nutritivo de derivados de levedura de cervejaria (*Saccharomyces* spp.). *Rev. Bras. Prod. Agroind.* **2013**, *15*, 7–17. [CrossRef]
43. León-González, M.E.; Gómez-Mejía, E.; Rosales-Conrado, N.; Madrid-Albarrán, Y. Residual brewing yeast as a source of polyphenols: Extraction, identification and quantification by chromatographic and chemometric tools. *Food Chem.* **2018**, *267*, 246–254. [CrossRef]
44. Alanazi, A.M.; Mostafa, G.A.E.; Al-Badr, A.A. *Profiles of Drug Substances, Excipients and Related Methodology*; Elsevier: Amsterdam, The Netherlands, 2015; Volume 40, pp. 43–158.
45. Shimizu, H.; Araki, K.; Shioya, S.; Suga, K.-I. Optimal production of glutathione by controlling the specific growth rate of yeast in fed-batch culture. *Biotechnol. Bioeng.* **1991**, *38*, 196–205. [CrossRef]
46. Liu, C.-H.; Hwang, C.-F.; Liao, C.-C. Medium optimization for glutathione production by *Saccharomyces cerevisiae*. *Process. Biochem.* **1999**, *34*, 17–23. [CrossRef]
47. Wohlt, J.; Corcione, T.; Zajac, P. Effect of Yeast on Feed Intake and Performance of Cows Fed Diets Based on Corn Silage during Early Lactation. *J. Dairy Sci.* **1998**, *81*, 1345–1352. [CrossRef]
48. Vohra, A.; Syal, P.; Madan, A. Probiotic yeasts in livestock sector. *Anim. Feed Sci. Technol.* **2016**, *219*, 31–47. [CrossRef]
49. Harlow, B.; Bryant, R.; Cohen, S.; O'Connell, S.; Flythe, M.D. Degradation of spent craft brewer's yeast by caprine rumen hyper ammonia-producing bacteria. *Lett. Appl. Microbiol.* **2016**, *63*, 307–312. [CrossRef] [PubMed]
50. Oliveira, R.L.; Oliveira, R.J.F.; Bezerra, L.; Nascimento, T.; de Pellegrini, C.B.; de Freitas Neto, M.D.; do Nascimento, N.G., Jr.; de Souza, W.F. Substitution of corn meal with dry brewer´s yeast in the diet of sheep. *Rev. Colomb. Cienc. Pecu.* **2016**, *29*, 99–107. [CrossRef]
51. Line, J.E.; Bailey, J.S.; Cox, N.A.; Stern, N.J.; Tompkins, T. Effect of yeast-supplemented feed on Salmonella and Campylobacter populations in broilers. *Poult. Sci.* **1998**, *77*, 405–410. [CrossRef] [PubMed]
52. Milewski, S.; Wójcik, R.; Zaleska, B.; Małaczewska, J.; Tański, Z.; Siwicki, A.K. Effect of *Saccharomyces cerevisiae* dried yeast on the meat performance traits and selected indicators of humoral immunity in lambs. *Acta Vet. Brno* **2013**, *82*, 147–151. [CrossRef]
53. Martin, D.S.; Orive, M.; Iñarra, B.; Castelo, J.; Estévez, A.; Nazzaro, J.; Iloro, I.; Elortza, F.; Zufía, J. Brewers' Spent Yeast and Grain Protein Hydrolysates as Second-Generation Feedstuff for Aquaculture Feed. *Waste Biomass Valoriz.* **2020**, *11*, 5307–5320. [CrossRef]

54. Nguyen, N.H.Y.; Trinh, L.T.; Chau, D.T.; Baruah, K.; Lundh, T.; Kiessling, A. Spent brewer's yeast as a replacement for fishmeal in diets for giant freshwater prawn (*Macrobrachium rosenbergii*), reared in either clear water or a biofloc environment. *Aquac. Nutr.* **2019**, *25*, 970–979. [CrossRef]
55. Castro, C.; Pérez-Jiménez, A.; Coutinho, F.; Pousão-Ferreira, P.; Brandão, T.M.; Oliva-Teles, A.; Peres, H. Digestive enzymes of meagre (*Argyrosomus regius*) and white seabream (*Diplodus sargus*). Effects of dietary brewer's spent yeast supplementation. *Aquaculture* **2013**, *416–417*, 322–327. [CrossRef]
56. Vieira, E.F.; Ferreira, I.M. Antioxidant and antihypertensive hydrolysates obtained from by-products of cannery sardine and brewing industries. *Int. J. Food Prop.* **2016**, *20*, 662–673. [CrossRef]
57. Vieira, E.F.; Pinho, O.; Ferreira, I.M. Bio-functional properties of sardine protein hydrolysates obtained by brewer's spent yeast and commercial proteases. *J. Sci. Food Agric.* **2017**, *97*, 5414–5422. [CrossRef] [PubMed]
58. Vieira, E.F.; Van Camp, J.; Ferreira, I.; Grootaert, C. Protein hydrolysate from canned sardine and brewing by-products improves TNF-α-induced inflammation in an intestinal–endothelial co-culture cell model. *Eur. J. Nutr.* **2017**, *57*, 2275–2286. [CrossRef] [PubMed]
59. Vieira, E.F.; Da Silva, D.D.; Carmo, H.; Ferreira, I. Protective ability against oxidative stress of brewers' spent grain protein hydrolysates. *Food Chem.* **2017**, *228*, 602–609. [CrossRef]
60. Mathias, T.R.D.S.; De Aguiar, P.F.; Silva, J.B.D.A.E.; De Mello, P.P.M.; Sérvulo, E.F.C. Brewery Waste Reuse for Protease Production by Lactic Acid Fermentation. *Food Technol. Biotechnol.* **2017**, *55*, 218–224. [CrossRef]
61. Djokoto, D.; Dzogbefia, V.; Oldham, J. Rapid Extraction of Pawpaw Juice with the Application of Locally Produced Pectic Enzymes from *Saccharomyces Cerevisiae* ATCC 51712. *Food Biotechnol.* **2006**, *20*, 31–41. [CrossRef]
62. Dzogbefia, V.P.; Amoke, E.; Oldham, J.H.; Ellis, W.O. Production and Use of Yeast Pectolytic Enzymes to Aid Pineapple Juice Extraction. *Food Biotechnol.* **2001**, *15*, 25–34. [CrossRef]
63. Martins, Z.E.; Pinho, O.; Ferreira, I. Impact of new ingredients obtained from brewer's spent yeast on bread characteristics. *J. Food Sci. Technol.* **2018**, *55*, 1–6. [CrossRef]
64. Pancrazio, G.; Cunha, S.C.; De Pinho, P.G.; Loureiro, M.; Meireles, S.; Ferreira, I.; Pinho, O. Spent brewer's yeast extract as an ingredient in cooked hams. *Meat Sci.* **2016**, *121*, 382–389. [CrossRef]
65. Wieser, H. Chemistry of gluten proteins. *Food Microbiol.* **2007**, *24*, 115–119. [CrossRef]
66. Yano, H. Comparison of Oxidized and Reduced Glutathione in the Bread-Making Qualities of Rice Batter. *J. Food Sci.* **2012**, *77*, C182–C188. [CrossRef] [PubMed]
67. Verheyen, C.; Albrecht, A.; Herrmann, J.; Strobl, M.; Jekle, M.; Becker, T. The contribution of glutathione to the destabilizing effect of yeast on wheat dough. *Food Chem.* **2015**, *173*, 243–249. [CrossRef] [PubMed]
68. Henrion, M.; Francey, C.; Lê, K.-A.; Lamothe, L. Cereal B-Glucans: The Impact of Processing and How It Affects Physiological Responses. *Nutrients* **2019**, *11*, 1729. [CrossRef] [PubMed]
69. Mantovani, M.S.; Bellini, M.F.; Angeli, J.P.F.; Oliveira, R.J.; Silva, A.F.; Ribeiro, L.R. β-Glucans in promoting health: Prevention against mutation and cancer. *Mutat. Res. Rev. Mutat. Res.* **2008**, *658*, 154–161. [CrossRef]
70. Waszkiewicz-Robak, B.; Bartnikowska, E. Effects of spent brewer's yeast and biological β-glucans on selected parameters of lipid metabolism in blood and liver in rats. *J. Anim. Feed Sci.* **2009**, *18*, 699–708. [CrossRef]
71. Bell, S.; Goldman, V.M.; Bistrian, B.R.; Arnold, A.H.; Ostroff, G.; Forse, R.A. Effect of β-Glucan from Oats and Yeast on Serum Lipids. *Crit. Rev. Food Sci. Nutr.* **1999**, *39*, 189–202. [CrossRef]
72. Zhu, F.; Du, B.; Xu, B. A critical review on production and industrial applications of beta-glucans. *Food Hydrocoll.* **2016**, *52*, 275–288. [CrossRef]
73. Salgado, M.; Santos, F.; Rodríguez-Rojo, S.; Reis, R.L.; Duarte, A.R.C.; Cocero, M.J. Development of barley and yeast β-glucan aerogels for drug delivery by supercritical fluids. *J. CO2 Util.* **2017**, *22*, 262–269. [CrossRef]
74. Marson, G.V.; Saturno, R.P.; Comunian, T.A.; Consoli, L.; Machado, M.T.D.C.; Hubinger, M.D. Maillard conjugates from spent brewer's yeast by-product as an innovative encapsulating material. *Food Res. Int.* **2020**, *136*, 109365. [CrossRef]
75. Guedes, J.D.S.; Pimentel, T.C.; Diniz-Silva, H.T.; Almeida, E.T.D.C.; Tavares, J.F.; De Souza, E.L.; Garcia, E.F.; Magnani, M. Protective effects of β-glucan extracted from spent brewer yeast during freeze-drying, storage and exposure to simulated gastrointestinal conditions of probiotic lactobacilli. *LWT* **2019**, *116*, 108496. [CrossRef]
76. Vieira, E.F.; Das Neves, J.; Vitorino, R.; Dias-Da-Silva, D.; Carmo, H.; Ferreira, I. Impact of in Vitro Gastrointestinal Digestion and Transepithelial Transport on Antioxidant and ACE-Inhibitory Activities of Brewer's Spent Yeast Autolysate. *J. Agric. Food Chem.* **2016**, *64*, 7335–7341. [CrossRef] [PubMed]

77. Vieira, E.F.; Melo, A.; Ferreira, I.M. Autolysis of intracellular content of Brewer's spent yeast to maximize ACE-inhibitory and antioxidant activities. *LWT* **2017**, *82*, 255–259. [CrossRef]
78. Marson, G.V.; Machado, M.T.D.C.; De Castro, R.J.S.; Hubinger, M.D. Sequential hydrolysis of spent brewer's yeast improved its physico-chemical characteristics and antioxidant properties: A strategy to transform waste into added-value biomolecules. *Process. Biochem.* **2019**, *84*, 91–102. [CrossRef]
79. Komorowska, A.; Sieliwanowicz, B.; Mrówka, E.; Stecka, K.; Ha, A. Studies on yeast extracts enriched in 5′ nucleotides flavour enhancers obtained from spent brewery yeast. *Electron. J. Polish Agric. Univ.* **2003**, *6*, 3.
80. Vieira, E.; Brandão, T.; Ferreira, I. Evaluation of Brewer's Spent Yeast to Produce Flavor Enhancer Nucleotides: Influence of Serial Repitching. *J. Agric. Food Chem.* **2013**, *61*, 8724–8729. [CrossRef]
81. Rakowska, R.; Sadowska, A.; Dybkowska, E.; Świderski, F. Spent yeast as natural source of functional food additives. *Rocz. Państwowego Zakładu Hig.* **2017**, *68*, 115–121.
82. Varavinit, S.; Shobsngob, S.; Bhidyachakorawat, M.; Suphantharika, M. Production of Meat-Like Flavor. *Sci. Asia* **2000**, *26*, 219–224. [CrossRef]
83. Alim, A.; Yang, C.; Song, H.; Liu, Y.; Zou, T.; Zhang, Y.; Zhang, S. The behavior of umami components in thermally treated yeast extract. *Food Res. Int.* **2019**, *120*, 534–543. [CrossRef]
84. Cameron, J.S.; Moro, F.; Simmonds, H.A. Gout, uric acid and purine metabolism in paediatric nephrology. *Pediatr. Nephrol.* **1993**, *7*, 105–118. [CrossRef]
85. Trevelyan, W.E. Chemical methods for the reduction of the purine content of baker's yeast, a form of single-cell protein. *J. Sci. Food Agric.* **1976**, *27*, 225–230. [CrossRef]
86. Canepa, A.; Pieber, M.; Romero, C.; Tohá, J.C. A method for large reduction of the nucleic acid content of yeast. *Biotechnol. Bioeng.* **1972**, *14*, 173–177. [CrossRef]
87. Wiebe, M.G. Myco-protein from Fusarium venenatum: A well-established product for human consumption. *Appl. Microbiol. Biotechnol.* **2002**, *58*, 421–427. [CrossRef] [PubMed]
88. Trevelyan, W.E. Autolytic methods for the reduction of the purine content of baker's yeast, a form of single-cell protein. *J. Sci. Food Agric.* **1976**, *27*, 753–762. [CrossRef] [PubMed]
89. Gibson, D.; Dwivedi, B. Production of Meat Substitutes from Spent Brewers' Yeast and Soy Protein. *Can. Inst. Food Technol. J.* **1970**, *3*, 113–115. [CrossRef]
90. Dwivedi, B.; Gibson, D.; Bell, J. Nutritive Value of Meat Substitute Prepared from Brewers' Yeast and Isolated Soy Protein. *Can. Inst. Food Sci. Technol. J.* **1972**, *5*, 155–158. [CrossRef]
91. Mathias, T.R.D.S.; Alexandre, V.M.F.; Cammarota, M.C.; De Mello, P.P.M.; Sérvulo, E.F.C. Characterization and determination of brewer's solid wastes composition. *J. Inst. Brew.* **2015**, *121*, 400–404. [CrossRef]
92. Radosavljević, M.; Pejin, J.; Pribić, M.; Kocić-Tanackov, S.; Mladenović, D.; Djukić-Vuković, A.; Mojović, L. Brewing and malting technology by-products as raw materials in L-(+)-lactic acid fermentation. *J. Chem. Technol. Biotechnol.* **2020**, *95*, 339–347. [CrossRef]
93. Sawisit, A.; Seesan, S.; Chan, S.; Kanchanatawee, S.; Jantama, S.S.; Jantama, K. Validation of Fermentative Parameters for Efficient Succinate Production in Batch Operation by Actinobacillus succinogenes 130ZT. *Adv. Mater. Res.* **2012**, *550*, 1448–1454. [CrossRef]
94. Neira, K.; Jeison, D. Anaerobic co-digestion of surplus yeast and wastewater to increase energy recovery in breweries. *Water Sci. Technol.* **2010**, *61*, 1129–1135. [CrossRef]
95. Sosa-Hernández, O.; Parameswaran, P.; Alemán-Nava, G.S.; Torres, C.I.; Parra-Saldívar, R. Evaluating biochemical methane production from brewer's spent yeast. *J. Ind. Microbiol. Biotechnol.* **2016**, *43*, 1195–1204. [CrossRef]
96. Bocher, B.T.; Agler, M.; Garcia, M.L.; Beers, A.R.; Angenent, L.T. Anaerobic digestion of secondary residuals from an anaerobic bioreactor at a brewery to enhance bioenergy generation. *J. Ind. Microbiol. Biotechnol.* **2008**, *35*, 321–329. [CrossRef] [PubMed]
97. Zupančič, G.D.; Škrjanec, I.; Logar, R.M. Anaerobic co-digestion of excess brewery yeast in a granular biomass reactor to enhance the production of biomethane. *Bioresour. Technol.* **2012**, *124*, 328–337. [CrossRef] [PubMed]
98. Šafaříková, M.; Ptáčková, L.; Kibriková, I.; Šafařík, I. Biosorption of water-soluble dyes on magnetically modified *Saccharomyces cerevisiae* subsp. uvarum cells. *Chemosphere* **2005**, *59*, 831–835. [CrossRef] [PubMed]
99. Ferraz, A.; Tavares, T.; Teixeira, J.A. Cr (III) removal and recovery from. *Chem. Eng. J.* **2004**, *105*, 11–20. [CrossRef]
100. Riordan, C.; Bustard, M.; Putt, R.; McHale, A.P. Removal of uranium from solution using residual brewery yeast: Combined biosorption and precipitation. *Biotechnol. Lett.* **1997**, *19*, 385–388. [CrossRef]

101. Machado, M.D.; Santos, M.S.F.; Gouveia, C.; Soares, H.M.V.M.; Soares, E.V. Removal of heavy metals using a brewer's yeast strain of *Saccharomyces cerevisiae*: The flocculation as a separation process. *Bioresour. Technol.* **2008**, *99*, 2107–2115. [CrossRef]
102. Laopaiboon, L.; Sridee, W.; Jaisil, P.; Laopaiboon, P. The use of dried spent yeast as a low-cost nitrogen supplement in ethanol fermentation from sweet sorghum juice under very high gravity conditions. *Electron. J. Biotechnol.* **2011**, *14*. [CrossRef]
103. Kawa-Rygielska, J.; Pietrzak, W. Ethanol fermentation of very high gravity (VHG) maize mashes by *Saccharomyces cerevisiae* with spent brewer's yeast supplementation. *Biomass Bioenergy* **2014**, *60*, 50–57. [CrossRef]

2

Yeast Cellular Stress: Impacts on Bioethanol Production

Joshua Eardley and David J. Timson *

School of Pharmacy and Biomolecular Sciences, University of Brighton, Huxley Building, Lewes Road, Brighton BN2 4GJ, UK; j.eardley@brighton.ac.uk
* Correspondence: d.timson@brighton.ac.uk

Abstract: Bioethanol is the largest biotechnology product and the most dominant biofuel globally. *Saccharomyces cerevisiae* is the most favored microorganism employed for its industrial production. However, obtaining maximum yields from an ethanol fermentation remains a technical challenge, since cellular stresses detrimentally impact on the efficiency of yeast cell growth and metabolism. Ethanol fermentation stresses potentially include osmotic, chaotropic, oxidative, and heat stress, as well as shifts in pH. Well-developed stress responses and tolerance mechanisms make *S. cerevisiae* industrious, with bioprocessing techniques also being deployed at industrial scale for the optimization of fermentation parameters and the effective management of inhibition issues. Overlap exists between yeast responses to different forms of stress. This review outlines yeast fermentation stresses and known mechanisms conferring stress tolerance, with their further elucidation and improvement possessing the potential to improve fermentation efficiency.

Keywords: bioethanol; chaotropicity; fermentation; biofuel yield; compatible solute; cell stress

1. Introduction

The biotechnological potential of *Saccharomyces cerevisiae* has been exploited traditionally for the purposes of baking, brewing, and wine making. This species has also been incredibly useful in basic biochemistry and genetics studies [1]. Yeasts can be found thriving in diverse sugar-rich habitats and can be considered to be microbial weed species due to their dominant, biocide producing nature [2]. Its fermentative nature evolved between 125 and 150 million years ago and is shared by many yeast species, but the acquisition of ethanol tolerance by *S. cerevisiae* is a more recent event occurring after the whole genome duplication event, which is believed to have happened 100 million years ago [3,4]. This ~6000 gene-containing unicellular eukaryote was the first organism to have its genome fully sequenced [5,6]. Today it remains an indispensable model system for experimental studies and a mainstay of biotechnology. The curated Saccharomyces Genome Database (SGD; www.yeastgenome.org) contains vast amounts of information for this microorganism, obtained through "omics" technologies, traditional biochemistry and molecular biology techniques [7]. Yeast continues to play a pivotal role in the study of fundamental biological processes including aging, cell cycle and stress, as a model organism and in emerging biotechnological areas including drug screening, computational and systems biology—see, for example [8–20]. The high genetic tractability of this microorganism allows for the manipulation of metabolic pathways; for example biofuel researchers often minimize carbon fluxes to natural metabolic by-products to maximize ethanol productivity (see also Section 3.1) [21,22].

The displacement of conventional transport fuels with biofuels including high octane number bioethanol increases fuel security, whilst also being an effective strategy for reducing transport sector emissions [23]. First-generation bioethanol remains the most dominant biofuel globally, yet policy now favors adoption of second-generation or advanced biofuels due to their higher sustainability rating [24]. First-generation bioethanol production generally involves the pre-processing of food-feedstocks, notably including sugarcane, corn, wheat, for the efficient extraction of contained fermentable sugars like glucose [25,26]. Sugarcane bioethanol production is particularly prevalent in Brazil, a global pioneer for transportation ethanol use [27]. However large-scale bioethanol production previously caused a spike in global food prices, triggering the so-called food-versus-fuel debate [28–30]. Non-food sources, including waste lignocellulosic biomass, are utilized for the production of second-generation bioethanol, yielding pentose and hexose sugars whilst possessing a higher sustainability rating. Higher associated cost and energy pre-processing steps currently render commercial production of second-generation bioethanol unviable [31]. Furthermore, pre-processing steps associated with second-generation bioethanol feedstocks may yield highly inhibitory (chaotropic) by-products including vanillin, not found or generated in first-generation bioethanol fermentations [32].

Due to its industrious, high ethanol productivity, high ethanol tolerance and ability to ferment a wide range of sugars, *S. cerevisiae* is the principal yeast employed for global bioethanol production [21,33]. Obtaining maximum ethanol yields remains a technological challenge however, with contaminant microorganisms, feedstock impurities, undesired side products and cellular stress impacting on yeast ethanol productivity [34,35]. *S. cerevisiae* is "stressed" sequentially and in concert throughout an ethanol fermentation [32]. On occasion, these inhibitory factors, notably ethanol-induced inhibition, but also contamination, can make an ethanol fermentation become "stuck" [36,37]. The provision of commercially available, higher ethanol tolerance yeasts to "restart" fermentation can be required, demonstrating the necessity for deploying a highly stress tolerant yeast for industrial fuel ethanol production [38].

Here, we provide an overview of the major stresses encountered by yeast cells during fermentations and the responses those cells make to mitigate these stresses. We pay particular attention to the production of ethanol by fermentations in the biofuel industry. Many of the processes are also relevant to other fermentations carried out by yeast, for example the production of other chemicals or the use of yeast as a host for the production of recombinant proteins. We also largely focus on the budding yeast *S. cerevisiae*. We do this partly because this organism is so well studied, partly due to its significant role in the biotechnology industry and also because studies in this organism tend to be generalizable to other unicellular fungi. The roles and functions of the proteins mentioned in this review are summarized in Table 1.

Table 1. The roles of proteins mentioned in this review.

Protein	Systematic Gene Name	Function	Role in Stress	References
Cmp2p	*YML057W*	Calcineurin catalytic subunit	Involved in sensing alkaline stress	[39,40]
Cna1p	*YLR433C*	Calcineurin catalytic subunit	Involved in sensing alkaline stress	[39,40]
Cnb1p	*YKL190W*	Calcineurin regulatory subunit	Involved in sensing alkaline stress	[41]
Crz1p	*YNL027W*	Transcription factor	Involved in responding to alkaline stress	[42]
Elo1p	*YJL196C*	Medium-chain fatty acyl elongase	Enables production of longer fatty acids in heat and chaotrope stress	[43]
Ena1p	*YDR040C*	Sodium ion pump	Involved in responding to alkaline stress	[44]
Hac1p	*YFL031W*	Transcription factor	Activates genes involved in the unfolded protein response	[45]
Hog1p	*YLR113W*	MAP kinase	Terminal kinase of the HOG pathway. Activates genes in response to osmotic and other stresses	[46]
Hsf1p	*YGL073W*	Transcription factor	Activates genes involved in the heat shock response	[47]
Ire1p	*YHR079C*	Protein kinase and nuclease	Cleaves *HAC1* RNA making it translationally competent	[48]
Kar2p	*YJL034W*	Chaperone	Detects and responds to the presence of unfolded proteins in the endoplasmic reticulum	[49]
Msn2p	*YMR037C*	Transcription factor	Activates STRE responsive genes	[50]
Msn4p	*YKL062W*	Transcription factor	Activates STRE responsive genes	[50]
Ole1p	*YGL055W*	Δ9 fatty acid desaturase	Enables production of saturated fatty acids in heat and chaotrope stress	[51]
Pma1p	*YGL008C*	Hydrogen ion pump	Pumps protons into the vacuole and extracellular medium in acid stress	[52,53]
Pro1p	*YDR300C*	γ-glutamyl kinase	Enables synthesis of proline which protects against heat and chaotrope stress	[54]
Pro2p	*YOR323C*	γ-glutamyl phosphate reductase	Enables synthesis of proline which protects against heat and chaotrope stress	[55]
Pro3p	*YER023W*	Δ^1-pyrroline-5-carboxylate reductase	Enables synthesis of proline which protects against heat and chaotrope stress	[55,56]
Rlm1p	*YPL089C*	Transcription factor	Activated by Slt2p. Controls expression of genes involved in cell wall maintenance and strengthening	[57,58]
Slt2p	*YHR030C*	MAP kinase	Terminal kinase of the CWI pathway. Activates genes in response to cell wall stress	[59]
Tps1p	*YBR126C*	Trehalose-6-phosphate synthase	Enables synthesis of trehalose in heat and chaotrope stress	[60]
Tps2p	*YDR074W*	Trehalose-6-phosphate phosphatase	Enables synthesis of trehalose in heat and chaotrope stress	[61]
Trl1p	*YJL087C*	RNA ligase	Ligates *HAC1* following cleavage by Ire1p	[62,63]

2. Yeast Stress Responses

Fluctuating environments affect all living organisms. Yeast's response to fermentation associated stress conditions includes the switching of growth and metabolic strategies, including the synthesis of stress proteins functioning to mitigate cell damage and stress. These responses are is coordinated by cell stress stimulus sensing and signal transduction, triggering downstream transcriptional and post-translational responses [64]. The stress response elements (STRE; consensus CCCCT) are activated during numerous stress conditions including oxidative, heat shock, nitrogen starvation, low external pH, weak lipophilic acids and ethanol [65–70]. Yeast when subjected to ethanol stress initially struggles to maintain energy production. This compromised energy production leads to the increased expression of genes associated with energy metabolism [71]. While stress is often perceived as detrimental, which it often is doing industrial fermentations, it should also be remembered that environmental challenges and stresses can improve evolutionary fitness and drive the selection of more robust traits, strains and species [72,73].

The genetics underlying stress tolerance have been extensively investigated in *S. cerevisiae* and genetic and metabolic improvements having been attempted with success [68–71,74–81]. It is common for genes found to be up-regulated in the presence of ethanol stress to be targets for genetic engineering approaches [82]. A common experimental approach for the eventual isolation of desirable industrial traits including high stress tolerance is directed evolution. Selective pressures imposed on microorganisms promote the eventual emergence of traits that provide a selective advantage [83–85]. Repetitive batch cultivations in the presence of lignocellulosic hydrolysates for example evolved strains displaying increased tolerance to all inhibitors present [86]. Evolutionary methods are essentially "blind", but they enable tolerance to develop through changes in many genes, including ones whose functions are not fully known. Alternatively, more rational methods might be employed. This would involve identifying systems which are known to be involved with stress tolerance and making specific changes which aim to enhance these functions. For example, ethanol tolerance improvement could be achieved through approaches including enhancing the unfolded protein response (UPR), or engineering the enzymes involved in the biosynthesis of cell membrane lipids [87–90]. Tools including genome shuffling and global transcription machinery engineering have successfully bred stress tolerant yeast [91–95]. The emergence of powerful technologies including CRISPR-Cas9 should also allow for the better engineering of industrially relevant phenotypic traits [96,97].

2.1. Compatible Solutes or Inert Osmolytes

Many microorganisms produce, or increase production of, small molecules in response to external stresses. These compatible solutes are typically hydrophilic and facilitate the protection of biological macromolecules and cells [98,99]. The main yeast compatible solutes are glycerol, proline and trehalose (Figure 1). With the exception of glycerol, these compounds are kosmotropic and promote the ordering of biological macromolecules and assemblies, see Section 3.3 [100]. The physical biochemistry underpinning the mechanism of stabilization has not been fully elucidated [101]. It is likely to involve both direct and indirect effects [102]. Some compatible solutes bind to, and stabilize, biological macromolecules [103]. Kosmotropes generally act indirectly by increasing the ordering of water molecules, i.e., reducing the entropy of the system [104,105]. This enables them to "neutralize" the effects of chaotropes such as ethanol and to protect biological macromolecules from thermal, acid, or alkali denaturation. In the case of osmotic stress, they increase the osmolarity of the cytoplasm and thus help equalize the osmotic pressure on both sides of the plasma membrane [106].

Glycerol is a three carbon, trihydroxy alcohol. It is a by-product of ethanol fermentation in *S. cerevisiae*, having a role in maintaining cellular redox balance, with imbalances leading to oxidative stress [107]. Glycerol production also prevents osmotic-induced oxidative stress, caused by the intracellular accumulation of reactive oxygen species (ROS) [108]. Aquaporins and glycerol facilitators allow for the rapid export of glycerol for the conditioning of the external cellular environment. It is important in the adaptation to low osmolarity, due to its stabilizing effects on cellular components [106].

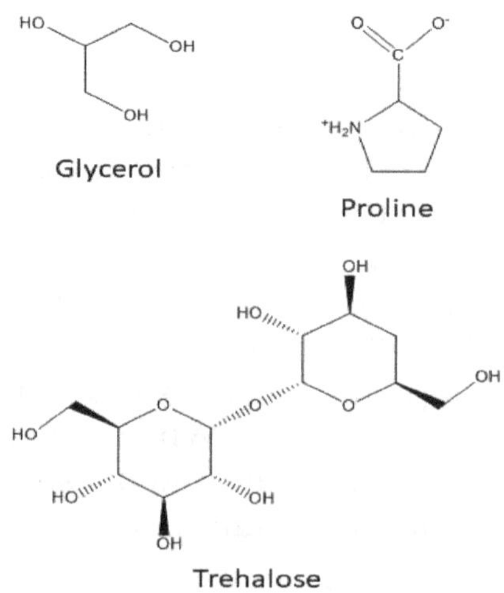

Figure 1. Structures of the three main compatible solutes deployed by yeast. Note that there is little structural similarity between these molecules, except that they all relatively small and hydrophilic.

Proline is an imino (secondary amino) acid which is synthesized in yeast from glutamate by the actions of the enzymes Pro1p, Pro2p and Pro3p [55]. Unlike glycerol, proline biosynthesis does not appear to be upregulated in response to stress. Natural levels may contribute to stress resistance and strains engineered to produce more, or consume less, proline, show greater tolerance to a range of stresses [109]. Trehelose is a disaccharide of two glucose subunits joined by an α (1$\rightarrow$1) glycosidic bond. It is synthesized in yeast from UDP-glucose and glucose 6-phosphate by the enzymes Tps1p and Tps2p [110]. The genes encoding these enzymes have stress response elements upstream ensuring then they are coregulated [111]. The transcription factors responsible for activating the genes encoding these enzymes, along with others regulated by STREs, are Msn2p and Msn4p [50]. These proteins translocate to the nucleus under stress conditions [74,112]. Trehalose itself assists in the activation of the heat shock response via the transcription factor Hsf1p [113].

2.2. The Unfolded Protein Response (UPR)

The unfolded protein response facilitates generic, cell wide responses to stresses which results in disruptions to "proteostasis" or the accumulation of unfolded proteins in the endoplasmic reticulum [114]. The UPR also affects general processes related to secretory pathway homeostasis and therefore plays a maintenance role in yeast cell wall integrity [115]. Its rapid initiation is ensured by cells constitutively expressing mRNA encoding Hac1p [116]. In the non-stressed state this mRNA is maintained in the unspliced form which is not translationally active. Increased levels of unfolded protein are sensed by the chaperone Kar2p. This causes it to stop binding Ire1p, a protein with both nuclease and kinase activities [117–120]. Ire1p can also directly detect unfolded proteins and damaged phospholipids [121,122]. *HAC1* mRNA is spliced by Ire1p and the RNA ligase Trl1p [116,120,123]. It is then translated to produce the transcription factor Hac1p which then translocates to the nucleus where it binds to the UPR response element. This sequence lies upstream from genes which are regulated by the UPR [124]. Hac1p then recruits RNA polymerase and other proteins required for transcriptional activation and increases the expression of a range of genes which code for proteins such as chaperones, and enzymes for phospholipid biosynthesis, cell wall biogenesis, DNA repair and aerobic metabolism [117,122,125–127]. They precise subset of genes upregulated depends partly on the conditions giving rise to stress [122,128,129]. Cell wall stress also activates the broad transcriptional response of the UPR through the mitogen-activated protein (MAP) kinase cascade, with ER stress also activating the CWI pathway, demonstrating that the UPR and CWI are co-ordinately regulated in *Saccharomyces cerevisiae* in order to protect cells against related stressors [130].

3. Yeast Stresses and Their Mitigation

3.1. Osmotic Stress

High initial fermentable sugar concentrations is particularly pertinent to first-generation bioethanol production, with high initial external osmolarity in an alcoholic fermentation facilitating the passive diffusion of cellular water down the concentration gradient, leading to hyperosmotic stress [106]. Effects include rapid decrease of cell volume and turgor pressure, with the concurrent generation of reactive-oxygen species (ROS), leading to redox state imbalances and oxidative stress. Oxidative stress and osmotic stress differ significantly but display overlapping responses with the robust yeast continuing to proliferate over a range of external water activities (a_W) [108]. Intracellular volume and water balance are tightly regulated through cell osmoregulation, for the proper maintenance and functioning of biochemical and biological processes [66].

Interest in the molecular mechanisms of yeast osmoadaptation originated from the need to improve the performance of yeast strains under industrial conditions. Hyperosmotic stress disrupts cytoskeleton structure, causes the remodeling of chromatin and can even lead to cell cycle arrest and apoptosis [131–133]. An intracellular shift in metabolism results, mediated by the high-osmolarity-glycerol (HOG) pathway and cell wall integrity (CWI) pathways, producing glycerol as a compatible solute and providing cell surface stability respectively, with these processes being co-coordinatively linked [66,115,134]. This cross talk between the two pathways suggests the evolution of highly coordinated stress responses.

The CWI pathway responds to damage to the cell wall. Extracellular domains of transmembrane receptors sense chemicals which may damage the cell wall [135]. The pathway also responds to osmotic, ethanol and pH stress. How these stresses are sensed is currently unknown [136]. Following sensing of cell wall associated stress, a signaling pathway connects the cell membrane with the MAP kinase Slt2p [136,137]. This kinase then activates two transcription factors, Swi4p/Swi6p and Rlm1p [57,138]. The first of these normally regulates cell cycle progression, but in this case regulates some genes involved in cell wall biosynthesis [139,140]. Rlm1p controls the expression of at least 80 genes including those which code for enzymes involved in the synthesis of cell wall components [141,142]. The HOG pathway also responds to cell surface receptors activating a signaling pathway which culminates in the activation of the MAP kinase Hog1p [46,66,143]. Phosphorylated Hog1p acts both in the nucleus and the cytoplasm. In the nucleus it phosphorylates and activates transcription factors which control the expression of over 600 genes [144,145]. This includes genes responsible for controlling the production of compatible solutes and the pausing of the cell cycle in G2 phase [146]. In the cytoplasm Hog1p directly reduces the export of glycerol thus increasing the cellular concentration of this compatible solute [147,148].

3.2. Heat Stress

Stress due to high temperatures can be encountered in ethanol fermentations, particularly in the early stages or if the external temperature is higher than the optimum for the fermentation [149,150]. Excess temperatures also increase sensitivity to ethanol [151]. A brief discussion of the effects of, and remedies for, heat stress is included here due to their similarity with chaotrope stress, see Section 3.3 [65]. High temperatures represent a general threat to living systems. Proteins, nucleic acids and phospholipid bilayers are denatured by increased temperature. Therefore, cells respond by counteracting this. Chaperone protein expression is upregulated In the so called heat shock response [152]. These proteins assist folding of nascent proteins and those which become partially unfolded as a result of denaturation by temperature and other stresses. Membrane composition is altered. Shorter and unsaturated fatty acids in phospholipids are replaced by longer and less saturated ones [153]. Both longer and less saturated fatty acids have higher melting temperatures and thus increase membrane rigidity. This is achieved by the expression of enzymes which extend the number of carbon atoms in fatty acids, e.g., Elo1p [43]. The amount of ergosterol in the membranes also increases [154]. Ergosterol is the predominant sterol in yeast membranes; increasing the mole fraction of this sterol in the membrane causes it to become more rigid and promotes the formation of lipid rafts [155–157]. Some strains which have been artificially evolved to have enhanced long term thermotolerance produce fecosterol in place of ergosterol which further rigidifies the membrane (Figure 2) [158]. In industrial fermentations, heat stress can be mitigated by thermostatic control of the reaction vessels. The use of thermotolerant strains also reduces the negative effects of heat stress [159–161]. Given the similarities between heat and chaotrope stress these strains are also likely to have improved tolerance to high ethanol concentrations [65,162,163]. Alternatively, additives to the fermentation mix, such as magnesium sulphate, can improve thermotolerance [164].

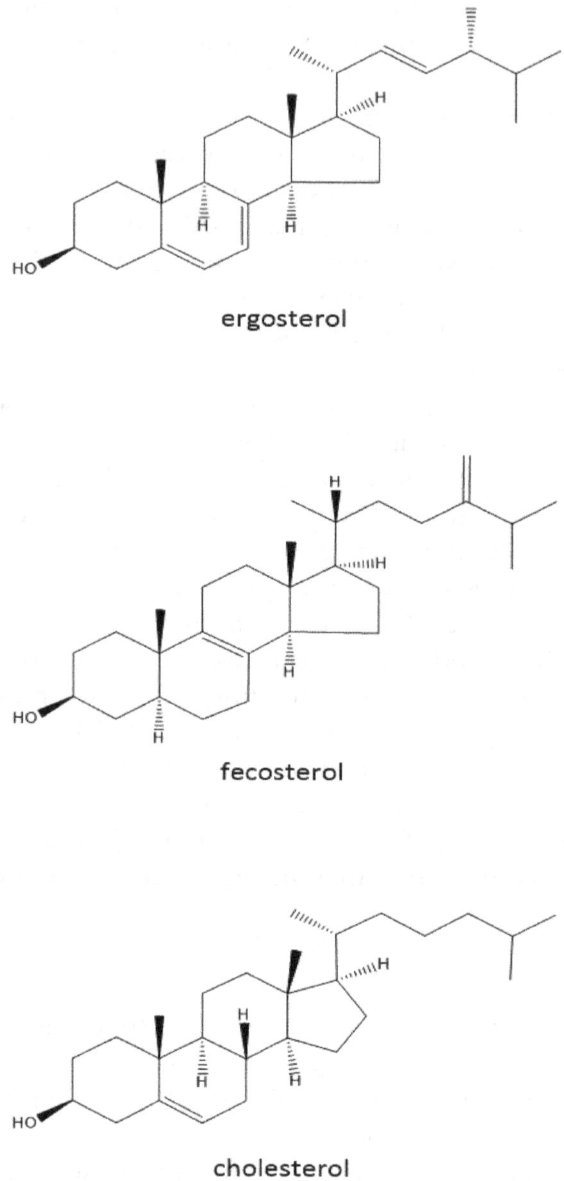

Figure 2. Structures of sterols found in yeast membranes. Cholesterol, which is commonly found in mammalian cell membranes, is shown for comparison.

3.3. *Chaotrope Stress*

Chaotropes are compounds which cause the disordering of biological macromolecules such as proteins and nucleic acids, as well as the disruption of biological assemblies held together by non-covalent interactions such as cellular membranes [100,165]. Many chaotropes do not to interact directly with the macromolecules and assemblies. Instead, they disrupt the hydrogen bonding networks in the solvent (water) thus increasing the overall entropy of the system [166,167]. However, the mechanism of chaotropicity is likely to vary between different compounds [165,168]. Since chaotropes disrupt critical cellular macromolecules and assemblies they exert a detrimental effect on biological systems. They affect key features of proteostasis and impose a dosage dependant "fitness cost" due to the "toxicity" of misfolded proteins [169,170]. The products of ethanol fermentations are chaotropic. Thus, the production of alcohol will ultimately be self-limiting as there will come a point where the concentration of the chaotropic compound exceeds the cell's ability to mitigate it [32]. These mitigations are similar to those seen in high temperatures stress [65]. The ER's quality control system, involving the unfolded protein response becomes activated [87]. This leads to increased expression of ER-located molecular chaperones whilst also aiding in the coordinated transition from fermentative metabolism to slower mitochondrial respiration (post-diauxic shift) during starvation [171]. The fatty acids in membrane phospholipids become longer and less saturated, catalyzed by Elo1p and Ole1p respectively. Compatible solutes such as glycerol and trehelose are produced [172].

Ethanol, a chaotropic solute, is utilized commercially as a protein denaturant, with chaotropic agents in general entropically disordering macromolecular systems, adversely affecting cell constituents including phospholipid bilayers, proteins, and other hydrated cell components [100]. The chaotropic activity of common fermentation products including ethanol (+5.93 kJ kg^{-1} mol^{-1}) and some by-products of lignocellulose pre-treatment including vanillin (+174 kJ kg^{-1} mol^{-1}), have been quantified using the agar gelation-temperature model system [100]. The growth rates susceptibility of *S. cerevisiae* to these substances has also been determined [173–175]. Chaotropes, including the aliphatic alcohols, cross and perturb cellular membranes in the high millimolar to molar range leading to a reduction in cell viability, slower growth/proliferation and lower obtainable ethanol yields [37,76,173].

Mitigation of chaotropicity in industrial settings is difficult to achieve. If a fermentation runs to completion, the alcohol poisons the cells. Furthermore, the production of glycerol, which is miscible with ethanol, presents an additional problem since the ethanol must be purified from this mixture. Kosmotropes are they opposite of chaotropes. They cause ordering and rigidity in biological macromolecules and phospholipid membranes. Typically, they reinforce the hydrogen bonding networks in the solvent, reducing the overall entropy of the system [176]. In theory, kosmotropes could be added to fermentation mixes to alleviate chaotrope stress. Indeed, the compatible solutes trehelose and proline are kosmotropic [100]. Glycerol presents an intriguing paradox. It is a compatible solute produced in response to stress by many microorganisms. However, in the agar gel setting assay it was shown to be mildly chaotropic [100]. Other studies suggest glycerol can enhance the hydrogen bonding network in water—a kosmotropic attribute [177,178].

There are a number of issues with using kosmotropes to neutralise chaotrope stress. First, some kosmotropes are themselves toxic to yeast cells and impair efficient growth and fermentation [175]. Second, there is currently no method of rapidly measuring or predicting the net solution chaotropicity. So, knowing the concentration of kosmotrope required to neutralize a given concentration of ethanol is currently impossible [179]. Third, the addition of chemicals to the fermentation mix will alter they osmolarity of the medium and may cause osmotic stress in the yeast. Fourth, the additional cost of purchasing kosmotropic compounds and, potentially, separating them from the ethanol at the end of the process may render this strategy nonviable.

3.4. Other Stresses

Stresses do not occur in isolation. Almost all instances of compounds which cause one form of stress also induce others [32,101]. For example, ethanol causes both osmotic and chaotropic stress [99]. In the context of ethanol fermentations, oxidative, and pH stress are less commonly encountered or occur as a consequence of other stresses.

Oxidative stress can be caused by chaotropic and osmotic stress during fermentations. Both can cause the buildup of reactive oxygen species. These free radicals cause non-specific damage to biomacromolecules and membrane lipids [180]. Uncorrected damage to DNA can cause hereditable mutations which are likely to be deleterious. Reactive oxygen species can also cause lipid peroxidation which damages cell membranes [181]. Yeast cells have a complex, integrated responses to free radical production [182]. This includes enzymes which repair double and single strand breaks in DNA, which reverse lipid peroxidation and which directly reduce reactive oxygen species [183–186].

Yeast cells normally condition the growth media to a slightly acidic pH. This requires the export of protons across the cell membrane by the plasma-membrane H^{+}-ATPase, Pma1p [52]. Thus, alkaline conditions represent an external stress for yeast with growth being slow or undetectable above pH 8.2 [187,188]. Multiple systems respond to alkaline stress [188]. One key pathway involves the calcium activated phosphatase calcineurin complex which is comprised of one of two catalytic subunits, Cna1p or Cmp2p, and one regulatory subunit, Cnb1p [40,41,189]. This activates the transcription factor Crz1p which controls the expression of stress related genes [190]. In addition, signaling through protein kinase A is transiently inhibited leading to the nuclear localization of the transcription factors Msn2p and Msn4p which bind to stress response elements, see above [191]. One consequence of alkaline stress is a reduction in mRNA transcription rates and in the stability of mRNA molecules [192]. A key response is the upregulation of the ATP dependent sodium ion transporter Ena1p [193]. This is regulated by both the calcineurin and protein kinase A pathways [194]. It is not immediately obvious why the export of sodium ions into the alkaline medium would mitigate the stress caused by excess hydroxide irons. It is generally considered that, unlike many mammalian sodium transporters, Ena1p does not cotransport protons although some early evidence suggests that it may do so [195]. The export of sodium ions may also assist in maintaining the membrane potential under conditions where protons are likely to be consumed.

Excessively acidic conditions also cause stress in yeast. Weak acids such as acetic acid, citric acid and boric acid can inhibit yeast growth and induce apoptosis [196]. These compounds are transported out of the cell by multidrug resistance transporters [197,198]. Pma1p activity is increased in order to transport excess protons out of the cell [199]. Acidification of the external medium results in a corresponding decrease in the pH of the vacuole suggesting that the cell transports excess protons to this compartment in order to maintain cytoplasmic pH [199].

In fermentations, the pH can be monitored and controlled by inline sensors and the addition of appropriate amounts of acid or base. The amounts can be determined either by feedback control or estimation using the Henderson Hasselbalch equation.

4. Conclusions

S. cerevisiae is considered a robust and versatile microorganism and has become an outstanding model system for the elucidation of Eukaryotic cell stress biology. Responses to each form of stress typically involve complex interactions of signaling pathways and transcription factors. This can affect the expression how many different genes. Although genomic and proteomic studies have revealed many candidate genes and proteins for the mitigation of various forms of stress, unravelling and understanding these data is challenging. In particular, understanding which genes are critical for response compared to those whose expression changes consequentially (e.g., many stress responses require energy and so ATP generating metabolic pathway enzymes are likely to be upregulated as a consequence). Distinguishing between those changes which make a minor contribution to stress responses compared to more significant contributions is critical. Here is should be noted that fold changes in expression levels do not provide a reliable guide as to which changes are the most important. This can only be determined by further biochemical experiments, for example assessing the response of strains with key genes deleted. The major players are likely to be those which act in systems which have been identified by traditional molecular biology approaches to be involved in stress remediation. The heat shock response, for example, produces coordinated sets of actions which mitigate the effects of heat on cellular components. Unsurprisingly, elements of this response are also used in adaptations to other forms of stress.

Understanding these responses is critical to enabling rational efforts to engineer strains of yeast which are better adapted to stressful conditions. Inevitably, there must be limits to the amount and duration of stress which yeast can endure. These limits will ultimately impose maximum yields in ethanol fermentations and determine whether, or not, it will be economically viable to transition away from fossil fuels towards sustainable biofuels.

Cellular stresses incurred by yeast throughout a bioethanol fermentation impact on cell growth and fermentative metabolism, reducing final obtainable ethanol titers. The plethora of general and stress-specific response and tolerance mechanisms makes yeast an industrially applicable microorganism. There are two main strategies for minimalizing the effects of stress on an ethanol fermentation. First, we can minimize the stress caused to the yeast. For example, temperatures and pH values of fermentations can be strictly control. To do this requires an understanding of the situations which cause the yeast cells to be stressed and thus reduce ethanol productivity. Second, we can select or engineer strains which are resistant to stresses. That many stress responses overlap assists us in this process. For example, yeast which are highly resistant to temperature stress are also likely to be able to resist high concentrations of ethanol. Again, this requires a knowledge of the biochemistry of stress sensing, stress responses and of unmitigated stresses. Therefore, it is clear that continued research efforts to understand the biochemistry of stress in yeast is vital for improving the economic and environmental sustainability of the biofuel industry.

Author Contributions: J.E.—initial draft of Section 1, Section 2, Section 2.1, Section 3.1, and Section 3.3; editing of entire manuscript; D.J.T.—initial draft of Section 2.2, Section 3.2, Section 3.4, and Section 4; editing of entire manuscript; figures and table. Both authors have read and agreed to the published version of the manuscript.

Acknowledgments: The authors thank MDPI for waiving the Open Access fee for this article. We thank Samantha Banford for her assistance with the revised version of the paper.

References

1. Barnett, J.A. Beginnings of microbiology and biochemistry: The contribution of yeast research. *Microbiology* **2003**, *149*, 557–567. [CrossRef] [PubMed]
2. Cray, J.A.; Bell, A.N.; Bhaganna, P.; Mswaka, A.Y.; Timson, D.J.; Hallsworth, J.E. The biology of habitat dominance; can microbes behave as weeds? *Microb. Biotechnol.* **2013**, *6*, 453–492. [CrossRef] [PubMed]
3. Hagman, A.; Säll, T.; Compagno, C.; Piskur, J. Yeast "make-accumulate-consume" life strategy evolved as a multi-step process that predates the whole genome duplication. *PLoS ONE* **2013**, *8*, e68734. [CrossRef] [PubMed]
4. Williams, K.M.; Liu, P.; Fay, J.C. Evolution of ecological dominance of yeast species in high-sugar environments. *Evolution* **2015**, *69*, 2079–2093. [CrossRef] [PubMed]
5. Goffeau, A.; Barrell, B.G.; Bussey, H.; Davis, R.W.; Dujon, B.; Feldmann, H.; Galibert, F.; Hoheisel, J.D.; Jacq, C.; Johnston, M.; et al. Life with 6000 genes. *Science* **1996**, *274*, 546–567. [CrossRef]
6. Engel, S.R.; Dietrich, F.S.; Fisk, D.G.; Binkley, G.; Balakrishnan, R.; Costanzo, M.C.; Dwight, S.S.; Hitz, B.C.; Karra, K.; Nash, R.S.; et al. The reference genome sequence of Saccharomyces cerevisiae: Then and now. *G3 Genes Genomes Genet.* **2014**, *4*, 389–398. [CrossRef]
7. Cherry, J.M.; Hong, E.L.; Amundsen, C.; Balakrishnan, R.; Binkley, G.; Chan, E.T.; Christie, K.R.; Costanzo, M.C.; Dwight, S.S.; Engel, S.R.; et al. Saccharomyces Genome Database: The genomics resource of budding yeast. *Nucleic Acids Res.* **2012**, *40*, D700–D705. [CrossRef] [PubMed]
8. Kaeberlein, M.; Burtner, C.R.; Kennedy, B.K. Recent developments in yeast aging. *PLoS Genet.* **2007**, *3*, e84. [CrossRef] [PubMed]
9. Barberis, M.; Todd, R.G.; van der Zee, L. Advances and challenges in logical modeling of cell cycle regulation: Perspective for multi-scale, integrative yeast cell models. *FEMS Yeast Res.* **2017**, *17*, fow103. [CrossRef] [PubMed]

10. Petranovic, D.; Vemuri, G.N. Impact of yeast systems biology on industrial biotechnology. *J. Biotechnol.* **2009**, *144*, 204–211. [CrossRef]
11. Bilsland, E.; Sparkes, A.; Williams, K.; Moss, H.J.; de Clare, M.; Pir, P.; Rowland, J.; Aubrey, W.; Pateman, R.; Young, M.; et al. Yeast-based automated high-throughput screens to identify anti-parasitic lead compounds. *Open Biol.* **2013**, *3*, 120158. [CrossRef] [PubMed]
12. Bilsland, E.; Pir, P.; Gutteridge, A.; Johns, A.; King, R.D.; Oliver, S.G. Functional expression of parasite drug targets and their human orthologs in yeast. *PLoS Negl. Trop. Dis.* **2011**, *5*, e1320. [CrossRef] [PubMed]
13. Ruetenik, A.; Barrientos, A. Exploiting post-mitotic yeast cultures to model neurodegeneration. *Front. Mol. Neurosci.* **2018**, *11*, 400. [CrossRef] [PubMed]
14. Lindström, M.; Liu, B. Yeast as a model to unravel mechanisms behind FUS toxicity in amyotrophic lateral sclerosis. *Front. Mol. Neurosci.* **2018**, *11*, 218. [CrossRef] [PubMed]
15. Nielsen, J. Yeast systems biology: Model organism and cell factory. *Biotechnol. J.* **2019**, *14*, 1800421. [CrossRef]
16. Seynnaeve, D.; Vecchio, M.D.; Fruhmann, G.; Verelst, J.; Cools, M.; Beckers, J.; Mulvihill, D.P.; Winderickx, J.; Franssens, V. Recent insights on Alzheimer's disease originating from yeast models. *Int. J. Mol. Sci.* **2018**, *19*, 1947. [CrossRef]
17. Raghavan, V.; Aquadro, C.F.; Alani, E. Baker's Yeast Clinical Isolates Provide a Model for How Pathogenic Yeasts Adapt to Stress. *Trends Genet.* **2019**, *35*, 804–817. [CrossRef]
18. Daignan-Fornier, B.; Pinson, B. Yeast to study human purine metabolism diseases. *Cells* **2019**, *8*, 67. [CrossRef]
19. Cazzanelli, G.; Pereira, F.; Alves, S.; Francisco, R.; Azevedo, L.; Dias Carvalho, P.; Almeida, A.; Côrte-Real, M.; Oliveira, M.J.; Lucas, C. The yeast Saccharomyces cerevisiae as a model for understanding RAS proteins and their role in human tumorigenesis. *Cells* **2018**, *7*, 14. [CrossRef]
20. Di Gregorio, S.E.; Duennwald, M.L. ALS yeast models—Past success stories and new opportunities. *Front. Mol. Neurosci.* **2018**, *11*, 394. [CrossRef]
21. Macedo, N.; Brigham, C.J. From beverages to biofuels: The journeys of ethanol-producing microorganisms. *Int. J. Biotechnol. Wellness Ind.* **2014**, *3*, 79–87.
22. Naghshbandi, M.P.; Tabatabaei, M.; Aghbashlo, M.; Gupta, V.K.; Sulaiman, A.; Karimi, K.; Moghimi, H.; Maleki, M. Progress toward improving ethanol production through decreased glycerol generation in Saccharomyces cerevisiae by metabolic and genetic engineering approaches. *Renew. Sustain. Energy Rev.* **2019**, *115*, 109353. [CrossRef]
23. Timilsina, G.R. Biofuels in the long-run global energy supply mix for transportation. *Philos. Trans. R. Soc. A Math. Phys. Eng. Sci.* **2014**, *372*, 20120323. [CrossRef] [PubMed]
24. Gasparatos, A.; Stromberg, P.; Takeuchi, K. Sustainability impacts of first-generation biofuels. *Anim. Front.* **2013**, *3*, 12–26. [CrossRef]
25. Lee, R.A.; Lavoie, J.-M. From first-to third-generation biofuels: Challenges of producing a commodity from a biomass of increasing complexity. *Anim. Front.* **2013**, *3*, 6–11. [CrossRef]
26. Das Neves, M.A.; Kimura, T.; Shimizu, N.; Nakajima, M. State of the art and future trends of bioethanol production. *Dynam. Biochem. Proc. Biotechnol. Mol. Biol.* **2007**, *1*, 1–14.
27. Paulino de Souza, J.; Dias do Prado, C.; Eleutherio, E.C.A.; Bonatto, D.; Malavazi, I.; Ferreira da Cunha, A. Improvement of Brazilian bioethanol production - Challenges and perspectives on the identification and genetic modification of new strains of *Saccharomyces cerevisiae* yeasts isolated during ethanol process. *Fungal Biol.* **2018**, *122*, 583–591. [CrossRef]
28. Banerjee, A. Food, feed, fuel: Transforming the competition for grains. *Dev. Chang.* **2011**, *42*, 529–557. [CrossRef]
29. Meyer, P.M.; Rodrigues, P.H.; Millen, D.D. Impact of biofuel production in Brazil on the economy, agriculture, and the environment. *Anim. Front.* **2013**, *3*, 28–37. [CrossRef]
30. Ewing, M.; Msangi, S. Biofuels production in developing countries: Assessing tradeoffs in welfare and food security. *Environ. Sci. Policy* **2009**, *12*, 520–528. [CrossRef]
31. Eisentraut, A. *Sustainable Production of Second-Generation Biofuels*; OECD/IEA: Paris, France, 2010.
32. Cray, J.A.; Stevenson, A.; Ball, P.; Bankar, S.B.; Eleutherio, E.C.; Ezeji, T.C.; Singhal, R.S.; Thevelein, J.M.; Timson, D.J.; Hallsworth, J.E. Chaotropicity: A key factor in product tolerance of biofuel-producing microorganisms. *Curr. Opin. Biotechnol.* **2015**, *33*, 228–259. [CrossRef] [PubMed]
33. Mohd Azhar, S.H.; Abdulla, R.; Jambo, S.A.; Marbawi, H.; Gansau, J.A.; Mohd Faik, A.A.; Rodrigues, K.F. Yeasts in sustainable bioethanol production: A review. *Biochem. Biophys. Rep.* **2017**, *10*, 52–61. [CrossRef] [PubMed]

34. Jönsson, L.J.; Martín, C. Pretreatment of lignocellulose: Formation of inhibitory by-products and strategies for minimizing their effects. *Bioresour. Technol.* **2016**, *199*, 103–112. [CrossRef] [PubMed]
35. Beckner, M.; Ivey, M.L.; Phister, T.G. Microbial contamination of fuel ethanol fermentations. *Lett. Appl. Microbiol.* **2011**, *53*, 387–394. [CrossRef] [PubMed]
36. Narendranath, N.; Hynes, S.; Thomas, K.; Ingledew, W. Effects of lactobacilli on yeast-catalyzed ethanol fermentations. *Appl. Environ. Microbiol.* **1997**, *63*, 4158–4163. [CrossRef] [PubMed]
37. Gibson, B.R.; Lawrence, S.J.; Leclaire, J.P.; Powell, C.D.; Smart, K.A. Yeast responses to stresses associated with industrial brewery handling. *FEMS Microbiol. Rev.* **2007**, *31*, 535–569. [CrossRef]
38. Maisonnave, P.; Sanchez, I.; Moine, V.; Dequin, S.; Galeote, V. Stuck fermentation: Development of a synthetic stuck wine and study of a restart procedure. *Int. J. Food Microbiol.* **2013**, *163*, 239–247. [CrossRef]
39. Liu, Y.; Ishii, S.; Tokai, M.; Tsutsumi, H.; Ohki, O.; Akada, R.; Tanaka, K.; Tsuchiya, E.; Fukui, S.; Miyakawa, T. The Saccharomyces cerevisiae genes (CMP1 and CMP2) encoding calmodulin-binding proteins homologous to the catalytic subunit of mammalian protein phosphatase 2B. *Mol. Gen. Genet.* **1991**, *227*, 52–59. [CrossRef]
40. Cyert, M.S.; Kunisawa, R.; Kaim, D.; Thorner, J. Yeast has homologs (CNA1 and CNA2 gene products) of mammalian calcineurin, a calmodulin-regulated phosphoprotein phosphatase. *Proc. Natl. Acad. Sci. USA* **1991**, *88*, 7376–7380. [CrossRef] [PubMed]
41. Cyert, M.S.; Thorner, J. Regulatory subunit (CNB1 gene product) of yeast Ca2+/calmodulin-dependent phosphoprotein phosphatases is required for adaptation to pheromone. *Mol. Cell. Biol.* **1992**, *12*, 3460–3469. [CrossRef]
42. Matheos, D.P.; Kingsbury, T.J.; Ahsan, U.S.; Cunningham, K.W. Tcn1p/Crz1p, a calcineurin-dependent transcription factor that differentially regulates gene expression in Saccharomyces cerevisiae. *Genes Dev.* **1997**, *11*, 3445–3458. [CrossRef] [PubMed]
43. Schneiter, R.; Tatzer, V.; Gogg, G.; Leitner, E.; Kohlwein, S.D. Elo1p-dependent carboxy-terminal elongation of C14:1Delta(9) to C16:1Delta(11) fatty acids in Saccharomyces cerevisiae. *J. Bacteriol.* **2000**, *182*, 3655–3660. [CrossRef] [PubMed]
44. Haro, R.; Garciadeblas, B.; Rodríguez-Navarro, A. A novel P-type ATPase from yeast involved in sodium transport. *FEBS Lett.* **1991**, *291*, 189–191. [CrossRef]
45. Mori, K.; Kawahara, T.; Yoshida, H.; Yanagi, H.; Yura, T. Signalling from endoplasmic reticulum to nucleus: Transcription factor with a basic-leucine zipper motif is required for the unfolded protein-response pathway. *Genes Cells* **1996**, *1*, 803–817. [CrossRef] [PubMed]
46. Brewster, J.L.; de Valoir, T.; Dwyer, N.D.; Winter, E.; Gustin, M.C. An osmosensing signal transduction pathway in yeast. *Science* **1993**, *259*, 1760–1763. [CrossRef]
47. Wiederrecht, G.; Seto, D.; Parker, C.S. Isolation of the gene encoding the S. cerevisiae heat shock transcription factor. *Cell* **1988**, *54*, 841–853. [CrossRef]
48. Nikawa, J.; Yamashita, S. IRE1 encodes a putative protein kinase containing a membrane-spanning domain and is required for inositol phototrophy in Saccharomyces cerevisiae. *Mol. Microbiol.* **1992**, *6*, 1441–1446. [CrossRef]
49. Tokunaga, M.; Kawamura, A.; Kohno, K. Purification and characterization of BiP/Kar2 protein from Saccharomyces cerevisiae. *J. Biol. Chem.* **1992**, *267*, 17553–17559.
50. Martinez-Pastor, M.T.; Marchler, G.; Schuller, C.; Marchler-Bauer, A.; Ruis, H.; Estruch, F. The Saccharomyces cerevisiae zinc finger proteins Msn2p and Msn4p are required for transcriptional induction through the stress response element (STRE). *EMBO J.* **1996**, *15*, 2227–2235. [CrossRef]
51. Stukey, J.E.; McDonough, V.M.; Martin, C.E. The OLE1 gene of Saccharomyces cerevisiae encodes the delta 9 fatty acid desaturase and can be functionally replaced by the rat stearoyl-CoA desaturase gene. *J. Biol. Chem.* **1990**, *265*, 20144–20149.
52. Serrano, R.; Kielland-Brandt, M.C.; Fink, G.R. Yeast plasma membrane ATPase is essential for growth and has homology with (Na+ + K+), K+- and Ca2+-ATPases. *Nature* **1986**, *319*, 689–693. [CrossRef] [PubMed]
53. Serrano, R. Characterization of the plasma membrane ATPase of Saccharomyces cerevisiae. *Mol. Cell. Biochem.* **1978**, *22*, 51–63. [CrossRef] [PubMed]
54. Li, W.; Brandriss, M.C. Proline biosynthesis in Saccharomyces cerevisiae: Molecular analysis of the PRO1 gene, which encodes gamma-glutamyl kinase. *J. Bacteriol.* **1992**, *174*, 4148–4156. [CrossRef] [PubMed]
55. Tomenchok, D.M.; Brandriss, M.C. Gene-enzyme relationships in the proline biosynthetic pathway of Saccharomyces cerevisiae. *J. Bacteriol.* **1987**, *169*, 5364–5372. [CrossRef] [PubMed]

56. Brandriss, M.C.; Falvey, D.A. Proline biosynthesis in Saccharomyces cerevisiae: Analysis of the PRO3 gene, which encodes delta 1-pyrroline-5-carboxylate reductase. *J. Bacteriol.* **1992**, *174*, 3782–3788. [CrossRef] [PubMed]

57. Watanabe, Y.; Takaesu, G.; Hagiwara, M.; Irie, K.; Matsumoto, K. Characterization of a serum response factor-like protein in Saccharomyces cerevisiae, Rlm1, which has transcriptional activity regulated by the Mpk1 (Slt2) mitogen-activated protein kinase pathway. *Mol. Cell. Biol.* **1997**, *17*, 2615–2623. [CrossRef] [PubMed]

58. Dodou, E.; Treisman, R. The Saccharomyces cerevisiae MADS-box transcription factor Rlm1 is a target for the Mpk1 mitogen-activated protein kinase pathway. *Mol. Cell. Biol.* **1997**, *17*, 1848–1859. [CrossRef] [PubMed]

59. Torres, L.; Martín, H.; García-Saez, M.I.; Arroyo, J.; Molina, M.; Sánchez, M.; Nombela, C. A protein kinase gene complements the lytic phenotype of Saccharomyces cerevisiae lyt2 mutants. *Mol. Microbiol.* **1991**, *5*, 2845–2854. [CrossRef]

60. Bell, W.; Klaassen, P.; Ohnacker, M.; Boller, T.; Herweijer, M.; Schoppink, P.; Van der Zee, P.; Wiemken, A. Characterization of the 56-kDa subunit of yeast trehalose-6-phosphate synthase and cloning of its gene reveal its identity with the product of CIF1, a regulator of carbon catabolite inactivation. *Eur. J. Biochem.* **1992**, *209*, 951–959. [CrossRef]

61. De Virgilio, C.; Bürckert, N.; Bell, W.; Jenö, P.; Boller, T.; Wiemken, A. Disruption of TPS2, the gene encoding the 100-kDa subunit of the trehalose-6-phosphate synthase/phosphatase complex in Saccharomyces cerevisiae, causes accumulation of trehalose-6-phosphate and loss of trehalose-6-phosphate phosphatase activity. *Eur. J. Biochem.* **1993**, *212*, 315–323. [CrossRef]

62. Phizicky, E.M.; Schwartz, R.C.; Abelson, J. Saccharomyces cerevisiae tRNA ligase. Purification of the protein and isolation of the structural gene. *J. Biol. Chem.* **1986**, *261*, 2978–2986. [PubMed]

63. Xu, Q.; Teplow, D.; Lee, T.D.; Abelson, J. Domain structure in yeast tRNA ligase. *Biochemistry* **1990**, *29*, 6132–6138. [CrossRef] [PubMed]

64. Causton, H.C.; Ren, B.; Koh, S.S.; Harbison, C.T.; Kanin, E.; Jennings, E.G.; Lee, T.I.; True, H.L.; Lander, E.S.; Young, R.A. Remodeling of yeast genome expression in response to environmental changes. *Mol. Biol. Cell.* **2001**, *12*, 323–337. [CrossRef] [PubMed]

65. Piper, P.W. The heat shock and ethanol stress responses of yeast exhibit extensive similarity and functional overlap. *FEMS Microbiol. Lett.* **1995**, *134*, 121–127. [CrossRef] [PubMed]

66. Schüller, C.; Brewster, J.; Alexander, M.; Gustin, M.; Ruis, H. The HOG pathway controls osmotic regulation of transcription via the stress response element (STRE) of the Saccharomyces cerevisiae CTT1 gene. *EMBO J.* **1994**, *13*, 4382–4389. [CrossRef] [PubMed]

67. Moskvina, E.; Imre, E.M.; Ruis, H. Stress factors acting at the level of the plasma membrane induce transcription via the stress response element (STRE) of the yeast Saccharomyces cerevisiae. *Mol. Microbiol.* **1999**, *32*, 1263–1272. [CrossRef]

68. Moskvina, E.; Schüller, C.; Maurer, C.; Mager, W.; Ruis, H. A search in the genome of Saccharomyces cerevisiae for genes regulated via stress response elements. *Yeast* **1998**, *14*, 1041–1050. [CrossRef]

69. Ding, J.; Huang, X.; Zhang, L.; Zhao, N.; Yang, D.; Zhang, K. Tolerance and stress response to ethanol in the yeast Saccharomyces cerevisiae. *Appl. Microbiol. Biotechnol.* **2009**, *85*, 253. [CrossRef]

70. Crawford, R.A.; Pavitt, G.D. Translational regulation in response to stress in Saccharomyces cerevisiae. *Yeast* **2019**, *36*, 5–21. [CrossRef]

71. Alexandre, H.; Ansanay-Galeote, V.; Dequin, S.; Blondin, B. Global gene expression during short-term ethanol stress in Saccharomyces cerevisiae. *FEBS Lett.* **2001**, *498*, 98–103. [CrossRef]

72. Hallsworth, J.E. Stress-free microbes lack vitality. *Fungal Biol.* **2018**, *122*, 379–385. [CrossRef] [PubMed]

73. Aertsen, A.; Michiels, C.W. Diversify or die: Generation of diversity in response to stress. *Crit. Rev. Microbiol.* **2005**, *31*, 69–78. [CrossRef] [PubMed]

74. Estruch, F. Stress-controlled transcription factors, stress-induced genes and stress tolerance in budding yeast. *FEMS Microbiol. Rev.* **2000**, *24*, 469–486. [CrossRef] [PubMed]

75. Rangel, D.E.N.; Finlay, R.D.; Hallsworth, J.E.; Dadachova, E.; Gadd, G.M. Fungal strategies for dealing with environment- and agriculture-induced stresses. *Fungal Biol.* **2018**, *122*, 602–612. [CrossRef] [PubMed]

76. Stanley, D.; Bandara, A.; Fraser, S.; Chambers, P.; Stanley, G.A. The ethanol stress response and ethanol tolerance of Saccharomyces cerevisiae. *J. Appl. Microbiol.* **2010**, *109*, 13–24. [CrossRef] [PubMed]

77. Endo, A.; Nakamura, T.; Ando, A.; Tokuyasu, K.; Shima, J. Genome-wide screening of the genes required for tolerance to vanillin, which is a potential inhibitor of bioethanol fermentation, in Saccharomyces cerevisiae. *Biotechnol. Biofuels* **2008**, *1*, 3. [CrossRef] [PubMed]
78. Swinnen, S.; Schaerlaekens, K.; Pais, T.; Claesen, J.; Hubmann, G.; Yang, Y.; Demeke, M.; Foulquié-Moreno, M.R.; Goovaerts, A.; Souvereyns, K.; et al. Identification of novel causative genes determining the complex trait of high ethanol tolerance in yeast using pooled-segregant whole-genome sequence analysis. *Genome Res.* **2012**, *22*, 975–984. [CrossRef] [PubMed]
79. Hu, X.; Wang, M.; Tan, T.; Li, J.; Yang, H.; Leach, L.; Zhang, R.; Luo, Z. Genetic dissection of ethanol tolerance in the budding yeast Saccharomyces cerevisiae. *Genetics* **2007**, *175*, 1479–1487. [CrossRef]
80. Haas, R.; Horev, G.; Lipkin, E.; Kesten, I.; Portnoy, M.; Buhnik-Rosenblau, K.; Soller, M.; Kashi, Y. Mapping Ethanol Tolerance in Budding Yeast Reveals High Genetic Variation in a Wild Isolate. *Front. Genet.* **2019**, *10*, 998. [CrossRef]
81. Liu, R.; Liang, L.; Choudhury, A.; Garst, A.D.; Eckert, C.A.; Oh, E.J.; Winkler, J.; Gill, R.T. Multiplex navigation of global regulatory networks (MINR) in yeast for improved ethanol tolerance and production. *Metab. Eng.* **2019**, *51*, 50–58. [CrossRef]
82. Kasavi, C.; Eraslan, S.; Arga, K.Y.; Oner, E.T.; Kirdar, B. A system based network approach to ethanol tolerance in Saccharomyces cerevisiae. *BMC Syst. Biol.* **2014**, *8*, 90. [CrossRef] [PubMed]
83. Ismail, A.; Ali, A. Selection of high ethanol-yielding Saccharomyces. *Folia Microbiol.* **1971**, *16*, 350–354. [CrossRef] [PubMed]
84. Brown, S.; Oliver, S. Isolation of ethanol-tolerant mutants of yeast by continuous selection. *Eur. J. Appl. Microbiol. Biotechnol.* **1982**, *16*, 119–122. [CrossRef]
85. Jiménez, J.; Benítez, T. Selection of ethanol-tolerant yeast hybrids in pH-regulated continuous culture. *Appl. Environ. Microbiol.* **1988**, *54*, 917–922. [CrossRef] [PubMed]
86. Madhavan, A.; Sindhu, R.; Arun, K.; Pandey, A.; Binod, P.; Gnansounou, E. Advances in Biofuel Production by Strain Development in Yeast from Lignocellulosic Biomass. *Bioprocess. Biomol. Prod.* **2019**, 289–302.
87. Miyagawa, K.-I.; Ishiwata-Kimata, Y.; Kohno, K.; Kimata, Y. Ethanol stress impairs protein folding in the endoplasmic reticulum and activates Ire1 in Saccharomyces cerevisiae. *Biosci. Biotechnol. Biochem.* **2014**, *78*, 1389–1391. [CrossRef] [PubMed]
88. Navarro-Tapia, E.; Nana, R.K.; Querol, A.; Pérez-Torrado, R. Ethanol cellular defense induce unfolded protein response in yeast. *Front. Microbiol.* **2016**, *7*, 189. [CrossRef]
89. Alexandre, H.; Rousseaux, I.; Charpentier, C. Relationship between ethanol tolerance, lipid composition and plasma membrane fluidity in Saccharomyces cerevisiae and Kloeckera apiculata. *FEMS Microbiol. Lett.* **1994**, *124*, 17–22. [CrossRef]
90. Vanegas, J.M.; Contreras, M.F.; Faller, R.; Longo, M.L. Role of unsaturated lipid and ergosterol in ethanol tolerance of model yeast biomembranes. *Biophys. J.* **2012**, *102*, 507–516. [CrossRef]
91. Shi, D.-J.; Wang, C.-L.; Wang, K.-M. Genome shuffling to improve thermotolerance, ethanol tolerance and ethanol productivity of Saccharomyces cerevisiae. *J. Ind. Microbiol. Biotechnol.* **2009**, *36*, 139–147. [CrossRef]
92. Alper, H.; Moxley, J.; Nevoigt, E.; Fink, G.R.; Stephanopoulos, G. Engineering yeast transcription machinery for improved ethanol tolerance and production. *Science* **2006**, *314*, 1565–1568. [CrossRef] [PubMed]
93. Snoek, T.; Nicolino, M.P.; Van den Bremt, S.; Mertens, S.; Saels, V.; Verplaetse, A.; Steensels, J.; Verstrepen, K.J. Large-scale robot-assisted genome shuffling yields industrial Saccharomyces cerevisiae yeasts with increased ethanol tolerance. *Biotechnol. Biofuels* **2015**, *8*, 32. [CrossRef] [PubMed]
94. Xue, T.; Chen, D.; Su, Q.; Yuan, X.; Liu, K.; Huang, L.; Fang, J.; Chen, J.; He, W.; Chen, Y. Improved ethanol tolerance and production of Saccharomyces cerevisiae by global transcription machinery engineering via directed evolution of the SPT8 gene. *Food Biotechnol.* **2019**, *33*, 155–173. [CrossRef]
95. Jetti, K.D.; GNS, R.R.; Garlapati, D.; Nammi, S.K. Improved ethanol productivity and ethanol tolerance through genome shuffling of Saccharomyces cerevisiae and Pichia stipitis. *Int. Microbiol.* **2019**, *22*, 247–254. [CrossRef] [PubMed]
96. Mitsui, R.; Yamada, R.; Ogino, H. Improved stress tolerance of Saccharomyces cerevisiae by CRISPR-Cas-mediated genome evolution. *Appl. Biochem. Biotechnol.* **2019**, *189*, 810–821. [CrossRef] [PubMed]
97. Liu, K.; Yuan, X.; Liang, L.; Fang, J.; Chen, Y.; He, W.; Xue, T. Using CRISPR/Cas9 for multiplex genome engineering to optimize the ethanol metabolic pathway in Saccharomyces cerevisiae. *Biochem. Eng. J.* **2019**, *145*, 120–126. [CrossRef]

98. Brown, A.D. Compatible solutes and extreme water stress in eukaryotic micro-organisms. *Adv. Microb. Physiol.* **1978**, *17*, 181–242.
99. Hallsworth, J.E.; Prior, B.A.; Nomura, Y.; Iwahara, M.; Timmis, K.N. Compatible solutes protect against chaotrope (ethanol)-induced, nonosmotic water stress. *Appl. Environ. Microbiol.* **2003**, *69*, 7032–7034. [CrossRef]
100. Cray, J.A.; Russell, J.T.; Timson, D.J.; Singhal, R.S.; Hallsworth, J.E. A universal measure of chaotropicity and kosmotropicity. *Environ. Microbiol.* **2013**, *15*, 287–296. [CrossRef]
101. Timson, D.J. The roles and applications of chaotropes and kosmotropes in industrial fermentation processes. *World J. Microbiol. Biotechnol.* **2020**, *36*, 89. [CrossRef]
102. McCammick, E.M.; Gomase, V.S.; McGenity, T.J.; Timson, D.J.; Hallsworth, J.E. Water-Hydrophobic Compound Interactions with the Microbial Cell. In *Handbook of Hydrocarbon and Lipid Microbiology*; Timmis, K.N., Ed.; Springer: Berlin/Heidelberg, Germany, 2010; pp. 1451–1466. [CrossRef]
103. Singer, M.A.; Lindquist, S. Multiple Effects of Trehalose on Protein Folding In Vitro and In Vivo. *Mol. Cell* **1998**, *1*, 639–648. [CrossRef]
104. Lever, M.; Blunt, J.; Maclagan, R. Some ways of looking at compensatory kosmotropes and different water environments. *Comp. Biochem. Physiol. Part A Mol. Integr. Physiol.* **2001**, *130*, 471–486. [CrossRef]
105. Moelbert, S.; Normand, B.; De Los Rios, P. Kosmotropes and chaotropes: Modelling preferential exclusion, binding and aggregate stability. *Biophys. Chem.* **2004**, *112*, 45–57. [CrossRef] [PubMed]
106. Hohmann, S. Osmotic stress signaling and osmoadaptation in yeasts. *Microbiol. Mol. Biol. Rev.* **2002**, *66*, 300–372. [CrossRef] [PubMed]
107. Nevoigt, E.; Stahl, U. Osmoregulation and glycerol metabolism in the yeast Saccharomyces cerevisiae. *FEMS Microbiol. Rev.* **1997**, *21*, 231–241. [CrossRef] [PubMed]
108. Mager, W.H.; de Boer, A.H.; Siderius, M.H.; Voss, H.-P. Cellular responses to oxidative and osmotic stress. *Cell Stress Chaperones* **2000**, *5*, 73. [CrossRef]
109. Takagi, H. Proline as a stress protectant in yeast: Physiological functions, metabolic regulations, and biotechnological applications. *Appl. Microbiol. Biotechnol.* **2008**, *81*, 211–223. [CrossRef]
110. Francois, J.; Parrou, J.L. Reserve carbohydrates metabolism in the yeast Saccharomyces cerevisiae. *FEMS Microbiol. Rev.* **2001**, *25*, 125–145. [CrossRef]
111. Winderickx, J.; de Winde, J.H.; Crauwels, M.; Hino, A.; Hohmann, S.; Van Dijck, P.; Thevelein, J.M. Regulation of genes encoding subunits of the trehalose synthase complex in Saccharomyces cerevisiae: Novel variations of STRE-mediated transcription control? *Mol. Gen. Genet.* **1996**, *252*, 470–482. [CrossRef]
112. Gorner, W.; Durchschlag, E.; Martinez-Pastor, M.T.; Estruch, F.; Ammerer, G.; Hamilton, B.; Ruis, H.; Schuller, C. Nuclear localization of the C2H2 zinc finger protein Msn2p is regulated by stress and protein kinase A activity. *Genes. Dev.* **1998**, *12*, 586–597. [CrossRef]
113. Conlin, L.K.; Nelson, H.C.M. The Natural Osmolyte Trehalose Is a Positive Regulator of the Heat-Induced Activity of Yeast Heat Shock Transcription Factor. *Mol. Cell. Biol.* **2007**, *27*, 1505–1515. [CrossRef] [PubMed]
114. Wu, H.; Ng, B.S.; Thibault, G. Endoplasmic reticulum stress response in yeast and humans. *Biosci. Rep.* **2014**, *34*. [CrossRef] [PubMed]
115. Levin, D.E. Cell wall integrity signaling in Saccharomyces cerevisiae. *Microbiol. Mol. Biol. Rev.* **2005**, *69*, 262–291. [CrossRef] [PubMed]
116. Gonzalez, T.N.; Sidrauski, C.; Dörfler, S.; Walter, P. Mechanism of non-spliceosomal mRNA splicing in the unfolded protein response pathway. *EMBO J.* **1999**, *18*, 3119–3132. [CrossRef] [PubMed]
117. Cox, J.S.; Shamu, C.E.; Walter, P. Transcriptional induction of genes encoding endoplasmic reticulum resident proteins requires a transmembrane protein kinase. *Cell* **1993**, *73*, 1197–1206. [CrossRef]
118. Kimata, Y.; Kimata, Y.I.; Shimizu, Y.; Abe, H.; Farcasanu, I.C.; Takeuchi, M.; Rose, M.D.; Kohno, K. Genetic evidence for a role of BiP/Kar2 that regulates Ire1 in response to accumulation of unfolded proteins. *Mol. Biol. Cell* **2003**, *14*, 2559–2569. [CrossRef]
119. Okamura, K.; Kimata, Y.; Higashio, H.; Tsuru, A.; Kohno, K. Dissociation of Kar2p/BiP from an ER sensory molecule, Ire1p, triggers the unfolded protein response in yeast. *Biochem. Biophys. Res. Commun.* **2000**, *279*, 445–450. [CrossRef]
120. Sidrauski, C.; Walter, P. The transmembrane kinase Ire1p is a site-specific endonuclease that initiates mRNA splicing in the unfolded protein response. *Cell* **1997**, *90*, 1031–1039. [CrossRef]

121. Gardner, B.M.; Walter, P. Unfolded proteins are Ire1-activating ligands that directly induce the unfolded protein response. *Science* **2011**, *333*, 1891–1894. [CrossRef]
122. Ho, N.; Yap, W.S.; Xu, J.; Wu, H.; Koh, J.H.; Goh, W.W.B.; George, B.; Chong, S.C.; Taubert, S.; Thibault, G. Stress sensor Ire1 deploys a divergent transcriptional program in response to lipid bilayer stress. *J. Cell Biol.* **2020**, *219*. [CrossRef]
123. Sidrauski, C.; Cox, J.S.; Walter, P. tRNA ligase is required for regulated mRNA splicing in the unfolded protein response. *Cell* **1996**, *87*, 405–413. [CrossRef]
124. Mori, K.; Ogawa, N.; Kawahara, T.; Yanagi, H.; Yura, T. Palindrome with spacer of one nucleotide is characteristic of the cis-acting unfolded protein response element in Saccharomyces cerevisiae. *J. Biol. Chem.* **1998**, *273*, 9912–9920. [CrossRef] [PubMed]
125. Welihinda, A.A.; Tirasophon, W.; Green, S.R.; Kaufman, R.J. Gene induction in response to unfolded protein in the endoplasmic reticulum is mediated through Ire1p kinase interaction with a transcriptional coactivator complex containing Ada5p. *Proc. Natl. Acad. Sci. USA* **1997**, *94*, 4289–4294. [CrossRef] [PubMed]
126. Moir, R.D.; Gross, D.A.; Silver, D.L.; Willis, I.M. SCS3 and YFT2 link transcription of phospholipid biosynthetic genes to ER stress and the UPR. *PLoS Genet.* **2012**, *8*, e1002890. [CrossRef] [PubMed]
127. Travers, K.J.; Patil, C.K.; Wodicka, L.; Lockhart, D.J.; Weissman, J.S.; Walter, P. Functional and genomic analyses reveal an essential coordination between the unfolded protein response and ER-associated degradation. *Cell* **2000**, *101*, 249–258. [CrossRef]
128. Thibault, G.; Ismail, N.; Ng, D.T. The unfolded protein response supports cellular robustness as a broad-spectrum compensatory pathway. *Proc. Natl. Acad. Sci. USA* **2011**, *108*, 20597–20602. [CrossRef]
129. Fun, X.H.; Thibault, G. Lipid bilayer stress and proteotoxic stress-induced unfolded protein response deploy divergent transcriptional and non-transcriptional programmes. *Biochim. Biophys. Acta (BBA)-Mol. Cell Biol. Lipids* **2020**, *1865*. [CrossRef]
130. Krysan, D.J. The cell wall and endoplasmic reticulum stress responses are coordinately regulated in Saccharomyces cerevisiae. *Commun. Integr. Biol.* **2009**, *2*, 233–235. [CrossRef]
131. Chowdhury, S.; Smith, K.W.; Gustin, M.C. Osmotic stress and the yeast cytoskeleton: Phenotype-specific suppression of an actin mutation. *J. Cell Biol.* **1992**, *118*, 561–571. [CrossRef]
132. Mas, G.; De Nadal, E.; Dechant, R.; De La Concepción, M.L.R.; Logie, C.; Jimeno-González, S.; Chávez, S.; Ammerer, G.; Posas, F. Recruitment of a chromatin remodelling complex by the Hog1 MAP kinase to stress genes. *EMBO J.* **2009**, *28*, 326–336. [CrossRef]
133. Silva, R.D.; Sotoca, R.; Johansson, B.; Ludovico, P.; Sansonetty, F.; Silva, M.T.; Peinado, J.M.; Côrte-Real, M. Hyperosmotic stress induces metacaspase-and mitochondria-dependent apoptosis in Saccharomyces cerevisiae. *Mol. Microbiol.* **2005**, *58*, 824–834. [CrossRef] [PubMed]
134. Rodríguez-Peña, J.M.; García, R.; Nombela, C.; Arroyo, J. The high-osmolarity glycerol (HOG) and cell wall integrity (CWI) signalling pathways interplay: A yeast dialogue between MAPK routes. *Yeast* **2010**, *27*, 495–502. [CrossRef] [PubMed]
135. Jendretzki, A.; Wittland, J.; Wilk, S.; Straede, A.; Heinisch, J.J. How do I begin? Sensing extracellular stress to maintain yeast cell wall integrity. *Eur. J. Cell Biol.* **2011**, *90*, 740–744. [CrossRef] [PubMed]
136. Sanz, A.B.; García, R.; Rodríguez-Peña, J.M.; Arroyo, J. The CWI pathway: Regulation of the transcriptional adaptive response to cell wall stress in yeast. *J. Fungi* **2018**, *4*, 1. [CrossRef] [PubMed]
137. Levin, D.E. Regulation of cell wall biogenesis in Saccharomyces cerevisiae: The cell wall integrity signaling pathway. *Genetics* **2011**, *189*, 1145–1175. [CrossRef]
138. Baetz, K.; Moffat, J.; Haynes, J.; Chang, M.; Andrews, B. Transcriptional coregulation by the cell integrity mitogen-activated protein kinase Slt2 and the cell cycle regulator Swi4. *Mol. Cell. Biol.* **2001**, *21*, 6515–6528. [CrossRef]
139. Kim, K.-Y.; Truman, A.W.; Levin, D.E. Yeast Mpk1 mitogen-activated protein kinase activates transcription through Swi4/Swi6 by a noncatalytic mechanism that requires upstream signal. *Mol. Cell. Biol.* **2008**, *28*, 2579–2589. [CrossRef]
140. Kim, K.-Y.; Truman, A.W.; Caesar, S.; Schlenstedt, G.; Levin, D.E. Yeast Mpk1 cell wall integrity mitogen-activated protein kinase regulates nucleocytoplasmic shuttling of the Swi6 transcriptional regulator. *Mol. Biol. Cell* **2010**, *21*, 1609–1619. [CrossRef]
141. Jung, U.S.; Levin, D.E. Genome-wide analysis of gene expression regulated by the yeast cell wall integrity signalling pathway. *Mol. Microbiol.* **1999**, *34*, 1049–1057. [CrossRef]

142. Lagorce, A.; Hauser, N.C.; Labourdette, D.; Rodriguez, C.; Martin-Yken, H.; Arroyo, J.; Hoheisel, J.D.; François, J. Genome-wide analysis of the response to cell wall mutations in the yeast Saccharomyces cerevisiae. *J. Biol. Chem.* **2003**, *278*, 20345–20357. [CrossRef]
143. O'Rourke, S.M.; Herskowitz, I.; O'Shea, E.K. Yeast go the whole HOG for the hyperosmotic response. *Trends Genet.* **2002**, *18*, 405–412. [CrossRef]
144. Posas, F.; Chambers, J.R.; Heyman, J.A.; Hoeffler, J.P.; de Nadal, E.; Ariño, J.n. The transcriptional response of yeast to saline stress. *J. Biol. Chem.* **2000**, *275*, 17249–17255. [CrossRef] [PubMed]
145. Rep, M.; Krantz, M.; Thevelein, J.M.; Hohmann, S. The transcriptional response of Saccharomyces cerevisiae to osmotic shock Hot1p and Msn2p/Msn4p are required for the induction of subsets of high osmolarity glycerol pathway-dependent genes. *J. Biol. Chem.* **2000**, *275*, 8290–8300. [CrossRef] [PubMed]
146. Alexander, M.R.; Tyers, M.; Perret, M.; Craig, B.M.; Fang, K.S.; Gustin, M.C. Regulation of cell cycle progression by Swe1p and Hog1p following hypertonic stress. *Mol. Biol. Cell* **2001**, *12*, 53–62. [CrossRef] [PubMed]
147. Beese, S.E.; Negishi, T.; Levin, D.E. Identification of positive regulators of the yeast fps1 glycerol channel. *PLoS Genet* **2009**, *5*, e1000738. [CrossRef]
148. Lee, J.; Reiter, W.; Dohnal, I.; Gregori, C.; Beese-Sims, S.; Kuchler, K.; Ammerer, G.; Levin, D.E. MAPK Hog1 closes the S. cerevisiae glycerol channel Fps1 by phosphorylating and displacing its positive regulators. *Genes Dev.* **2013**, *27*, 2590–2601. [CrossRef] [PubMed]
149. Auesukaree, C. Molecular mechanisms of the yeast adaptive response and tolerance to stresses encountered during ethanol fermentation. *J. Biosci. Bioeng.* **2017**, *124*, 133–142. [CrossRef]
150. Fleet, G.M. Yeasts-growth during fermentation. In *Wine Microbiology & Biotechnology*; Harword Academic Publishers: Amsterdam, The Netherlands, 1993; pp. 27–54.
151. Van Uden, N. Effect of alcohols on the temperature relations of growth and death in yeasts. In *Alcohol Toxicity in Yeast and Bacteria*; CRC Press: Boca Raton, FL, USA, 1989; pp. 77–88.
152. Verghese, J.; Abrams, J.; Wang, Y.; Morano, K.A. Biology of the heat shock response and protein chaperones: Budding yeast (Saccharomyces cerevisiae) as a model system. *Microbiol. Mol. Biol. Rev.* **2012**, *76*, 115–158. [CrossRef]
153. Swan, T.M.; Watson, K. Stress tolerance in a yeast lipid mutant: Membrane lipids influence tolerance to heat and ethanol independently of heat shock proteins and trehalose. *Can. J. Microbiol.* **1999**, *45*, 472–479. [CrossRef]
154. Swan, T.M.; Watson, K. Stress tolerance in a yeast sterol auxotroph: Role of ergosterol, heat shock proteins and trehalose. *FEMS Microbiol. Lett.* **1998**, *169*, 191–197. [CrossRef]
155. Endress, E.; Bayerl, S.; Prechtel, K.; Maier, C.; Merkel, R.; Bayerl, T.M. The effect of cholesterol, lanosterol, and ergosterol on lecithin bilayer mechanical properties at molecular and microscopic dimensions: A solid-state NMR and micropipet study. *Langmuir* **2002**, *18*, 3293–3299. [CrossRef]
156. Hsueh, Y.-W.; Gilbert, K.; Trandum, C.; Zuckermann, M.; Thewalt, J. The Effect of Ergosterol on Dipalmitoylphosphatidylcholine Bilayers: A Deuterium NMR and Calorimetric Study. *Biophys. J.* **2005**, *88*, 1799–1808. [CrossRef] [PubMed]
157. Xu, X.; Bittman, R.; Duportail, G.; Heissler, D.; Vilcheze, C.; London, E. Effect of the structure of natural sterols and sphingolipids on the formation of ordered sphingolipid/sterol domains (rafts) Comparison of cholesterol to plant, fungal, and disease-associated sterols and comparison of sphingomyelin, cerebrosides, and ceramide. *J. Biol. Chem.* **2001**, *276*, 33540–33546. [PubMed]
158. Caspeta, L.; Chen, Y.; Ghiaci, P.; Feizi, A.; Buskov, S.; Hallstrom, B.M.; Petranovic, D.; Nielsen, J. Biofuels. Altered sterol composition renders yeast thermotolerant. *Science* **2014**, *346*, 75–78. [CrossRef] [PubMed]
159. Abdel-Fattah, W.; Fadil, M.; Nigam, P.; Banat, I. Isolation of thermotolerant ethanologenic yeasts and use of selected strains in industrial scale fermentation in an Egyptian distillery. *Biotechnol. Bioeng.* **2000**, *68*, 531–535. [CrossRef]
160. Rajoka, M.; Ferhan, M.; Khalid, A. Kinetics and thermodynamics of ethanol production by a thermotolerant mutant of Saccharomyces cerevisiae in a microprocessor-controlled bioreactor. *Lett. Appl. Microbiol.* **2005**, *40*, 316–321. [CrossRef] [PubMed]
161. Banat, I.; Marchant, R. Characterization and potential industrial applications of five novel, thermotolerant, fermentative, yeast strains. *World J. Microbiol. Biotechnol.* **1995**, *11*, 304–306. [CrossRef]

162. Piper, P.W.; Talreja, K.; Panaretou, B.; Moradas-Ferreira, P.; Byrne, K.; Praekelt, U.M.; Meacock, P.; Récnacq, M.; Boucherie, H. Induction of major heat-shock proteins of Saccharomyces cerevisiae, including plasma membrane Hsp30, by ethanol levels above a critical threshold. *Microbiology* **1994**, *140*, 3031–3038. [CrossRef]
163. Benjaphokee, S.; Hasegawa, D.; Yokota, D.; Asvarak, T.; Auesukaree, C.; Sugiyama, M.; Kaneko, Y.; Boonchird, C.; Harashima, S. Highly efficient bioethanol production by a Saccharomyces cerevisiae strain with multiple stress tolerance to high temperature, acid and ethanol. *New Biotechnol.* **2012**, *29*, 379–386. [CrossRef]
164. Birch, R.M.; Walker, G.M. Influence of magnesium ions on heat shock and ethanol stress responses of Saccharomyces cerevisiae. *Enzyme Microb. Technol.* **2000**, *26*, 678–687. [CrossRef]
165. Ball, P.; Hallsworth, J.E. Water structure and chaotropicity: Their uses, abuses and biological implications. *Phys. Chem. Chem. Phys.* **2015**, *17*, 8297–8305. [CrossRef] [PubMed]
166. Abu-Hamdiyyah, M. The Effect of Urea on the Structure of Water and Hydrophobic Bonding1. *J. Phys. Chem.* **1965**, *69*, 2720–2725. [CrossRef]
167. Sahle, C.J.; Schroer, M.A.; Juurinen, I.; Niskanen, J. Influence of TMAO and urea on the structure of water studied by inelastic X-ray scattering. *Phys. Chem. Chem. Phys.* **2016**, *18*, 16518–16526. [CrossRef] [PubMed]
168. Timson, D.J. Four Challenges for Better Biocatalysts. *Fermentation* **2019**, *5*, 39. [CrossRef]
169. Geiler-Samerotte, K.A.; Dion, M.F.; Budnik, B.A.; Wang, S.M.; Hartl, D.L.; Drummond, D.A. Misfolded proteins impose a dosage-dependent fitness cost and trigger a cytosolic unfolded protein response in yeast. *Proc. Natl. Acad. Sci. USA* **2011**, *108*, 680–685. [CrossRef] [PubMed]
170. Telini, B.P.; Menoncin, M.; Bonatto, D. Does Inter-Organellar Proteostasis Impact Yeast Quality and Performance During Beer Fermentation? *Front. Genet.* **2020**, *11*, 2. [CrossRef]
171. Tran, D.M.; Ishiwata-Kimata, Y.; Mai, T.C.; Kubo, M.; Kimata, Y. The unfolded protein response alongside the diauxic shift of yeast cells and its involvement in mitochondria enlargement. *Sci. Rep.* **2019**, *9*, 12780. [CrossRef] [PubMed]
172. Hallsworth, J.E. Ethanol-induced water stress in yeast. *J. Ferment. Bioeng.* **1998**, *85*, 125–137. [CrossRef]
173. De Klerk, C.; Fosso-Kankeu, E.; Du Plessis, L.; Marx, S. Assessment of the viability of Saccharomyces cerevisiae in response to synergetic inhibition during bioethanol production. *Curr. Sci.* **2018**, *115*. [CrossRef]
174. Fosso-Kankeu, E.; Marx, S.; Meyer, A. Simulated inhibitory effects of typical byproducts of biomass pretreatment process on the viability of Saccharomyces cerevisiae and bioethanol production yield. *Afr. J. Biotechnol.* **2015**, *14*, 2383–2394.
175. Eardley, J.; Dedi, C.; Dymond, M.; Hallsworth, J.E.; Timson, D.J. Evidence for chaotropicity/kosmotropicity offset in a yeast growth model. *Biotechnol. Lett.* **2019**, *41*, 1309–1318. [CrossRef] [PubMed]
176. Arakawa, T.; Timasheff, S.N. The stabilization of proteins by osmolytes. *Biophys. J.* **1985**, *47*, 411–414. [CrossRef]
177. Gekko, K.; Timasheff, S.N. Thermodynamic and kinetic examination of protein stabilization by glycerol. *Biochemistry* **1981**, *20*, 4677–4686. [CrossRef] [PubMed]
178. Gekko, K. Calorimetric Study on Thermal Denaturation of Lysozyme in Polyol-Water Mixtures. *J. Biochem.* **1982**, *91*, 1197–1204. [CrossRef] [PubMed]
179. Timson, D.J.; Eardley, J. Destressing yeast for higher biofuel yields: Can Excess Chaotropicity Be Mitigated? *Appl. Biochem. Biotechnol.* **2020**, *192*, 1368–1375. [CrossRef] [PubMed]
180. Jamieson, D.J. Oxidative stress responses of the yeast Saccharomyces cerevisiae. *Yeast* **1998**, *14*, 1511–1527. [CrossRef]
181. Garre, E.; Raginel, F.; Palacios, A.; Julien, A.; Matallana, E. Oxidative stress responses and lipid peroxidation damage are induced during dehydration in the production of dry active wine yeasts. *Int. J. Food Microbiol.* **2010**, *136*, 295–303. [CrossRef]
182. Ikner, A.; Shiozaki, K. Yeast signaling pathways in the oxidative stress response. *Mutat. Res./Fundam. Mol. Mech. Mutagenesis* **2005**, *569*, 13–27. [CrossRef]
183. Moradas-Ferreira, P.; Costa, V.; Piper, P.; Mager, W. The molecular defences against reactive oxygen species in yeast. *Mol. Microbiol.* **1996**, *19*, 651–658. [CrossRef]
184. Girard, P.; Boiteux, S. Repair of oxidized DNA bases in the yeast Saccharomyces cerevisiae. *Biochimie* **1997**, *79*, 559–566. [CrossRef]
185. Pan, X.; Ye, P.; Yuan, D.S.; Wang, X.; Bader, J.S.; Boeke, J.D. A DNA integrity network in the yeast Saccharomyces cerevisiae. *Cell* **2006**, *124*, 1069–1081. [CrossRef] [PubMed]
186. Herrero, E.; Ros, J.; Bellí, G.; Cabiscol, E. Redox control and oxidative stress in yeast cells. *Biochim. Biophys. Acta (BBA)-Gen. Subj.* **2008**, *1780*, 1217–1235. [CrossRef] [PubMed]

187. Ariño, J. Integrative responses to high pH stress in S. cerevisiae. *Omics* **2010**, *14*, 517–523. [CrossRef] [PubMed]
188. Serra-Cardona, A.; Canadell, D.; Ariño, J. Coordinate responses to alkaline pH stress in budding yeast. *Microb. Cell* **2015**, *2*, 182–196. [CrossRef] [PubMed]
189. Viladevall, L.; Serrano, R.; Ruiz, A.; Domenech, G.; Giraldo, J.; Barcelo, A.; Arino, J. Characterization of the calcium-mediated response to alkaline stress in Saccharomyces cerevisiae. *J. Biol. Chem.* **2004**, *279*, 43614–43624. [CrossRef]
190. Stathopoulos, A.M.; Cyert, M.S. Calcineurin acts through the CRZ1/TCN1-encoded transcription factor to regulate gene expression in yeast. *Genes. Dev.* **1997**, *11*, 3432–3444. [CrossRef]
191. Casado, C.; González, A.; Platara, M.; Ruiz, A.; Ariño, J. The role of the protein kinase A pathway in the response to alkaline pH stress in yeast. *Biochem. J.* **2011**, *438*, 523–533. [CrossRef]
192. Canadell, D.; García-Martínez, J.; Alepuz, P.; Pérez-Ortín, J.E.; Ariño, J. Impact of high pH stress on yeast gene expression: A comprehensive analysis of mRNA turnover during stress responses. *Biochim. Biophys. Acta (BBA)-Gene Regul. Mech.* **2015**, *1849*, 653–664. [CrossRef]
193. Ariño, J.; Ramos, J.; Sychrova, H. Monovalent cation transporters at the plasma membrane in yeasts. *Yeast* **2019**, *36*, 177–193. [CrossRef]
194. Platara, M.; Ruiz, A.; Serrano, R.; Palomino, A.; Moreno, F.; Ariño, J. The Transcriptional Response of the Yeast Na+-ATPase ENA1 Gene to Alkaline Stress Involves Three Main Signaling Pathways. *J. Biol. Chem.* **2006**, *281*, 36632–36642. [CrossRef]
195. Benito, B.; Quintero, F.J.; Rodríguez-Navarro, A. Overexpression of the sodium ATPase of Saccharomyces cerevisiae: Conditions for phosphorylation from ATP and Pi. *Biochim. Biophys. Acta (BBA)-Biomembr.* **1997**, *1328*, 214–225. [CrossRef]
196. Giannattasio, S.; Guaragnella, N.; Corte-Real, M.; Passarella, S.; Marra, E. Acid stress adaptation protects Saccharomyces cerevisiae from acetic acid-induced programmed cell death. *Gene* **2005**, *354*, 93–98. [CrossRef] [PubMed]
197. Tenreiro, S.; Nunes, P.c.A.; Viegas, C.A.; Neves, M.S.; Teixeira, M.C.; Cabral, M.G.; Sá-Correia, I. AQR1 gene (ORF YNL065w) encodes a plasma membrane transporter of the major facilitator superfamily that confers resistance to short-chain monocarboxylic acids and quinidine in Saccharomyces cerevisiae. *Biochem. Biophys. Res. Commun.* **2002**, *292*, 741–748. [CrossRef] [PubMed]
198. Tenreiro, S.; Rosa, P.C.; Viegas, C.A.; Sá-Correia, I. Expression of the AZR1 gene (ORF YGR224w), encoding a plasma membrane transporter of the major facilitator superfamily, is required for adaptation to acetic acid and resistance to azoles in Saccharomyces cerevisiae. *Yeast* **2000**, *16*, 1469–1481. [CrossRef]
199. Carmelo, V.; Santos, H.; Sá-Correia, I. Effect of extracellular acidification on the activity of plasma membrane ATPase and on the cytosolic and vacuolar pH of Saccharomyces cerevisiae. *Biochim. Biophys. Acta (BBA)-Biomembr.* **1997**, *1325*, 63–70. [CrossRef]

3

Aromatic Potential of Diverse Non-Conventional Yeast Species for Winemaking and Brewing

Amparo Gamero [1,2,*], Annereinou Dijkstra [1], Bart Smit [1,3] and Catrienus de Jong [1,4]

1 NIZO, P.O. Box 20, 6710 BA Ede, The Netherlands; annereinou@hotmail.com (A.D.); bart@fermentor.nl (B.S.); catrienus.dejong@wur.nl (C.d.J.)
2 Food Technology Area, Faculty of Pharmacy, Universitat de València, Avda. Vicent Andrés Estellés s/n, Burjassot, 46100 Valencia, Spain
3 FerMentor, Ede, 6718 ZB Ede, The Netherlands
4 Wageningen Food & Biobased Research (WFBR), Wageningen University & Research, Bornse Weilanden 9, 6708 WG Wageningen, The Netherlands
* Correspondence: amparo.gamero@uv.es

Abstract: Traditionally, *Saccharomyces* species are those used to conduct industrial alcoholic fermentations. Recently, an increasing interest has arisen with respect to the potential of so-called non-conventional yeasts to improve wine and beer aroma profiles, keeping the particular *terroir* of each region or for the development of craft beers. In this study, the potential of diverse non-conventional yeasts to improve aroma in winemaking and brewing was investigated, testing several pure and mixed culture combinations. In addition, a comparison between microscale and labscale was carried out in order to assess the value of microwine and microbeer as screening tools. The results indicated that non-*Saccharomyces* yeasts were good candidates to enhance or diversify aroma profiles in alcoholic beverages, especially regarding acetate ester yield and fruity aromas. However, mixed cultures with *Saccharomyces* spp. are normally required to achieve a successful fermentation. The adjustment of pithing ratios is crucial for this purpose. Microscale is presented as an effective and efficient screening tool to compare different culture combinations, although scaling-up will always be necessary in order to get results closer to real winemaking or brewing processes.

Keywords: non-conventional yeasts; aromas; bioflavouring; wine; beer; microscale; alcoholic fermentation; screening

1. Introduction

The most common starter cultures employed in alcoholic fermentation are *Saccharomyces* species. These species combine several desirable characteristics, such as efficiency in fermenting sugars and producing ethanol, synthesis of aromas, and high ethanol tolerance [1]. However, in the early stages of wine fermentation, as well as in spontaneously fermented wines, several non-*Saccharomyces* species also participate in the fermentation process. These species may have a positive impact on desirable organoleptic characteristics of the wine such as improved mouthfeel and more complex flavour profiles [2].

Recently there has been an increasing interest in both winemaking and brewing towards non-*Saccharomyces* yeasts for their potential for aroma enhancement while maintaining the particular *terroir* of each region or for the development of craft beers.

In the case of wine, application studies with a selection of these so-called non-conventional yeasts indeed improve wine aroma; for example, the sequential fermentation of Ribolla Gialla grape juice employing commercial *Saccharomyces cerevisiae* Lalvin T73 and *Zygosaccharomyces kombuchaensis*

CBS8849 or *Kazachstania gamospora* CBS 10400 resulted in positive organoleptic properties and suitable fermentation dynamics, rapid sugar consumption and industrial strain compatibility [3]. In addition, Jood et al. [4] showed that mixed-culture fermentations involving *S. cerevisiae* and different *Kazachstania* species have very distinct aroma profiles, whereas Varela et al. [5] showed that wines fermented with a combination of *Metschnikowia pulcherrima* and *S. uvarum* were characterised by increased concentrations of 2-phenylethanol and 2-phenylethyl acetate, both associated with positive sensory attributes. In simultaneous and sequential inoculations of selected *Rhodotorula mucilaginosa* and *S.cerevisiae* strains, both the variety and levels of fermentative aromacompounds were improved, especially that of (Z)-3-hexene-1-ol, nerol oxide, certain acetates and ethyl esters. These changes are associated with an improved perception of citrus, sweet fruit, acid fruit, berry, and floral aroma [6].

In a similar approach, the use of non-*Saccharomyces* yeasts in brewing has been explored [7]. *Torulaspora delbrueckii*, for example, was studied in single-strain fermentations for the production of low-alcohol beer and for enhancing beer flavour profiles. Some strains fermented all the wort sugars, whereas others did not, but all strains tested displayed improved flavour-forming properties [8]. In another study, the use of pure cultures of *T. delbrueckii* and mixed cultures of *T. delbrueckii* and *S. cerevisiae* for brewing resulted in a well noticeable change in flavour profile due to *T. delbrueckii*. Distinctive sensory profiles with pure cultures of *T. Delbrueckii* were characterised by 'fruit/citric' notes and 'full-bodied' attributes whereas some mixed cultures enhanced the 'fruity/ester' and 'hop' aromas. At the same time, these beers showed enhanced clarity and a more persistent and compact foam [2]. Another approach involving non-*Saccharomyces* yeasts to produce low-alcohol beers was carried out by van Rijswijck et al. [9]. These authors found that the non-conventional yeasts *Cyberlindnera fabianii* and *Pichia kudriavzevii* both in pure and mixed cultures with *S. cerevisiae* yielded beer with lower ethanol content and higher levels of esters. *Brettanomyces custersii*, isolated from Lambic beer, showed higher β-glucosidase activity than *S. cerevisiae*, thus a higher potential for aroma release from precursors. Fermentation with pure cultures of a *Brettanomyces custersii* strain or in a co-culture with a *S. cerevisiae* resulted in enhanced release of volatiles from hop glycosides [2]. In a more recent study, Holt et al. [10] screened 17 non-conventional yeast species in sequential brewing fermentation, followed by inoculation of *S. cerevisiae*. In these fermentations, it was possible to enhance banana flavour (isoamyl acetate) by *Pichia kluyverii*, spicy notes (phenolic compounds) by *Brettanomyces* species and clove-like aroma (4-vinylguaiacol) by *T. delbrueckii.*

Furthermore, Gutiérrez et al. [11] explored the potential of 99 non-*Saccharomyces* yeasts to produce pleasant fruity aromas in three industrial media (beer, wine and cider) through an olfactory assay. Of these, 21 yeasts were further evaluated for their aroma profile and fermentation capacity using wort, grape and apple juice. These authors reached the conclusion that the choice of yeast has more impact than the medium composition. The yeasts with the best opportunities to tune beer flavour belong to the species *Galactomyces geotrichum, Kazachstania zonata, Kluyveromyces lactis, Lindnera meyerae, Pichia kluyveri, Starmera caribaea, Yarrowia lipolytica* and *Saccharomycodes ludwigii*. Finally, Ravasio et al. [12] explored the potential of 60 non-conventional yeasts to improve beer aroma and found that the species *Wickerhamomyces anomalus* was a good candidate to combine with lager yeast to increase fruity aroma.

In this paper, we analyse the fermentation performance and the aroma potential in microscale of 10 non-*Saccharomyces* species employed as pure and mixed cultures co-inoculated in different ratios with one *Saccharomyces* strain on both wort and wine must, showing that variation in pitching rates is a valid option to control diversification as well as the additional value in analysing large amounts of culture combinations using microwine and microbeer. In addition, a comparison between microscale and labscale was carried out to analyse the correspondence in fermentation performance and aroma profiles.

2. Materials and Methods

2.1. Strains

Ten non-*Saccharomyces* strains were included in this study and are listed in Table 1 (Westerdijk Fungal Biodiversity Institute-CBS-KNAW, Utrecht, The Netherlands). In addition, two commercial *Saccharomyces* species were included as references: *S.cerevisiae* Lalvin T73 (Lallemand) in wine and *S. Pastorianus* WS34/70 (Fermentis, Lesaffre) in beer.

Table 1. Non-*Saccharomyces* yeasts used in this study.

Strain	Species	Substrate	Origin
CBS 772.71	*Galactomyces geotrichum*	Soil	Puerto Rico
CBS 2567	*Hanseniaspora guilliermondii*	Juice	Israel
CBS 5552	*Wickerhamomyces subpelliculosus*	Sugar	Unknown
CBS 6641	*Torulaspora microellipsoides*	Plant	USA
CBS 7692	*Starmera caribaea*	Plant	West Indies
CBS 8031	*Hanseniaspora vineae*	Plant	Canada
CBS 8849	*Zygosaccharomyces kombuchaensis*	Fungus	Russia
CBS 8860	*Barnettozyma californica*	Fruit	Russia
CBS 9716	*Zygosaccharomyces rouxii*	Insect	Germany
CBS 10400	*Kazachstania gamospora*	Unknown	Japan

2.2. Microscale Fermentations

Strains were pre-cultured overnight in GPY medium at 30 °C. After centrifugation (1000× *g*, 5 min), cells were resuspended in sterile Syrah must (Sofralab, Magenta, France) or enzymatically treated glucose wort (14.5°P) with no hops (Carlsberg, Copenhagen, Denmark). Wine must contained 250 g/L of sugars and was sterilized overnight using 1 mL/L dimethyl dicarbonate (Sigma-Aldrich, St. Louis, MO, USA) whereas wort was autoclaved.

Fermentations were carried out in triplicate employing 24-well microplates (MP), where each well contained 5 mL of must or wort. The MP were covered with a breathable seal (Sigma-Aldrich, St. Louis, MO, USA) and incubated at 25 °C for 10 days in anaerobic jars. The strains were inoculated in MP as pure cultures ($OD_{600\ nm}$ of 0.2) or mixed cultures in combination with *S. cerevisiae* Lalvin T73 (microwine) or WS34/70 (microbeer) in proportion 1:1 (each strain $OD_{600\ nm}$ of 0.1). In addition, extra microbeer experiments were carried out employing $OD_{600\ nm}$ of 0.2 of non-*Saccharomyces* yeasts and $OD_{600\ nm}$ of 0.04 and 0.01 of WS34/70 (5:1 and 20:1 proportions, respectively). These extra experiments were done due to cultures mixed in 1:1 ratio resulted in microbeers similar as those produced by pure *S. cerevisiae* culture, indicating the outcompeting growth rate of the commercial strain.

2.3. Residual Sugars and Ethanol Analysis by High-Performance Liquid Chromatography (HPLC)

Mono/di-saccharides (glucose, fructose, maltose, sucrose, maltotriose) and ethanol were determined by HPLC (Thermo Fisher Scientific, USA) equipped with a refractive index detector Shodex RI 101. Prior to analyses, proteins and fat in the samples were precipitated by Grimbely Biggs reagents and removed after centrifugation.

The HPLC system was equipped with a Sil-10 AD vp injector, a LC-10ATvp pump and a CTO-6A oven. Sample volume of 25 µL was injected onto the mixed bed guard AG50W-X4 400 mesh (20 × 9 mm, packed with H^+, Bio-Rad, Hercules, CA, USA) and AG3-X4A, 200–400 mesh (Bio-Rad, Hercules, CA, USA). The column (Aminex HPX-87P, 300 × 7.8 mm; BioRad, Hercules, CA, USA), was held at 80 °C. Mono/disaccharides and ethanol were eluted isocratically with miliQ water, flowing at a rate of 0.4 mL/min. The autosampler temperature was 10 °C. The detection was carried out by a refractive index detector ERC-7510 (Erma, Japan). Data analysis was conducted with Chromeleon software version 6.60 (Thermo Fisher Scientific, Waltham, MA, USA). External standards (97%–99% purity) for the indicated mono/di-saccharides and ethanol were obtained by Sigma-Aldrich (St. Louis, MO, USA) and a 4-point calibration curve was constructed.

2.4. Aroma Analysis by GC-MS

Aroma compounds, higher alcohols, esters and vicinal diketones (Table 2) were extracted by headspace solid-phase microextraction (HS-SPME) employing a grey fibre DVB/CAR/PDMS (Divinylbenzene/Carboxen/Polydimethylsiloxane) and determined by GC-MS. The fibre was exposed to the headspace of the samples during 15 min at 60 °C and desorbed during 2 min in the GC injector at 250 °C. The injection device was a Combi PAL autosampler (CTC Analytics, Zwingen, Switzerland). The injected compounds were refocused at the beginning of the GC column by cryo-trapping at −110 °C (ThermoQuest Ce Instruments, Wigan, UK). Subsequently, the cryo-trap was heated to 200 °C at a rate of 50 °C/s.GC–MS analyses were carried out with a Finnigan Trace GC/MS (Thermo Fisher Scientific, Waltham, MA, USA) equipped with a 30 m × 0.25 μm VF-wax ms, df = 0.5 μm, capillary column (Varian, Palo Alto, CA, USA). The chromatographic oven was initially set at 40 °C for 1 min, then raised to 250 °C at 20 °C/min and then kept at 250 °C for 3 min. The carrier gas Helium was kept at a constant flow of 1.5 mL/min. The total runtime was 14.5 min. Mass spectral data was collected over a range of m/z 35–250 in full scan mode (scan time 0.25 s). Quantification of the compounds was done using calibration curves employing external standards of each aroma compound purchased from Sigma-Aldrich (St. Louis, MO, USA). Data analysis was performed by the Xcalibur 2.1 software (Thermo Fisher Scientific, Inc., Waltham, MA, USA).

Table 2. Sensory descriptors and thresholds of the aroma compounds investigated in the microfermentations.

Compound	Aroma	ST Wine * (mg/L)	ST Beer ** (mg/L)
Vicinal diketones			
Diacetyl	buttery	-	0.10–0.15
2,3-Pentanedione	toffee	-	0.90
Aldehydes			
Acetaldehyde	herbaceous, sherry, nutty, bruised apple	100	20–25
Higheralcohols			
Isobutyl alcohol	solvent, fusel, spiritous	40	100
Isoamyl alcohol	harsh, nail polish, banana	30	50–65
Phenylethyl alcohol	floral, rose, sweet	10	40
Acetateesters			
Ethyl acetate	nail polish, fruity	7.50	30–33
Isoamyl acetate	banana, pear	0.03	1.60–2.00
Phenylethyl acetate	flowery, rose, fruity, honey	0.25	3.80
Ethyl Esters			
Ethyl caproate	green apple, aniseed	0.05	0.23
Ethyl caprylate	sour apple, sweet soap	0.02	0.50–0.90
Ethyl caprate	apple, floral, soap	0.20	1.50

* ST wine: sensory thresholds in wine (10% ethanol) extracted from [13]; ** ST beer: sensory thresholds in beer extracted from [14] and [15].

2.5. Comparison of Microscale to Labscale

S. cerevisiae NIZO 900 was pre-cultured overnight at 30 °C in GPY (2% glucose, ScharlauChemie, Spain; 1% bacteriological peptone, Oxoid, UK; 0.5% yeast extract, Oxoid, UK). This strain was selected due to its good performance in fermenting alcoholic beverages. Afterwards, yeast cultures were centrifuged at 1000× *g* for 5 min, the medium was decanted, and the cells were subsequently re-suspended in triplicate in either 200 mL (bottles-lab scale) or 5 mL (microplates, MTPs-microscale) of Syrah must (Sofralab, Magenta, France) at a final $OD_{600\ nm}$ of 0.1. Priortoinoculation, wine must was sterilized overnight with 1 mL/L dimethyl dicarbonate (Sigma-Aldrich, St. Louis, MO, USA). Samples were taken after 1, 3, 7 and 10 days of fermentation, centrifuged (4 °C, 1000× *g*, 5 min) and frozen immediately at −20 °C, except for OD measurements.

2.5.1. Biomass Analysis

Optical density measurement at 600 nm ($OD_{600\ nm}$) was carried out in a UV/visible spectrophotometer (Ultra-spec 2000, Amersham Pharmacia Biotech, Little Chalfont, UK). To correct for non-linearity of the assay, samples were pre-diluted in medium to obtain reads between 0.2 and 0.4.

2.5.2. Fermentation Kinetics and Aroma Profiles

Residual sugars and ethanol in the resulting fermentations were analysed by HPLC as described in Section 2.3.

Aroma profiles were determined by headspace solid-phase dynamic extraction (HS-SPDE) in combination with GC-MS. The HS-SPDE extraction was carried out using a 2.5 mL HS-syringe with a polydimethylsiloxane active charcoal coated needle (PDMS/AC). After the equilibration of the samples at 60 °C for 15 min, 1 mL of the headspace was injected into the GC column (1 min at 250 °C in splitless mode). GC and MS conditions were the same as described in Section 2.4.

2.6. Statistical Analysis

Multifactorial and one-way ANOVA were performed at 95% of confidence level (p value ≤ 0.05) with the results of aroma production and fermentation kinetics in all the experiments. Average comparison was done using Fisher (microwine) or Tukey's (microbeer) tests. The statistical analyses were carried out by XLSTAT (Microsoft, Redmond, WA, USA).

3. Results and Discussion

3.1. Microwine

Microwine fermentations were carried out by 10 non-*Saccharomyces* strains (Table 1) in pure and mixed culture together with *S. cerevisiae* Lalvin T73 in proportion 1:1. Their fermentation performance and aroma potential were evaluated.

3.1.1. Fermentation Performance

Multifactorial ANOVA shows that non-conventional strain and type of culture, pure or mixed, were significant factors determining the residual sugar amount. However, just the type of culture was significantly influencing ethanol yield, indicating that ethanol synthesis was similar among the non-conventional yeasts employed in this study. Residual sugars and ethanol production of the resulting microwines are presented in Table 3.

One-way ANOVA points out that, in general, fermentations carried out by pure cultures of non-conventional yeasts left significantly higher amounts of residual sugars and produced lower quantities of ethanol compared to their corresponding mixed cultures and reference *S. cerevisiae* Lalvin T73. Ethanol was not even detected in case of *Starmera caribaea* CBS 7692 and *Barnettozyma california* CBS 8860 as practically no fermentation of the sugars occurred. One exception to this pattern was *Wickerhamomyces subpelliculosus* CBS 5552, which showed a fermentation performance as good as the reference strain even when used as a pure culture, leaving hardly any residual sugars and producing high ethanol levels.

Table 3. Residual sugars and ethanol in the resulting microwines.

Strain 1	Strain 2	Ratio	Sugars (g/L)	Ethanol% (v/v)
Lalvin T73			11 ± 4.4 [f]	20 ± 0.0 [b]
CBS 772.71			163 ± 7.8 [a–d]	13 ± 0.3 [d]
CBS 772.71	Lalvin T73	1:1	63 ± 1.7 [d–f]	17 ± 1.1 [b,c]
CBS 2567			170 ± 1.4 [a–c]	11 ± 0.1 [d,e]
CBS 2567	Lalvin T73	1:1	30 ± 1.3 [e,f]	21 ± 0.8 [a,b]
CBS 5552			20 ± 2.1 [f]	18 ± 1.1 [b,c]
CBS 5552	Lalvin T73	1:1	32 ± 0.6 [e,f]	18 ± 0.3 [b,c]
CBS 6641			157 ± 2.7 [a–d]	12 ± 0.7 [d]
CBS 6641	Lalvin T73	1:1	34 ± 1.4 [e,f]	24 ± 0.2 [a]
CBS 7692			235 ± 20 [a]	nd
CBS 7692	Lalvin T73	1:1	59 ± 7.3 [c–f]	19 ± 0.5 [b,c]
CBS 8031			173 ± 5.7 [a,b]	13 ± 0.3 [d]
CBS 8031	Lalvin T73	1:1	73 ± 2.3 [b–f]	18 ± 0.8 [b,c]
CBS 8849			135 ± 1.7 [b–f]	13 ± 0.0 [d]
CBS 8849	Lalvin T73	1:1	30 ± 3.2 [e,f]	21 ± 1.2 [a,b]
CBS 8860			250 ± 6.2 [a]	nd
CBS 8860	Lalvin T73	1:1	75 ± 5.4 [b–f]	22 ± 0.3 [a,b]
CBS 9716			138 ± 24.3 [a–e]	15 ± 1.0 [c,d]
CBS 9716	Lalvin T73	1:1	78 ± 2.0 [b–f]	22 ± 0.6 [a,b]
CBS 10400			170 ± 4.5 [a,b]	11 ± 0.4 [d,e]
CBS 10400	Lalvin T73	1:1	74 ± 2.0 [b–f]	18 ± 0.1 [b,c]

Letters indicate grouping according to ANOVA.

Its fermentation efficiency was in accordance with [16], who observed similar ethanol production by this strain compared to *S. cerevisiae* baker's yeast when bread fermentations were carried out and may indicate that this strain presented an intense Crabtree effect similar to *S. cerevisiae*.

The difference in ethanol production between pure and mixed cultures can be explained by the fact that it has been demonstrated that yeast in mixed cultures do not simply coexist, but rather interact, and as a result, oenological traits can be modulated [17]. The finding that single cultures of non-*Saccharomyces* yeasts are, in general, poor fermenters and sensitive to ethanol is in accordance to several authors [17–19]. *S. cerevisiae,* as consequence of Crabtree effect, prioritises fermentative metabolism over aerobic respiration when sugar levels are higher than 10 g/L. Nevertheless, many non-*Saccharomyces* yeasts do not show this effect and are able to grow using oxygen regardless of sugar concentration and thus, divert carbon into other metabolites, which reduces ethanol production [20]. In this way, fermentation performance improves when non-*Saccharomyces* yeasts are together with *S. cerevisiae,* with some exceptions. For instance, Dashko et al. [3] presented *Kazachstania gamospora* CBS 10400, the strain employed in this study, as Crabtree positive, observing a fermentation performance similar to *S. cerevisiae* in terms of residual sugars and ethanol yield when employed in sequential inoculation together to *S. cerevisiae.* In our study, CBS 10400 when used in mixed fermentation resulted in less residual sugars and much higher ethanol content, approaching that produced by *S. cerevisiae*.

On the other hand, the limited ethanol synthesis can also be considered an advantage for the wine sensory profile, since excessive ethanol amounts can increase the perception of bitterness and astringency and mask some volatile compounds [20]. Furthermore, this approach leads to the development of healthier wines and has been proposed as a strategy for counteracting climatic change and the resulting ethanol increase in wines [20]. Another interesting aspect is that it has been demonstrated that non-*Saccharomyces* strains that show poor efficiency in fermenting sugars may produce higher glycerol quantities, which is a highly desirable trait in wines, especially in red [11].

On the other hand, mixed cultures were similar to the reference strain in terms of residual sugars and ethanol yield. *Torulaspora microellipsoides* CBS 6641 in mixed culture even produced significantly more ethanol than the reference strain, being two-fold that produced in their corresponding pure culture. However, previous literature indicates that this strain produced low ethanol levels in sequential fermentations in wine, as well as when fermenting other substrates such as cashew apple juice [3,21]. Nevertheless, sequential and mixed fermentations can result in a different outcome and cashew apple juice present high quantity of phenolic compounds, which can exert microbial inhibitory effects.

3.1.2. Aroma Profiles

Table 4 shows the aroma profile of resulting microwines. The impact on aroma synthesis of non-*Saccharomyces* strains as a pure and mixed culture was studied employing multifactorial ANOVA. The type of culture, either pure or mixed, affected the production of all the aroma compounds except for ethyl caprate. In addition, the impact of non-conventional strains on acetate esters, phenylethyl alcohol and ethyl caprylate was significant.

Higher Alcohols

Higher alcohols contribute to floral, fruity or herbal aromas depending on their synergistic effects with other flavour-active compounds. In higher concentrations they could also influence flavour release and contribute to mouthfeel. Some of the most important higher alcohols in alcoholic beverages are isobutyl alcohol (solvent, fusel, spiritous), isoamyl alcohol (harsh, nail polish, banana) and phenylethyl alcohol (floral, rose, sweet), presenting, respectively, 40, 30 and 10 mg/L sensory thresholds in wine [13]. Excessive amounts of these compounds can be detrimental to wine quality; however, moderate amounts increase wine complexity [18].

Multifactorial ANOVA showed that the type of culture (pure or mixed) was a significant factor regarding higher alcohol yield. Contrarily, the non-*Saccharomyces* yeast participating in the fermentation did not influence the quantity of higher alcohols, except in the case of phenylethyl ethanol, where differences of even 8-fold were observed.

Isobutyl alcohol was above its sensory threshold in all of the mixed fermentations whereas isoamyl alcohol only in some of them (Table 4). In all these cases, the amount of the two alcohols were higher than in the corresponding pure non-*Saccharomyces* cultures and even higher than in the reference microwine, where the amounts of these alcohols were below their sensory thresholds (Table 4). Mixed fermentations carried out with *Hanseniaspora guilliermondii* CBS 2567, *T. microellipsoides* CBS 6641, *Zygosaccharomyces kombuchaensis* CBS 8849 and *Zygosaccharomyces rouxii* CBS 9716 had much higher isobutyl alcohol levels and *Barnettozyma californica* CBS 8860 had higher isobutyl and isoamyl alcohols amounts compared to the reference culture (Table 4). In previous works, intensive production of isobutyl and isoamyl alcohols was already seen in some of the non-*Saccharomyces* strains tested here, such as *H. Guilliermondii* CBS 2567 and *T. Microellipsoides* CBS 6641 when used as pure cultures in standard growth medium [22]. In this previous work, *H. guilliermondii* yielded 27 and 2.3 times the average quantity of isobutyl alcohol compared to *Saccharomyces* strains and *W. subpelliculosus* CBS 5552 produced 8 times the average quantity of isoamyl alcohol compared to *Saccharomyces* strains. Furthermore, the high production of alcohols by *Z. kombuchaensis* CBS 8849 has already been observed by other authors in pilot-scale sequential fermentations [3], suggesting a proper match between microscale and pilot scale regarding aroma profile.

It is worth mentioning the synergistic effect between reference strain and strains *Starmera caribaea* CBS 7692 and *B. californica* CBS 8860 regarding isoamyl alcohol production, since the amount of this alcohol in mixed fermentations was higher than the addition of quantities yielded by each strain separately (Table 5).

On the contrary, phenylethyl alcohol levels were higher in pure cultures, being above its sensory threshold in all cases except for *S. caribaea* CBS 7692 and with these differences being significant in the case of *Galactomyces geotrichum* CBS 772.71, *W. Subpelliculosus* CBS 5552, *T. microellipsoides* CBS 6641, *S. caribaea* CBS 7692 and *K. gamospora* CBS 10400 (Table 5). In mixed cultures, the amounts of this alcohol were below its sensory threshold with the exception of CBS 8031 (*Hanseniaspora vineae*). However, the reference microwine contained the highest phenylethyl alcohol level (Table 5).

Table 4. Aroma compounds (mg/L) in the resulting microwines.

Strain 1	Strain 2	Ratio	Isobutyl Alcohol	Isoamyl Alcohol	Phenylethyl Alcohol	Ethyl acetate	Isoamyl Acetate	Phenylethyl Acetate	Ethyl Caproate	Ethyl Caprylate	Ethyl Caprate
Lalvin T73			26 $^{a-c}$	24 $^{a-c}$	**33** ^{i}	3 ^{a}	**1.54** ^{bc}	**1.98** ^{a}	**0.55** ^{a,b}	**1.43** ^{e,f}	**0.57** ^{c,d}
CBS 772.71			12 ^{a}	23 $^{a-c}$	**17** ^{f,g}	4 $^{a-c}$	**0.81** ^{ab}	**23.79** ^{c}	**0.12** ^{a,b}	**0.22** ^{a,b}	**0.26** $^{a-d}$
CBS 772.71	Lalvin T73	1:1	**78** $^{b-f}$	**30** $^{a-d}$	3 ^{a,b}	nd	**0.09** ^{a}	**2.49** ^{a}	**0.87** ^{b}	**0.29** ^{a,b}	**0.42** $^{a-d}$
CBS 2567			18 ^{a,b}	11 ^{a}	**10** $^{b-e}$	**11** ^{c,d}	**1.55** ^{bc}	**13.35** ^{b}	**0.06** ^{a}	**0.10** ^{a}	**0.08** ^{a}
CBS 2567	Lalvin T73	1:1	**126** $^{e-g}$	21 $^{a-c}$	6 $^{a-d}$	nd	nd	**1.41** ^{a}	**0.53** ^{a,b}	**0.21** ^{a,b}	**0.30** $^{a-d}$
CBS 5552			13 ^{a}	19 $^{a-c}$	**22** ^{g,h}	**10** $^{b-d}$	**0.44** ^{ab}	**11.88** ^{b}	**0.22** ^{a,b}	**1.26** $^{d-f}$	**0.60** ^{d}
CBS 5552	Lalvin T73	1:1	**82** $^{c-g}$	24 $^{a-d}$	6 $^{a-d}$	nd	nd	**1.36** ^{a}	**0.71** ^{a,b}	**0.21** ^{a,b}	**0.25** $^{a-d}$
CBS 6641			10 ^{a}	16 ^{a,b}	**19** ^{g,h}	3 ^{a,b}	**0.25** ^{a}	**0.95** ^{a}	**0.14** ^{a,b}	**0.36** $^{a-c}$	**0.30** $^{a-d}$
CBS 6641	Lalvin T73	1:1	**141** ^{g}	**39** $^{b-d}$	3 ^{a}	nd	nd	**2.19** ^{a}	**0.78** ^{a,b}	**0.23** ^{a,b}	**0.40** $^{a-d}$
CBS 7692			14 ^{a}	12 ^{a}	9 $^{a-d}$	**20** ^{e}	**4.87** ^{d}	**28.53** ^{d}	**0.25** ^{a,b}	**0.75** $^{b-d}$	**0.11** ^{a,b}
CBS 7692	Lalvin T73	1:1	**76** $^{b-e}$	**41** ^{c,d}	3 ^{a}	nd	nd	**1.99** ^{a}	**0.79** ^{a,b}	**0.37** $^{a-c}$	**0.52** $^{a-d}$
CBS 8031			11 ^{a}	22 $^{a-c}$	**20** ^{g,h}	4 ^{a,b}	**0.76** ^{ab}	**26.24** ^{c,d}	**0.11** ^{a,b}	**0.14** ^{a}	**0.29** $^{a-d}$
CBS 8031	Lalvin T73	1:1	**75** $^{b-e}$	29 $^{a-d}$	**17** ^{f,g}	nd	nd	**1.72** ^{a}	**0.88** ^{b}	**0.29** ^{a,b}	**0.54** $^{b-d}$
CBS 8849			18 ^{a,b}	**25** $^{a-d}$	25 ^{h}	5 $^{a-c}$	**0.88** ^{ab}	**28.76** ^{d}	**0.10** ^{a,b}	**0.11** ^{a}	**0.12** $^{a-c}$
CBS 8849	Lalvin T73	1:1	**138** ^{f,g}	26 $^{a-d}$	5 $^{a-d}$	nd	nd	**2.42** ^{a}	**0.67** ^{a,b}	**0.22** ^{a,b}	**0.29** $^{a-d}$
CBS 8860			20 ^{a,b}	14 ^{a}	**11** $^{c-f}$	4 ^{a,b}	**2.25** ^{c}	**1.53** ^{a}	**0.29** ^{a,b}	**0.60** $^{a-c}$	**0.23** $^{a-d}$
CBS 8860	Lalvin T73	1:1	**125** $^{e-g}$	**48** ^{d}	4 $^{a-c}$	nd	nd	**2.35** ^{a}	**0.88** ^{b}	**0.44** $^{a-c}$	**0.53** $^{a-d}$
CBS 9716			29 $^{a-d}$	25 $^{a-d}$	**11** $^{d-f}$	2 ^{a}	**1.56** ^{b,c}	**1.49** ^{a}	**0.39** ^{a,b}	**1.61** ^{f}	**0.69** ^{d}
CBS 9716	Lalvin T73	1:1	**87** $^{d-g}$	**33** $^{a-d}$	3 ^{a}	nd	nd	**1.85** ^{a}	**0.81** ^{a,b}	**0.28** ^{a,b}	**0.59** ^{d}
CBS 10400			10 ^{a}	14 ^{a}	**16** $^{e-g}$	**14** ^{d}	**3.85** ^{d}	33.71 ^{e}	**0.21** ^{a,b}	**0.89** $^{c-e}$	**0.41** $^{a-d}$
CBS 10400	Lalvin T73	1:1	**76** $^{b-e}$	**32** $^{a-d}$	7 $^{a-d}$	nd	**0.06** ^{a}	**2.71** ^{a}	**0.83** ^{a,b}	**0.29** ^{a,b}	**0.58** ^{c,d}

Results in bold, above their sensory thresholds; letters indicate grouping according to ANOVA; standard deviations were below 20% in all of the cases.

Table 5. Residual sugars and ethanol in the resulting microbeers.

Strain 1	Strain 2	Ratio	Sugars (g/L)	Ethanol% (v/v)
WS34/70			nd [d]	5.9 ± 0.6 [a]
CBS 772.71	WS34/70	5:1	nd [d]	5.6 ± 1.4 [a]
CBS 772.71	WS34/70	20:1	nd [d]	6.5 ± 0.4 [a]
CBS 2567	WS34/70	5:1	5.4 ± 1.4 [a–c]	6.2 ± 1.1 [a]
CBS 2567	WS34/70	20:1	2.4 ± 0.7 [c,d]	6.0 ± 1.4 [a]
CBS 5552	WS34/70	5:1	nd [d]	6.4 ± 0.3 [a]
CBS 5552	WS34/70	20:1	nd [d]	5.4 ± 1.3 [a]
CBS 6641	WS34/70	5:1	2.2 ± 0.4 [c,d]	6.0 ± 0.6 [a]
CBS 6641	WS34/70	20:1	5.2 ± 1.0 [a–c]	6.4 ± 0.5 [a]
CBS 7692	WS34/70	5:1	0.9 ± 1.2 [d]	6.8 ± 0.3 [a]
CBS 7692	WS34/70	20:1	3.5 ± 1.9 [b–d]	6.2 ± 1.4 [a]
CBS 8031	WS34/70	5:1	nd [d]	6.9 ± 1.2 [a]
CBS 8031	WS34/70	20:1	1.0 ± 1.4 [d]	6.2 ± 0.6 [a]
CBS 8849	WS34/70	5:1	7.1 ± 0.6 [a,b]	5.5 ± 1.2 [a]
CBS 8849	WS34/70	20:1	nd [d]	6.5 ± 0.6 [a]
CBS 8860	WS34/70	5:1	nd [d]	6.0 ± 0.4 [a]
CBS 8860	WS34/70	20:1	0.8 ± 1.1 [d]	6.8 ± 0.1 [a]
CBS 9716	WS34/70	5:1	7.7 ± 0.4 [a]	5.7 ± 1.1 [a]
CBS 9716	WS34/70	20:1	3.9 ± 1.6 [a–d]	6.6 ± 0.3 [a]
CBS 10400	WS34/70	5:1	nd [d]	7.3 ± 0.4 [a]
CBS 10400	WS34/70	20:1	1.6 ± 0.5 [c,d]	6.6 ± 0.4 [a]

Letters indicate grouping according to ANOVA.

Acetate Esters

Acetate esters are desirable aroma compounds presenting a wide range of fruity flavours and are of special interest inthe case of young wines produced from neutral grape varieties [18]. Among the most important acetate esters present in alcoholic beverages are ethyl acetate (nail polish, fruity; sensory threshold: 7.5 mg/L), isoamyl acetate (banana, pear; sensory threshold: 0.03mg/L) and phenylethyl acetate (flowery, rose, fruity, honey; sensory threshold: 0.25 mg/L) [13].

Multifactorial ANOVA points out that non-*Saccharomyces* strain and type of culture, mixed or pure, significantly affected acetate ester yields.

Reference microwine did not excel in the amounts of any acetate ester and isoamyl acetate was missing in most of the mixed cultures (Table 4), indicating a negative interaction between *Saccharomyces* and non-*Saccharomyces* strains regarding this aroma compound.

Despite being the main acetate ester in wine, ethyl acetate only appeared above its sensory threshold in four pure cultures: *H. guilliermondii* CBS 2567, *W. subpelliculosus* CBS 5552, *S. caribaea* CBS 7692 and *K. gamospora* CBS 10400 (Table 5). Excessive acetate ester production, a trait traditionally linked to non-*Saccharomyces* strains, can negatively affect the aroma profile, giving "oxidised" notes [17,18,23]. However, in our study this seems not to be the case.

Similarly, in the case of isoamyl and phenylethyl acetate, pure cultures yielded significantly higher amounts (Table 4), indicating high activity of the acetyltransferases, which must be inhibited to a certain degree in mixed cultures. A remarkable example in this sense is the strain *S. caribaea* CBS 7692, since its huge acetate ester production was not shown when combined with *S. cerevisiae* (Table 4). This extraordinary acetate ester synthesis of *S. caribaea* CBS 7692 when used as pure culture was already described in previous works, where this strain and *H. guilliermondii* CBS 2567 produced from 44- to 84-fold the levels found in reference *S. cerevisiae* pure culture [11,22]. *H. guilliermondii* CBS 2567 also excelled in the production of phenylethyl acetate and the precursor alcohol when fermenting cashew apple juice [21]. In our study, acetate ester production by *H. guilliermondii* CBS 2567 was not so huge, but still remarkable (Table 4).

It is worth mentioning the high isoamyl acetate production by pure cultures of *S. caribaea* CBS 7692 and *K. gamospora* CBS 10400, which probably resultsin a strong banana flavour, as already observed in bread fermented by the latter [16]. These strains also present high synthesis of phenylethyl acetate, which was already seen by Dashko et al. [3] in sequential pilot scale fermentations. The strains *G. geotrichum* CBS 772.72, *Hanseniaspora vineae* CBS 8031 and *Z. kombuchaensis* CBS 8849 also showed high production of phenylethyl acetate (Table 4), which probably resulted in a remarkable rose aroma. Jolly et al. [18] described high production of this ester in *H. vineae* and *H. guilliermondii*, which was in accordance with our results, especially in case of the former (see CBS 8031 in Table 4).

Ethyl Esters

Ethyl esters, as acetate esters, are desirable aroma compounds presenting fruity and flowery flavours. Some of the most important ethyl esters in alcoholic beverages are ethyl caproate (green apple, aniseed aroma; sensory threshold: 0.05 mg/L), ethyl caprylate (sour apple, sweet soap aroma; sensory threshold: 0.02 mg/L) and ethyl caprate (apple, floral, soap aroma; sensory threshold: 0.20 mg/L) [13].

Multifactorial ANOVA shows that the type of culture carrying out the fermentation, pure or mixed, significantly influenced ethyl caproate and ethyl caprylate yield, being the latter also affected by the non-*Saccharomyces* strain. On the contrary, none of the factors were significantly important to determine ethyl caprate levels, which even appeared below its sensory threshold in some pure cultures.

Despite not being a significant difference probably due to high standard deviations linked to the low amounts detected, the following trend could be observed: ethyl caproate levels were equal or higher in mixed cultures compared to their corresponding pure cultures and the reference microwine (Table 4).

Regarding ethyl caprylate and ethyl caprate, only the microwine produced by pure culture of *Z. rouxii* CBS 9716 presented higher amounts compared to reference microwine (Table 4). Pure culture of *W. subpelliculosus* CBS 5552 also showed high ethyl caprylate quantity (Table 4). The relatively high ethyl ester production by *S. cerevisiae* compared to other *Saccharomyces* and non-*Saccharomyces* strains is a typical trait extensively described in the literature [3,22,24].

3.2. *Microbeer*

Microbeer fermentations were carried out with ten non-*Saccharomyces* strains (Table 1) in pure and mixed cultures together with *S. pastorianus* WS34/70 in proportions 1:1, 5:1 and 20:1 (non-*Saccharomyces*: *S. pastorianus*). Cultures mixed in 1:1 ratio, resulted in fermentation characteristics matching a pure *Saccharomyces* culture, indicating the outcompeting growth rate of the commercial strain. Thus, only the results of the 5:1 and 20:1 fermentations are shown.

3.2.1. Fermentation Performance

Multifactorial ANOVA shows that the quantity of residual sugars is affected by the adjunct strains, but the levels of ethanol were similar in all fermentations (Table 5). In more detail, most of the yeast combinations showed good fermentation performance, consuming practically all the sugars present in wort (Table 5). However, there were some exceptions to this pattern: small sugar amounts (5–8 g/L) were left by non-conventional yeasts *H. guilliermondii* CBS 2567, *Z. kombuchaensis* CBS 8849 and *Z. rouxii* CBS 9716 when used in combination to WS34/70 in proportion 5:1 and *T. microellipsoides* CBS 6641 in proportion 20:1 (Table 5). Different sugar consumption efficiency was already seen in previous works by these strains [11,21], especially by *Z. rouxii*, which is not able to properly ferment maltose [7]. Surprisingly, *T. microellipsoides* CBS 6641 left more residual sugars when used in higher proportion, indicating an interaction between yeasts regarding their fermentation abilities already seen in mixed fermentations by other authors [17]. These differences in sugar consumption were not translated into significant differences in ethanol yield, which was around 6% in all the fermentations (Table 5). This fact could be explained by the bifurcation of metabolic flux to glycerol production by certain strains which would use part of the sugars to this end, thus reducing the expected ethanol yield and as a strategy to cope with hyperosmotic stressful conditions, which has been extensively studied in *Saccharomyces* species [25,26].

3.2.2. Aroma Profiles

Table 6 shows the aroma profiles of the different microbeers. In this study, the impact of two factors on the aroma of the resulting beer were investigated, non-*Saccharomyces* strain and the pitching ratio. Statistical analyses showed that the strain employed is affecting the yield of all the aroma compounds except for isoamyl acetate whereas the pitching ratio especially impacted on vicinal diketones and certain esters, phenylethyl acetate, ethyl caproate and ethyl caprylate. Finally, strain and pitching ratio combined significantly influenced the amounts of vicinal diketones, phenylethyl acetate and ethyl caprate in line with previous studies [27,28].

Vicinal Diketones

Vicinal diketones are carbonyl compounds that are undesirable when they appear in high concentrations in beer. Diacetyl adds a buttery flavour to beer if found above its sensory threshold (0.10–0.15 mg/L) and 2,3-pentanedione its known for its toffee flavour (sensory threshold 0.90 mg/L) [14]. However, it is known that other aroma compounds could strongly influence the sensory threshold and perception of diacetyl in alcoholic beverages [29].

Both the adjunct strains, as well as dosage, significantly influenced the final quantity of vicinal diketones at the end of fermentation according to multifactorial ANOVA. The interaction of these two factors made a significant impact, as well. Diacetyl appears above its sensory threshold in most of the resulting microbeers (even in the reference microbeer), contrarily to the case of 2,3-pentanedione, whose quantity was below its sensory threshold in all of the cases (Table 6). However, it has to be taken into account that microbeers did not undergo maturation in this study. In the maturation process, yeast take up diacetyl and reduce it to 2,3-butanediol, which does not have an undesirable flavour [14].

Most of the 5:1 fermentations presented diacetyl levels near its threshold and similar to reference microbeer with the exception of *H. guilliermondii* CBS 2567, *W. subpelliculosus* CBS 5552, *S. caribaea* CBS 7692 and *B. californica* CBS 8860, which showed higher levels (Table 6). This fact was already observed in some of these non-conventional yeasts in previous works [22].

On the other hand, diacetyl levels, on average, increase with increasing dosage of adjunct strain, indicating that these non-conventional yeasts can produce more diacetyl, thus making necessary longer maturation times if used industrially. However, a few strains form an exception to the above and result in lower (*W. subpelliculosus* CBS 5552) or similar (*H. guilliermondii* CBS 2567, *Z. rouxii* CBS 9716 and *K. gamospora* CBS 10400) amounts of diacetyl when higher proportion of non-conventional yeast is employed.

Table 6. Aroma compounds (mg/L) in the resulting microbeers.

Strain 1	Strain 2	Ratio	Diacetyl	2,3-Pentadione	Isobutyl Alcohol	Isoamyl Alcohol	Phenylethyl Alcohol	Ethyl Acetate	Isoamyl Acetate	Phenylethyl Acetate	Ethyl Caproate	Ethyl Caprylate	Ethyl Caprate
WS34/70			**0.14** ^{a}	0.02 ^{b,c}	32 ^{d}	**74** ^{e}	27 $^{d–f}$	27 ^{a}	0.52 ^{a}	0.84 $^{a–e}$	0.04 ^{d}	0.12 ^{g}	0.06 ^{e,f}
CBS 772.71	WS34/70	5:1	**0.13** ^{a}	0.02 ^{a,b}	29 $^{a–d}$	**67** $^{a–e}$	23 $^{a–e}$	23 ^{a}	0.41 ^{a}	0.86 $^{a–e}$	0.04 ^{c,d}	0.11 ^{f,g}	0.05 $^{c–f}$
CBS 772.71	WS34/70	20:1	**0.41** ^{c}	0.10 ^{h}	29 $^{a–d}$	**68** $^{b–e}$	20 $^{a–c}$	20 ^{a}	0.37 ^{a}	1.22 $^{e–g}$	0.03 $^{a–d}$	0.10 $^{e–g}$	0.04 $^{b–e}$
CBS 2567	WS34/70	5:1	**0.65** ^{d}	0.06 ^{f,g}	27 $^{a–c}$	**62** $^{a–d}$	19 $^{a–c}$	19 ^{a}	0.35 ^{a}	1.04 $^{b–f}$	0.03 $^{a–c}$	0.03 $^{a–c}$	0.02 $^{a–d}$
CBS 2567	WS34/70	20:1	**0.62** ^{d}	0.05 ^{e,f}	25 ^{a}	**60** ^{a,b}	18 ^{a,b}	18 ^{a}	0.38 ^{a}	1.53 ^{f,g}	0.02 ^{a,b}	0.02 ^{a}	0.01 ^{a}
CBS 5552	WS34/70	5:1	**0.30** ^{b}	0.04 ^{d}	27 $^{a–c}$	**64** $^{a–d}$	34 ^{g,h}	**34** ^{a}	0.45 ^{a}	0.59 ^{a,b}	0.03 $^{b–d}$	0.09 $^{d–g}$	0.02 $^{a–d}$
CBS 5552	WS34/70	20:1	0.08 ^{a}	0.01 ^{a,b}	30 $^{b–d}$	**63** $^{a–d}$	34 ^{g,h}	**34** ^{a}	0.33 ^{a}	1.68 ^{g}	0.03 $^{a–c}$	0.04 $^{a–d}$	0.01 ^{a,b}
CBS 6641	WS34/70	5:1	0.09 ^{a}	0.01 ^{a,b}	28 $^{a–d}$	**67** $^{a–e}$	16 ^{a}	16 ^{a}	0.62 ^{a}	0.61 $^{a–c}$	0.04 ^{c,d}	0.12 ^{g,h}	0.05 $^{d–f}$
CBS 6641	WS34/70	20:1	**0.33** ^{b,c}	0.03 ^{d}	28 $^{a–d}$	**66** $^{a–e}$	19 $^{a–c}$	19 ^{a}	0.47 ^{a}	1.20 $^{d–g}$	0.03 $^{a–c}$	0.03 $^{a–c}$	0.01 ^{a,b}
CBS 7692	WS34/70	5:1	**0.41** ^{c}	0.04 ^{d,e}	25 ^{a,b}	**62** $^{a–d}$	25 $^{b–e}$	25 ^{a}	0.36 ^{a}	1.01 $^{b–e}$	0.03 $^{a–d}$	0.09 $^{e–g}$	0.03 $^{a–e}$
CBS 7692	WS34/70	20:1	**0.60** ^{d}	0.07 ^{g}	27 $^{a–c}$	**61** $^{a–c}$	19 ^{a,b}	19 ^{a}	0.53 ^{a}	1.12 $^{c–f}$	0.03 $^{a–d}$	0.04 $^{a–c}$	0.02 $^{a–d}$
CBS 8031	WS34/70	5:1	0.08 ^{a}	0.02 ^{a,b}	26 $^{a–c}$	**60** ^{a,b}	29 $^{e–g}$	29 ^{a}	0.43 ^{a}	2.25 ^{h}	0.03 $^{a–c}$	0.02 ^{ab}	0.01 ^{abc}
CBS 8031	WS34/70	20:1	**0.30** ^{b}	0.03 ^{c,d}	26 $^{a–c}$	**60** ^{a,b}	32 $^{f–h}$	**32** ^{a}	0.23 ^{a}	3.03 ^{i}	0.02 ^{a}	0.01 ^{a}	0.01 ^{a,b}
CBS 8849	WS34/70	5:1	0.08 ^{a}	0.02 ^{a,b}	30 $^{b–d}$	**68** $^{b–e}$	27 $^{c–f}$	27 ^{a}	0.44 ^{a}	0.72 $^{a–e}$	0.04 ^{c,d}	0.10 $^{e–g}$	0.04 $^{b–e}$
CBS 8849	WS34/70	20:1	**0.56** ^{d}	0.05 ^{f}	29 $^{a–d}$	**69** $^{c–e}$	22 $^{a–d}$	22 ^{a}	0.31 ^{a}	0.43 ^{a}	0.03 $^{a–d}$	0.04 $^{a–c}$	0.01 ^{a,b}
CBS 8860	WS34/70	5:1	**0.27** ^{b}	0.03 ^{d}	29 $^{a–d}$	**69** $^{c–e}$	26 $^{c–f}$	26 ^{a}	0.37 ^{a}	0.74 $^{a–e}$	0.04 ^{d}	0.16 ^{h}	0.07 ^{f}
CBS 8860	WS34/70	20:1	**0.42** ^{c}	0.11 ^{h}	31 ^{c,d}	**68** $^{b–e}$	24 $^{b–e}$	24 ^{a}	0.43 ^{a}	0.80 $^{a–e}$	0.03 $^{b–d}$	0.10 $^{e–g}$	0.06 ^{e,f}
CBS 9716	WS34/70	5:1	0.05 ^{a}	0.01 ^{a,b}	29 $^{a–d}$	**70** ^{d,e}	35 ^{g,h}	**35** ^{a}	0.40 ^{a}	0.60 ^{a,b}	0.04 ^{c,d}	0.10 $^{e–g}$	0.02 $^{a–d}$
CBS 9716	WS34/70	20:1	**0.10** ^{a}	0.02 ^{a,b}	28 $^{a–d}$	**69** $^{b–e}$	38 ^{h}	**38** ^{a}	0.38 ^{a}	0.66 $^{a–d}$	0.03 $^{a–d}$	0.07 $^{b–e}$	0.01 ^{a,b}
CBS 10400	WS34/70	5:1	**0.10** ^{a}	0.02 ^{a,b}	27 $^{a–d}$	**61** $^{a–d}$	20 $^{a–c}$	20 ^{a}	0.24 ^{a}	1.06 $^{b–f}$	0.03 $^{a–d}$	0.06 $^{b–e}$	0.05 $^{d–f}$
CBS 10400	WS34/70	20:1	0.05 ^{a}	0.01 ^{a}	25 ^{a,b}	**59** ^{a}	25 $^{b–f}$	25 ^{a}	0.28 ^{a}	1.66 ^{g}	0.03 $^{a–d}$	0.07 $^{c–f}$	0.12 ^{g}

Results in bold, above sensory thresholds; letters indicate grouping according to ANOVA; standard deviations were below 20% in all of the cases.

Higher Alcohols

Multifactorial ANOVA showed that the non-*Saccharomyces* strain participating in the fermentation was a significant factor regarding higher alcohol yield. Contrarily, the pitching ratio did not influence the quantity of higher alcohols, which was in accordance to [30], who compared the brewing performance of *Brettanomyces bruxellensis* as pure culture at different pitching ratios and the only aroma compound presenting different levels was 4-vinylguaiacol. This pattern was just the opposite as in the case of microwine, evidencing the importance of the fermentative substrate employed to carry out the fermentation process.

In general, adjunct cultures resulted in equal or lower concentration of higher alcohols compared to the reference microbeer (Table 6). In all microbeers, isoamyl alcohol was the only alcohol present in concentrations above its sensory threshold (Table 6). Ravasio et al. [12] obtained similar results when screening 60 non-conventional yeast for aroma production in brewing as they only detected isoamyl alcohol and 2-phenylethanol as higher alcohols produced by all the yeasts. In our data set, strains *Z. rouxii* CBS 9716 and *W. subpelliculosus* CBS 5552 resulted in 2-phenylethyl alcohol levels higher than that found in the reference microbeer (Table 6). The high production of alcohols in by *Z. rouxii* has already been reported by other authors for beer as well as for other food products such as soy source [14].

Acetate Esters

Except for ethyl acetate, none of the acetate esters investigated reached its sensory threshold (Table 6).

In general, levels of ethyl acetate and isoamyl acetate did not vary between the reference microbeer and the adjunct strains nor between pitching ratios, except for ethyl acetate, which appeared slightly above its sensory threshold in just two strains, *W. subpelliculosus* CBS 5552 and *Z. rouxii* CBS 9716, not being significant differences between pitching ratios (Table 6). High ethyl acetate synthesis by *Z. rouxii* was observed in soy sauce fermentation and beer [7,14], whereas high levels of acetate esters, especially ethyl acetate, were previously described in other species of the genus *Wickerhamomyces, W. anomalus* [7,12].

In contrast to ethyl acetate and isoamyl acetate, both adjunct strain and pitching ratio were crucial in case of 2-phenylethyl acetate as multifactorial ANOVA showed. In some mixed fermentations, higher amounts of 2-phenylethyl acetate than the reference strain *S. pastorianus* WS34/70 were observed and higher levels corresponded with pitching ratio. These strains were *H. guilliermondii* CBS 2567, *W. subpelliculosus* CBS 5552, *K. gamospora* CBS 10400 and *H. vineae* CBS 8031, being the latter able to synthesize higher amounts of this compound even in 5:1 fermentation (Table 6). Gamero et al. [22] observed high acetate ester synthesis by *H. guilliermondii*, whereas Gutierrez et al. [11] described a strong fruity flavour with *W. subpelliculosus*, which was even able to mask a high production of phenolic compounds.

Ethyl Esters

Multifactorial ANOVA showed that the non-*Saccharomyces* strains as well as their pitching ratio were significant factors regarding ethyl ester yield with the exception of ethyl caprate, which was not affected by the strain proportion. However, none of the ethyl esters studied reached its sensory threshold in any of the resulting microbeers. However, together they could be perceived if they have nearly the same characteristics and receptors as a result of additive or synergistic effects.

In general, ethyl ester levels decrease with increasing pitching ratios of the non-conventional yeasts (Table 6). This is in line with several studies where *Saccharomyces* yeasts have pointed out their huge capacity to synthesise ethyl esters, which has been positively correlated with the ethanol yield and fermentation capacity [10,22]. The exceptionsto this pattern were microbeers involving

B. californica CBS 8860 (5:1) and *K. gamospora* CBS 10400 (20:1), where a significantly higher amount of ethyl caprylate or ethyl caprate was found, respectively (Table 6).

3.3. Comparison Microscale vs.Labscale

To assess the effect of scaling-up, we compare microscale and labscale fermentations. Specifically, comparison of yeast growth, fermentation performance and aroma profiles were carried out between microwine and wine produced at labscale.

Similar trends were observed regarding yeast growth and fermentation kinetics despite appearing significant differences in the absolute values, which were expected as a consequence of the differences in oxygen levels, resulting of the differences in volume. The lower the volume, the less anaerobic environment is created. Similar behaviour would happen when scaling-up between lab and pilot scales or between pilot and industrial scales.

Sugar consumption and ethanol production (Figure 1) were similar up to day 3, and thereafter the microscale was marginally faster, resulting in a full sugar depletion at the end of fermentation. This is in line with the slightly higher biomass in the microwine system (Figure 2) and was according to [31], who observed accelerated sugar consumption in fermentations carried out in MTPs compared to those in Erlenmeyer flasks. However, these authors experienced lower ethanol levels in MTPs at the end of fermentation due to evaporation loss and despite being the same levels at early fermentation stages. In our study, this fact did not occur (Figure 1).

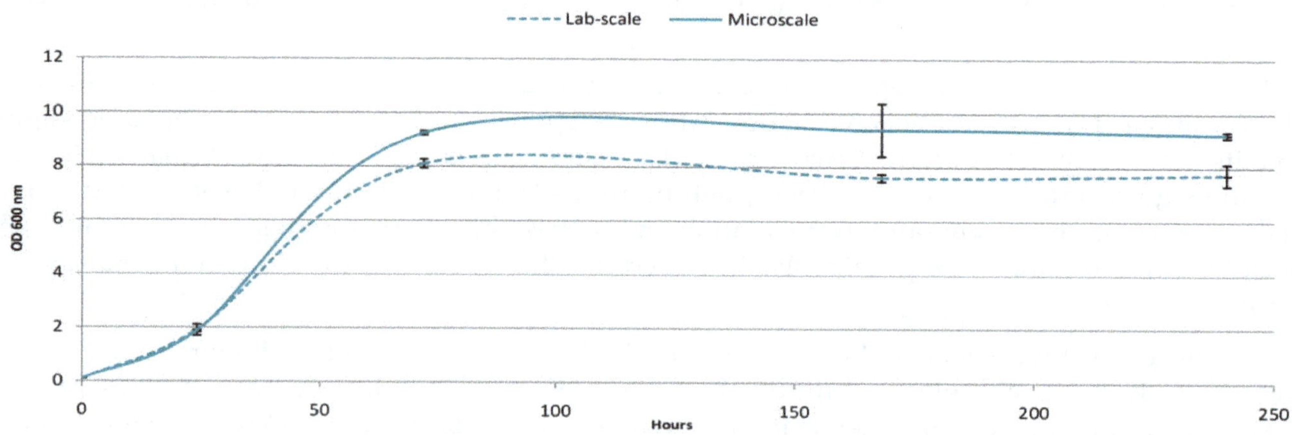

Figure 1. Fermentation performance in labscale wine and microwine.

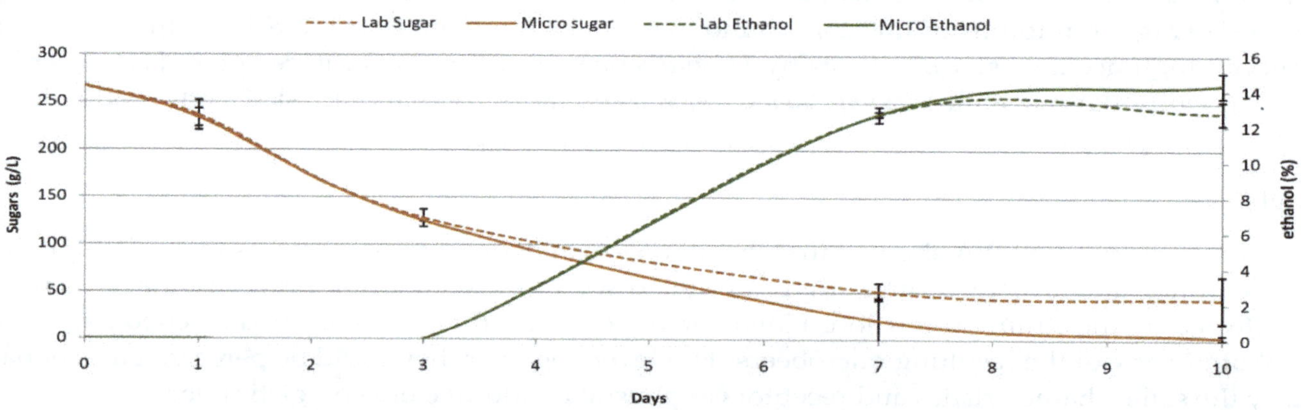

Figure 2. Yeast growth in labscale wine and microwine.

Most relevant levels of aroma compounds are presented in Figure 3. The significant differences that appeared in practically all the aroma compounds could also be explained by the levels of oxygen present in each scale. It has been demonstrated that differences in aeration causes differences in the yield of aroma compounds [32]. The initial oxygenation of the musts causes, in general, higher levels of alcohols and esters [32], which could explain the higher amounts of these aromas found in microscale compared to labscale. This was the case of isobutyl alcohol, isoamyl alcohol, phenylethanol, isoamyl acetate, ethyl caproate and ethyl caprylate. For instance, in the case of higher alcohols, these differences range from 4- to 20-fold [32]. One exception is ethyl acetate, which seems not to be affected by the levels of oxygen according to [32]. In our study, just slightly lower quantities of this compound were found in microscale. Regarding acetaldehyde and diacetyl, the former increased due to the higher presence of oxygen whereas the latter was not affected at all, being the same in both scales.

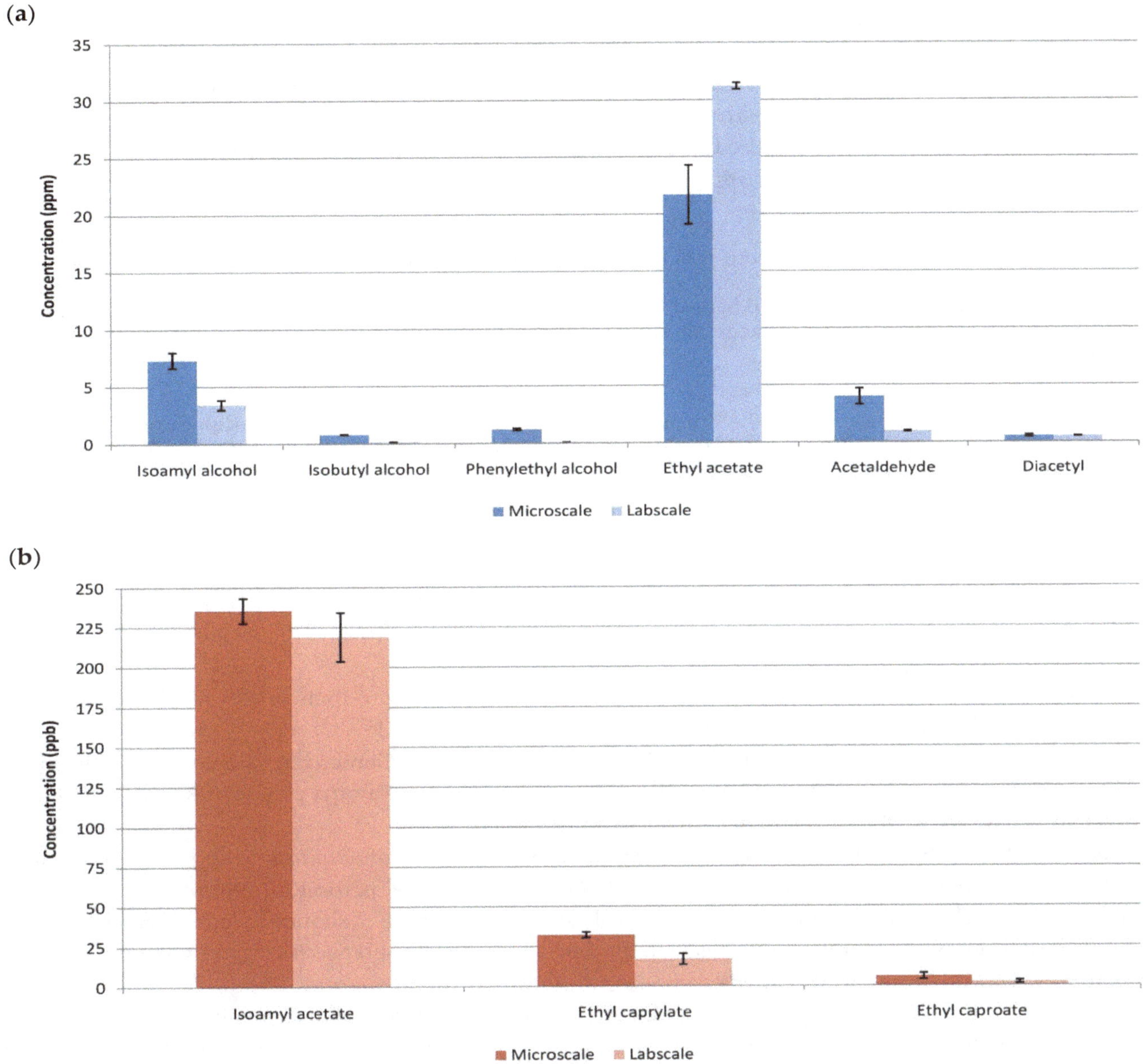

Figure 3. Aroma profiles at microwine and labscale. (**a**) Major compounds; (**b**) Minor compounds.

So the actual flavour levels in the microsystem were systematically higher than in the lab system. Besides oxygen, this matches with the slightly higher biomass levels and sugar consumption and does not compromise the screening power of the microsystem, although upscaling will always require fine tuning. Our results indicate that microsystems provide a useful and economical platform for the screening of interesting traits in industrial strains, as [31] affirmed, since the results are valid for comparison purposes. However, scaling-up will always be necessary in order to get closer to real winemaking or brewing processes.

4. Conclusions

Non-conventional yeasts show an enormous potential to improve and diversify aroma profiles of fermented beverages as wine and beer, especially regarding acetate ester yields and fruity aromas However, it has to be taken into account that mixed cultures of these species together to fermentative strains, *Saccharomyces* spp., is normally required to achieve successful fermentations. The adjustment of pitching ratio is crucial in these cases to avoid *Saccharomyces* spp. to take over and allow non-*Saccharomyces* yeasts to exploit all their aromatic potential.

Microsystems have proven their value as a tool for screening yeasts for key aromas in both wine and beer, giving the possibility of testing multiple combinations in a fast and simple way. However, scaling-up with the strains showing the highest potential will always be necessary in order to get closer to real winemaking or brewing outputs.

Author Contributions: Conceptualization, A.G., A.D., B.S. and C.d.J.; methodology, A.G., A.D., B.S. and C.d.J.; formal analysis, A.G.; investigation, A.G. and A.D.; resources, C.d.J.; writing—original draft preparation, A.G.; writing—review & editing, A.G., A.D., B.S. and C.d.J.; visualization, A.G.; supervision, A.G. and C.d.J.; project administration, A.G. and C.d.J.; funding acquisition, A.G., C.d.J. All authors have read and agreed to the published version of the manuscript.

Acknowledgments: The authors thank the European Comission for the funding; Westerdijk Fungal Biodiversity Institute-CBS-KNAW for providing most of the strains; Amparo Querol (IATA-CSIC) for providing commercial wine strain and Carlsberg Research Laboratory for providing wort for microbeer experiments.

References

1. Aslankoohi, E.; Herrera-Malaver, B.; Rezaei, M.N.; Steensels, J.; Courtin, C.M.; Verstrepen, K.J. Non-Conventional Yeast Strains Increase the Aroma Complexity of Bread. *PLoS ONE* **2016**, *11*, e0165126. [CrossRef] [PubMed]
2. Varela, C. The impact of non-*Saccharomyces* yeasts in the production of alcoholic beverages. *Appl. Microbiol. Biotechnol.* **2016**, *100*, 9861–9874. [CrossRef] [PubMed]
3. Dashko, S.; Zhou, N.; Tinta, T.; Sivilotti, P.; Lemut, M.S.; Trost, K.; Gamero, A.; Boekhout, T.; Butinar, L.; Vrhovsek, U.; et al. Use of non-conventional yeast improves the wine aroma profile of Ribolla Gialla. *J. Ind. Microbiol. Biotechnol.* **2015**, *42*, 997–1010. [CrossRef] [PubMed]
4. Jood, I.; Hoff, J.W.; Setati, M.E. Evaluating fermentation characteristics of *Kazachstania* spp. and their potential influence on wine quality. *World J. Microbiol. Biotechnol.* **2017**, *33*, 129. [CrossRef] [PubMed]
5. Varela, C.; Sengler, F.; Solomon, M.; Curtin, C. Volatile flavour profile of reduced alcohol wines fermented with the non-conventional yeast species *Metschnikowia pulcherrima* and *Saccharomyces uvarum*. *Food Chem.* **2016**, *209*, 57–64. [CrossRef] [PubMed]
6. Wang, X.-C.; Li, A.-H.; Dizy, M.; Ullah, N.; Sun, W.-X.; Tao, Y. Evaluation of aroma enhancement for "Ecolly" dry white wines by mixed inoculation of selected *Rhodotorula mucilaginosa* and *Saccharomyces cerevisiae*. *Food Chem.* **2017**, *228*, 550–559. [CrossRef]
7. Capece, A.; Romaniello, R.; Siesto, G.; Romano, P. Conventional and Non-Conventional Yeasts in Beer Production. *Fermentation* **2018**, *4*, 38. [CrossRef]
8. Michel, M.; Kopecká, J.; Meier-Dörnberg, T.; Zarnkow, M.; Jacob, F.; Hutzler, M. Screening for new brewing yeasts in the non-*Saccharomyces* sector with *Torulaspora delbrueckii* as model. *Yeast* **2016**, *33*, 129–144. [CrossRef]
9. Rijswijck, I.M.H.; Rooijackers, J.C.M.W.; Abee, T.; Smid, E.J. Performance of non-conventional yeasts in co-culture with brewers' yeast for steering ethanol and aroma production. *Microb. Biotechnol.* **2017**, *10*, 1591–1602. [CrossRef]

10. Holt, S.; Mukherjee, V.; Lievens, B.; Verstrepen, K.J.; Thevelein, J.M. Bioflavoring by non-conventional yeasts in sequential beer fermentations. *Food Microbiol.* **2018**, *72*, 55–66. [CrossRef]
11. Gutiérrez, A.; Boekhout, T.; Gojkovic, Z.; Katz, M. Evaluation of non-*Saccharomyces* yeasts in the fermentation of wine, beer and cider for the development of new beverages. *J. Inst. Brew.* **2018**, *124*, 389–402. [CrossRef]
12. Ravasio, D.; Carlin, S.; Boekhout, T.; Groenewald, M.; Vrhovsek, U.; Walther, A.; Wendland, J. Adding Flavor to Beverages with Non-Conventional Yeasts. *Fermentation* **2018**, *4*, 15. [CrossRef]
13. Swiegers, J.H.; Bartowsky, E.; Henschke, P.; Pretorius, I. Yeast and bacterial modulation of wine aroma and flavour. *Aust. J. Grape Wine Res.* **2005**, *11*, 139–173. [CrossRef]
14. Michel, M.; Meier-Dörnberg, T.; Jacob, F.; Methner, F.J.; Wagner, S.; Hutzler, M. Review: Pure non-*Saccharomyces* starter cultures for beer fermentation with a focus on secondary metabolites and practical applications. *J. Inst. Brew.* **2016**, *122*, 569–587. [CrossRef]
15. Šmogrovičová, D.; Dömény, Z. Beer volatile by-product formation at different fermentation temperature using immobilised yeasts. *Process. Biochem.* **1999**, *34*, 785–794. [CrossRef]
16. Zhou, N.; Schifferdecker, A.J.; Gamero, A.; Compagno, C.; Boekhout, T.; Piškur, J.; Knecht, W. *Kazachstania gamospora* and *Wickerhamomyces subpelliculosus*: Two alternative baker's yeasts in the modern bakery. *Int. J. Food Microbiol.* **2017**, *250*, 45–58. [CrossRef]
17. Comitini, F.; Gobbi, M.; Domizio, P.; Romani, C.; Lencioni, L.; Mannazzu, I.; Ciani, M. Selected non-*Saccharomyces* wine yeasts in controlled multistarter fermentations with *Saccharomyces cerevisiae*. *Food Microbiol.* **2011**, *28*, 873–882. [CrossRef]
18. Jolly, N.P.; Varela, C.; Pretorius, I.S. Not your ordinary yeast: Non-*Saccharomyces* yeasts in wine production uncovered. *FEMS Yeast Res.* **2013**, *14*, 215–237. [CrossRef]
19. Minnaar, P.; Du Plessis, H.; Jolly, N.; Van Der Rijst, M.; Du Toit, M. Non-*Saccharomyces* yeast and lactic acid bacteria in co-inoculated fermentations with two *Saccharomyces cerevisiae* yeast strains: A strategy to improve the phenolic content of Syrah wine. *Food Chem.* **2019**, *4*, 100070. [CrossRef]
20. Canonico, L.; Solomon, M.; Comitini, F.; Ciani, M.; Varela, C. Volatile profile of reduced alcohol wines fermented with selected non-*Saccharomyces* yeasts under different aeration conditions. *Food Microbiol.* **2019**, *84*, 103247. [CrossRef]
21. Gamero, A.; Ren, X.; Lamboni, Y.; de Jong, C.; Smid, E.J.; Linnemann, A.R. Development of a low-alcoholic fermented beverage employing cashew apple Juice and non-conventional yeasts. *Fermentation* **2019**, *5*, 71. [CrossRef]
22. Gamero, A.; Quintilla, R.; Groenewald, M.; Alkema, W.; Boekhout, T.; Hazelwood, L. High-throughput screening of a large collection of non-conventional yeasts reveals their potential for aroma formation in food fermentation. *Food Microbiol.* **2016**, *60*, 147–159. [CrossRef] [PubMed]
23. Sabel, A.; Martens, S.; Petri, A.; König, H.; Claus, H. *Wickerhamomyces anomalus* AS1: A new strain with potential to improve wine aroma. *Ann. Microbiol.* **2013**, *64*, 483–491. [CrossRef]
24. Gamero, A.; Tronchoni, J.; Querol, A.; Belloch, C. Production of aroma compounds by cryotolerant *Saccharomyces* species and hybrids at low and moderate fermentation temperatures. *J. Appl. Microbiol.* **2013**, *114*, 1405–1414. [CrossRef] [PubMed]
25. Arroyo-Lopez, F.N.; Torrado, R.P.; Querol, A.; Barrio, E. Modulation of the glycerol and ethanol syntheses in the yeast *Saccharomyces kudriavzevii* differs from that exhibited by *Saccharomyces cerevisiae* and their hybrid. *Food Microbiol.* **2010**, *27*, 628–637. [CrossRef]
26. Torrado, R.P.; Oliveira, B.M.; Zemančíková, J.; Sychrova, H.; Querol, A. Alternative Glycerol Balance Strategies among *Saccharomyces* Species in Response to Winemaking Stress. *Front. Microbiol.* **2016**, *7*, 628. [CrossRef]
27. Erten, H.; Tangüler, H.; Cakiroz, H. The Effect of Pitching Rate on Fermentation and Flavour Compounds in High Gravity Brewing. *J. Inst. Brew.* **2007**, *113*, 75–79. [CrossRef]
28. Verbelen, P.J.; Dekoninck, T.M.L.; Saerens, S.M.G.; Van Mulders, S.E.; Thevelein, J.M.; Delvaux, F.R. Impact of pitching rate on yeast fermentation performance and beer flavour. *Appl. Microbiol. Biotechnol.* **2009**, *82*, 155–167. [CrossRef]
29. Bartowsky, E.J.; Henschke, P.A. The 'buttery' attribute of wine—Diacetyl—Desirability, spoilage and beyond. *Int. J. Food Microbiol.* **2004**, *96*, 235–252. [CrossRef]
30. Yakobson, C. Pure Culture Fermentation Characteristics of *Brettanomyces* Yeast Species and Their Use in the Brewing Industry. Ph.D. Thesis, Heriot-Watt University, Edinburgh, UK, 2010.

31. Liccioli, T.; Tran, T.M.T.; Cozzolino, D.; Jiranek, V.; Chambers, P.J.; Schmidt, S.A. Microvinification—How small can we go? *Appl. Microbiol. Biotechnol.* **2010**, *89*, 1621–1628. [CrossRef]
32. Valero, E.; Moyano, L.; Millan, M.; Medina, M.; Ortega, J. Higher alcohols and esters production by *Saccharomyces cerevisiae*. Influence of the initial oxygenation of the grape must. *Food Chem.* **2002**, *78*, 57–61. [CrossRef]

4

Modulating Wine Pleasantness throughout Wine-Yeast Co-Inoculation or Sequential Inoculation

Alice Vilela

Department of Biology and Environment, School of Life Sciences and Environment, University of Trás-os-Montes and Alto Douro, CQ-VR, Chemistry Research Centre, 5001-801 Vila Real, Portugal; avimoura@utad.pt

Abstract: Wine sensory experience includes flavor, aroma, color, and (for some) even acoustic traits, which impact consumer acceptance. The quality of the wine can be negatively impacted by the presence of off-flavors and aromas, or dubious colors, or sediments present in the bottle or glass, after pouring (coloring matter that precipitates or calcium bitartrate crystals). Flavor profiles of wines are the result of a vast number of variations in vineyard and winery production, including grape selection, winemaker's knowledge and technique, and tools used to produce wines with a specific flavor. Wine color, besides being provided by the grape varieties, can also be manipulated during the winemaking. One of the most important "tools" for modulating flavor and color in wines is the choice of the yeasts. During alcoholic fermentation, the wine yeasts extract and metabolize compounds from the grape must by modifying grape-derived molecules, producing flavor-active compounds, and promoting the formation of stable pigments by the production and release of fermentative metabolites that affect the formation of vitisin A and B type pyranoanthocyanins. This review covers the role of *Saccharomyces* and non-*Saccharomyces* yeasts, as well as lactic acid bacteria, on the perceived flavor and color of wines and the choice that winemakers can make by choosing to perform co-inoculation or sequential inoculation, a choice that will help them to achieve the best performance in enhancing these wine sensory qualities, avoiding spoilage and the production of defective flavor or color compounds.

Keywords: wine yeasts; lactic acid bacteria; co-inoculation; sequence inoculation; flavor compounds; color pigments

---

1. Introduction

1.1. The Human Senses in Wine Evaluation

Five senses are involved in perceiving wine sensory quality: sight, taste, hearing, touch, and smell. Color perception results from the stimulus of the retina by light (wavelengths 380 to 760 nm). In wine, color and appearance are the first attributes by which quality is assessed. According to Spence [1], color is the most important product-intrinsic indicator used by consumers when searching, purchasing, and subsequently consuming food or a libation. Color, clarity, and hue affect the perception of other attributes such as flavor due to the association with color. For example, a yellow/green beverage is expected to have a lemon flavor and an acidic taste.

Taste is a chemical sense and happens when taste stimuli contact with the taste receptors located on the tongue, called taste buds. Humans can distinguish six basic tastes: sweet, sour, salty, bitter, umami, and fatty [2,3]. Between 20 and 30 levels of intensity can be distinguished for each taste, and each taste quality represents different nutritional or physiological requirements, or a potential dietary risk [4].

Sound (waves which strike the eardrum, causing it to vibrate [5]) is also important when judging a wine. For instances, when we hear a champagne cork popping, it is a sign that the wine has an enjoyable gas.

Texture in wine can be defined as the total sum of kinesthetic sensations derived from oral manipulation. It encompasses mouthfeel, masticatory properties, residual properties, and even visual and auditory properties [6].

Aroma and flavor are chemical senses stimulated by the chemical properties of odor molecules which must reach the olfactory bulb to interact with olfactory cells in the olfactory mucosa [7]; therefore, to smell, molecules must be airborne (i.e., volatile). The sensory term which we call "flavor" is a mingled experience based on human judgment, built on personal differences in perception thresholds.

In conclusion, and as reported by Swiegers et al. [8], all of the senses play a key role in wine/flavor development—color, aroma, mouthfeel, sound, and, ultimately, taste. Altogether, these sensory perceptions are very complex. Wine contains many flavor and aroma-active compounds. Terpenes, methoxypyrazines, esters, ethanol and other alcohols and aldehydes impart distinct flavors and aromas (floral, pepper, fruit, woody and vinylic flavors, among others) to wine [9–11]. The taste of wine can be described as sweet, sour, salty, umami, bitter, and, to a lesser extent, fat [12]. These properties are the result of the presence of sugars, polyols, salts, polyphenols, flavonoid compounds, amino acids, and fatty acids. Compounds such as glycerol, polysaccharides, and mannoproteins contribute to the viscosity and mouthfeel of wines [13]; grape anthocyanins contribute to the color [14], and ethanol, by sheer mass, also carries other alcohols along, promoting a mouth-warming effect [15].

1.2. Main Wine Aroma and Flavor Compounds from the Fermentative Origin

Yeast and bacteria are vital to the development of wine flavor. Many biosynthetic pathways, in wine yeast and malolactic bacteria, are responsible for the formation of wine aroma and flavor. However, we cannot discard the other factors that can also influence the wine chemical composition, such as viticultural practices, grape-must composition, pH, fermentation temperature, and technological aspects associated with the vinification process [8]. So, depending on their origin, wine aroma and flavor compounds can be named varietal aromas (originating from the grapes), fermentative aromas (originating during alcoholic and malolactic fermentations), and aging aromas (developed during the reductive or oxidative wine-aging that depends on storage conditions) [16].

Most of the wine aroma and flavor compounds are produced or released during wine fermentation due to microbial activities of *Saccharomyces* and non-*Saccharomyces* yeast genera (*Brettanomyces, Candida, Debaryomyces, Hanseniaspora, Hansenula, Kloeckera, Kluyveromyces, Lachancea, Metschnikowia, Pichia, Saccharomycodes, Schizosaccharomyces, Torulaspora*, and *Zygosaccharomyces*). Both in spontaneous and inoculated wine fermentations, non-*Saccharomyces* are important in early stages of the fermentation, before *Saccharomyces* becomes dominant in the culture, and contribute meaningfully to the global aroma profile of wines by producing flavor-active compounds [17,18].

A group of aroma compounds has been directly linked to specific varietal flavors and aromas in wines [19,20]. Most of these compounds are present at low concentrations in both grapes and fermented wine. These aroma compounds are found in grapes in the form of non-odorant precursors that, due to the metabolic activity of *Saccharomyces* and non-*Saccharomyces* yeast during fermentation, are transformed to aromas and flavor that are of great relevance in the sensory perception of wines [20] (Table 1).

Table 1. Main odorants contributing to varietal aromas of some monovarietal wines.

Compounds	Main Cultivars	Odour Descriptor	Ref.
Geraniol	Muscat, Gewurztraminer	Citrus, floral, geranium	[21]
Linalool	Muscat, Gewurztraminer	Floral, lavender	[21]
Nerol	Muscat	Floral	[21]
Tetrahydro-4-methyl-2-(2-methyl-1-propenyl)-2,5-cis-2h-pyran (cis-rose oxide)	Gewurztraminer	Geranium oil	[22]
3,6-Dimethyl-3a,4,5,7a-Tetrahydro-3h-1-Benzofuran-2-One	Gewurztraminer	Coconut, woody, sweet	[23]
3-Isobutyl-2-Methoxypyrazines	Sauvignon blanc	Bell pepper	[24,25]
4-Methyl-4-Mercaptopentan-2-One	Sauvignon blanc	Black currant	[24,25]
3-Mercapto-1-Hexanol (R Isomer)	Sauvignon blanc	Grapefruit, citrus peel	[24,25]
1,1,6-Trimethyl-1,2-Dihydronaphthalene	Riesling	Kerosene	[26]
3-Mercapto-1-Hexanol (S Isomer)	Semillon	Passion fruit	[27]
Rotundone	Shiraz	Black pepper	[28]

During alcoholic fermentation, some yeast, mainly non-*Saccharomyces* yeasts, can release β-glucosidases that hydrolyze the glycosidic bonds of the odorless non-volatile glycoside linked forms of monoterpenes (geraniol, linalool, nerol, among others), releasing the odor compounds to the wine [29]. Volatile thiols that give Sauvignon blanc wines their characteristic aroma (bell pepper, black currant, grapefruit, and citrus peel) are not present in grape juice but occur in grape must as odorless, non-volatile, cysteine-bound conjugates. During fermentation, the wine yeasts are responsible for the cleaving of the thiol from the precursor [30].

However, the major groups of aromas and flavor compounds from the fermentative origin are ethanol, higher alcohols or fusel alcohols, and esters. The biosynthetic pathways responsible for the formation of higher alcohols, the Ehrlich pathway, or the enzymes responsible for the formation of esters, have been studied in wine yeasts [31].

Higher alcohols are derived from amino acid catabolism via a pathway that was first described by Ehrlich [32] and later revised by Neubauer and Fromherz in 1911 [33]. Amino acids that are assimilated by the Ehrlich pathway (valine, leucine, isoleucine, methionine, and phenylalanine), present in grape must are metabolized by yeasts, sequentially, throughout the fermentation. Figure 1 shows the metabolism of phenylalanine with the production of 2-phenylethanol and, after oxidation of phenylacetaldehyde, the formation of phenylacetate. Both compounds possess a pleasant rose-like aroma/flavor.

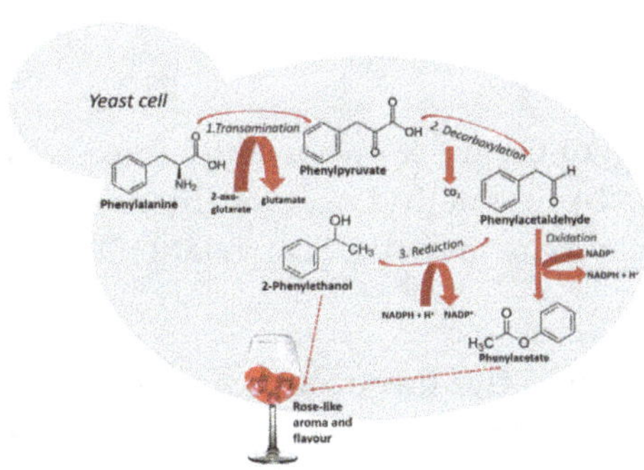

Figure 1. Schematic representation of the Ehrlich pathway for the catabolism of the aromatic amino acid, phenylalanine leading to the formation of 2-phenylethanol [34]. This biosynthetic pathway consists of three steps (reactions 1, 2 and 3): first, amino acids are deaminated to the corresponding α-ketoacids, in reactions catalyzed by transaminases. In a second step, α-ketoacids are decarboxylated and converted to their corresponding aldehydes (five decarboxylases are involved in this process), in a third step, alcohol dehydrogenases (Adh1p to Adh6p and Sfa1p) catalyze the reduction of aldehydes to their corresponding higher alcohols [35].

Studies have shown that profiles and concentrations of higher alcohols produced vary by yeast species, even when the fermentation conditions are similar, which indicates that the mechanisms that regulate the Ehrlich pathway are diverse in non-*Saccharomyces* yeasts compared to *Saccharomyces* [16,36]. So, Ehrlich pathway mechanisms should be explored in detail in non-*Saccharomyces* yeasts as it contributes to the formation of important and flavorful wine aromas [36].

The most important esters are synthetized by yeasts during alcoholic fermentation as a detoxification mechanism since they are less toxic than their correspondent alcohol or acidic precursors. Moreover, their synthesis serves as a mechanism for the regeneration of free CoA from its conjugates [16,37].

Esters (Figure 2) that contribute to wine aroma, derived from fermentation, belong to two categories: the acetate esters of higher alcohols and the ethyl esters of medium-chain fatty acids (MCFA). Acetate esters are formed inside the yeast cell, and in *S. cerevisiae* the reaction is metabolized by two alcohol acetyltransferases, AATase I and AATase II (encoded by genes *ATF1* and *ATF2* [35,38]). Eat1p is responsible for the production of acetate and propanoate esters [39,40]. Most medium-chain fatty acid ethyl ester biosynthesis during fermentation is catalyzed by two enzymes, Eht1p and Eeb1p [38,41].

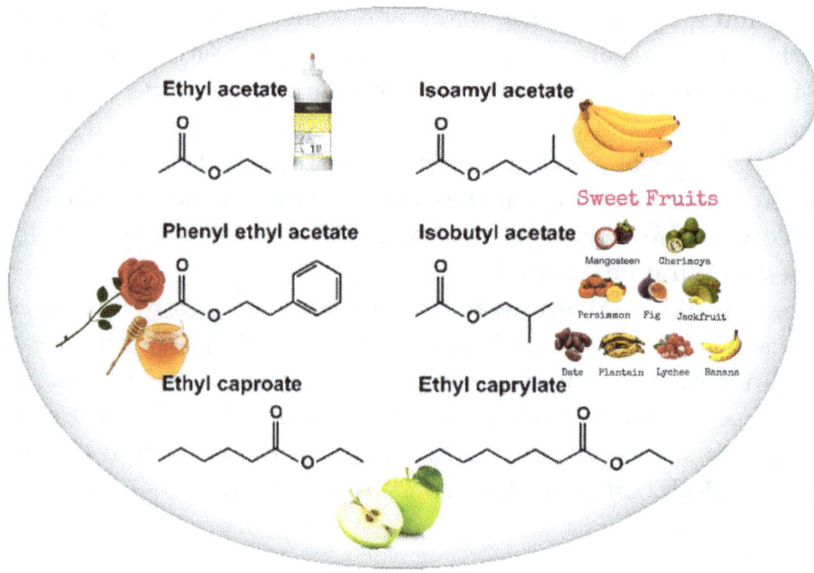

Figure 2. Schematic representation of the most important wine esters: ethyl acetate (glue-like aroma), isoamyl acetate (banana aroma), 2-phenylethyl acetate (roses and honey aromas), isobutyl acetate (sweet-fruits aromas), and ethyl caproate and ethyl caprylate with a sour-apple aroma [38].

Volatile fatty acids also contribute to the flavor and aroma of the wine. During yeast fermentation, long-chain fatty acids (LCFAs) are also formed via the fatty-acid synthesis pathway from acetyl-CoA in concentrations varying from ng/L to g/L [42]. Medium-chain fatty acids (MCFAs (C_6 to C_{12})) are produced primarily by yeasts as intermediates in the biosynthesis of LCFAs that are prematurely released from the fatty acid synthase complex. These acids (Table 2) directly contribute to the flavor of wine or serve as substrates that participate in the formation of ethyl acetates [43]. As most have unpleasant aromas (see Table 2), their formation should be minimized.

Table 2. Main medium-chain fatty acids (MCFAs (C_6 to C_{12})), produced by yeasts during alcoholic fermentation.

Fatty Acid	Associated Aroma	Odor Threshold (μg/L) [1]
Butanoic Acid	Rancid butter or baby vomit aroma	173
Hexanoic Acid	Sour, fatty, sweat, cheese	420
Octanoic Acid	Fatty, waxy, rancid oily, vegetable, cheesy	500
Decanoic Acid	Unpleasant rancid, sour, fatty, citrus	1000
2-Methylpropanoic Acid	Acidic sour, cheese, dairy, buttery, rancid	2300
2- and 3-Methylbutyric Acid	Sour, stinky feet, sweaty, cheese, tropical	33

[1] Measured in model wine, water/ethanol (90 + 10, *w/w*) [44].

Sulfur-containing compounds can also be formed by yeasts during alcoholic fermentation. They are usually perceived as off-flavors. The sulfur-containing compounds can be derived from the grape and the metabolic activities of yeast and bacteria. They can also occur due to the chemical reactions during the wine aging and storage and also due to environmental contamination [45]. They can be formed by enzymatic mechanism as the products of metabolic and fermentative pathways whose substrates are both amino acids and some sulfur-containing pesticides. When wine microorganisms metabolize these thiols, the sulfur compounds formed are considered off-flavors [46] which convey negative notes such as cabbage, garlic, onion, rotten eggs, rubber, and sulfur to wines [47]. However, there are some volatile thiols that may confer enjoyable aromatic notes at trace levels, such as 4-mercapto-4-methylpentan-2-one (4MMP), 3-mercaptohexan-1-ol (3MH), already mentioned in Table 1, and 3-mercaptohexyl acetate (3MHA), important for the characterization of the typical Sauvignon Blanc wine aroma [24,25,48].

Finally, another important family of aromatic compounds present in wines are the carbonyl compounds. In this group we may include acetaldehyde, acrolein, ethyl carbamate, formaldehyde, and furfural [49]. Several factors may contribute to the presence of carbonyl compounds in wines, including the fermentation of over-ripe grapes and increasing the maceration time, probably due to increased concentration of the precursors like amino acids and glucose in the must [50]. Due to their carbonyl group, carbonyl compounds present a high reactivity with the nucleophile's cellular constituents [51] and may cause cell damage. So, these compounds are toxic, and their formation should be avoided.

2. Yeast Modulation of Wine Aroma and Flavor Compounds

2.1. *Non-Saccharomyces and Saccharomyces Co-inoculation vs. Sequential Inoculation*

The wine industry attempts to diversify producing wines with distinctive characteristics and creating high-quality new products. A true test for winemakers is to blend several grapes, grown on different soil and climate conditions (terroir), with a developing science of yeast and bacterial metabolism, to produce the most enjoyable wine [52]. Many winemakers today use commercial yeast and bacteria starter cultures for alcoholic and malolactic fermentation, respectively. The selection of a "fit-for-purpose" starter strain has a pivotal role in optimizing flavor and aroma.

There is no consensus on the impact of indigenous yeasts on wine sensory properties; while some researchers show a positive effect, others show negative effects on the wine chemical composition and sensory properties [53,54]. For example, Varela et al. [55] showed an increase in the concentration of some higher alcohols and esters in wines produced with autochthonous yeasts compared to the wines produced with commercial yeasts. Moreover, some non-*Saccharomyces* yeast may increase the concentration of biogenic amines in wines [56].

In red winemaking, significant increases in the concentrations of desirable compounds such as ethyl lactate (sweet, fruity, acidic, ethereal with a brown nuance aroma), 2,3-butanediol (buttery aroma), 2-phenylethanol, and 2-phenylethyl acetate (both with a rose-like scent) can be obtained when non-*Saccharomyces* yeasts are introduced into the fermentation process [57,58].

Another selection criterion for non-*Saccharomyces* yeasts, when aiming to improve the wine aroma, is the presence of β-glucosidase activity that favors the hydrolysis of the non-volatile aromatic precursors from the grape [59,60]. Non-*Saccharomyces* species display superior β-glucosidase activity to that of *Saccharomyces* species, which has been defined as intracellular and strain-dependent [23].

Budic-Leto et al. [61] found that Prosek, a traditional Croatian dessert wine, produced with native and inoculated yeasts differed in its volatile compounds. Using descriptive sensory analysis, it was shown that the sensory properties of the wines were significantly different depending on the type of fermentation, namely determined for the attributes strawberry jam aroma, and fullness. So, recently, non-*Saccharomyces* yeast species have been suggested for winemaking as they could contribute to the improvement of wine quality mostly in terms of aromatic characteristics [62,63]. Thus, starter cultures composed of non-*Saccharomyces* yeasts together with *S. cerevisiae* have been used for co-, or sequential wine fermentations [64].

Co-inoculation involving *S. cerevisiae* and non-*Saccharomyces* yeasts species typically results in the disappearance (or the presence in relative low amounts) and loss of viability of non-*Saccharomyces* [65,66]. Though *S. cerevisiae* dominance can be explained by the depletion of sugar and nutrients from the grape must followed by ethanol production and lack of oxygen, some direct mechanisms for yeast species antagonism have also been described: (i) killer factors (so-called killer toxins or killer proteins), which are secreted peptides, encoded by extrachromosomal elements of *S. cerevisiae* that affect other yeast species [67]; (ii) similar compounds have also been described for *Torulaspora delbrueckii* species [68] and for the genera *Pichia, Kluyveromyces, Lachancea, Candida, Cryptococcus, Debaryomyces, Hanseniaspora, Hansenula, Kluyveromyces, Metschnikowia, Torulopsis, Ustilago, Williopsis,* and *Zygosaccharomyces,* indicating that the killer phenomenon is indeed widespread among yeasts. [69]; (iii) *S. cerevisiae* are also able to secret antimicrobial peptides (AMPs) during alcoholic fermentation that are active against wine-related yeasts (e.g., *Dekkera bruxellensis*) and bacteria (e.g., *Oenococcus oeni*). These AMPs correspond to fragments of the *S. cerevisiae* glyceraldehyde 3-phosphate dehydrogenase (GAPDH) protein [70].

Several authors agree that the sequential culture is better than the mixed culture, especially because it allows for a greater expression of the metabolism of non-*Saccharomyces* yeasts at the beginning of fermentation [71,72]. However, as recently reported by Loira et al. [73], the winemaker selection criteria for performing co-inoculation or sequential inoculation with the appropriate non-*Saccharomyces* is dependent on the characteristics of the wine to be produced, including the desired sensory properties. The ratio of inoculation (non-*Saccharomyces* vs. *Saccharomyces*) is also a subject that must be considered. Moreover, the contribution of the inoculation of non-*Saccharomyces* strains to wine fermentation can be direct or indirect, through biological interactions with *S. cerevisiae*. Recently, Renault et al. [74,75] described a synergic interaction between *S. cerevisiae* and *T. delbrueckii* resulting in increased levels of 3-sulfanylhexan-1-ol a compound that presents a sulphurous aroma and an initially fruity flavor. However, with over-aging, the aroma/flavor evolves to savory and chicken meaty with roasted coffee shades and a hint of fruitiness [76].

Not long ago, García et al. [63] performed small-scale fermentations where they studied the oenological characterization of five non-*Saccharomyces* native yeast species under several co-culture conditions in combination with a selected strain of *S. cerevisiae*, aiming to improve the sensory characteristics of the Malvar wines. Sequential inoculations were elaborated with *S. cerevisiae* CLI 889 in combination with several non-*Saccharomyces*: (i) *T. delbrueckii* CLI 918, which produced wines with a lower ethanol content and higher fruity and floral aroma; (ii) *C. stellata* CLI 920, which augmented the aroma complexity and glycerol content; (iii) *L. thermotolerans* 9-6C, which produced an increase in acidity and floral and ripe-fruit aroma and a lower volatile acidity; (iv) *Schizosaccharomyces pombe*, which produced wines with fruity aromas; and (v) *M. pulcherrima*, which produced wines with lower volatile acidity and an increase of glycerol and ripe-fruit aroma.

Continuing their work on Malvar wines, García et al. [77] performed fermentations at the pilot scale, using sequential-inoculation strategies which resulted in wines that tasters were able to distinguish from the controls. Moreover, the wines were most appreciated, namely, those produced in sequential cultures with *T. delbrueckii* CLI 918/*S. cerevisiae* CLI 889 and *C. stellata* CLI 920/*S. cerevisiae* CLI 889 and, also, with mixed and sequential cultures of *L. thermotolerans* 9-6C/*S. cerevisiae* CLI 889 strains. Studies have shown that sequential cultures can produce more different wines, when compared with the controls, providing sensory properties associated with the non-*Saccharomyces* strains. Some strains of *T. delbrueckii* in sequential fermentation with *S. cerevisiae* can produce significant amounts of 3-ethoxy-1-propanol [78], with a fruity-like aroma with a low perception threshold, 0.1 mg/L [79].

However, sequential inoculation is not only favorable for positive aromas sequential. It is a fermentation technique that can be used to prevent or diminish the production of some undesirable compounds, augmenting the production of others. Viana et al. [80] reported a decrease in the higher alcohol's concentration (considered as possessing a fuel-like aroma) from 452.5 mg/L (control) to 306.2 mg/L when carrying out mixed fermentations with *H. osmophila* and *S. cerevisiae*. Moreover, higher concentrations of 2-phenylethyl acetate (rose-like aroma) were obtained. A higher intensity of fruitiness was also detected in these wines when compared to the control wine, obtained throughout the fermentation of a pure *S. cerevisiae* culture.

2.2. Saccharomyces and Lactic Acid Bacteria co inoculation vs. Sequential Inoculation

The vinification involves different microbiological processes, mainly alcoholic fermentation (AF) conducted by *Saccharomyces cerevisiae* and malolactic fermentation (MLF) conducted by lactic acid bacteria (*Oenococcus oeni*). These two distinctive fermentation processes represent an essential step in the improvement of the quality of red wines [81].

MLF naturally occurs after AF, however, the timing of the start of MLF depends on several parameters like temperature, pH, alcoholic degree and the concentration of sulfur dioxide (SO_2), as well as on certain yeast metabolites available, such as medium-chain fatty acids and peptidic fractions [82,83]. The success of spontaneous MLF is not always guaranteed, and the addition of starter culture can improve its viability. To overcome this issue, the winemakers may carry out traditional LAB inoculation after alcoholic fermentation (sequential inoculation), or simultaneous inoculation in the must with yeast (co-inoculation). The co-inoculation has, in the last several years, been adopted by some winemakers, particularly in warm climates with higher temperatures, where high concentrations of ethanol can inhibit LAB growth [84].

There are not many studies focusing on the impact of the co-inoculation technique on the aromatic and biochemical profile of wines. Abrahamse and Bartowsky [85] and Knoll et al. [86] have shown that the timing of inoculation with LAB in both white and red wines could influence the profile of aromatic yeast-derived compounds such as higher alcohols, terpene, esters, and fatty acids. However, these works were performed at the lab scale, and no sensory evaluation was performed on the wines.

Antalick and collaborators [87] demonstrated the impact of the timing of inoculation with LAB on the metabolic profile of wines manufactured at production scale. They have clearly shown that this technique has an impact on the aromatic profile of the wines, mainly in the presence of lactic and fruity notes. Co-inoculation can modulate the intensity of these descriptors, due to the production/degradation of metabolites or by the development of an aromatic mask over the short and long term. Moreover, they discuss that the metabolic and aromatic changes that occur with co-inoculation depend strongly on the yeast and LAB strains, as well as on the composition of the must. Co-inoculation of musts at the beginning of vinification can also lead to a faster vinification process without an excessive increase in volatile acidity [88].

Due to the importance of the interaction between yeasts (*Saccharomyces* and non-*Saccharomyces*) and bacteria during wine processing, Berbegal et al. [89] applied a next-generation sequencing (NGS) analysis to several fermentation modalities: uninoculated must, pied-de-cuve, *S. cerevisiae*, *S. cerevisiae*, and *Torulaspora delbrueckii* co-inoculated and sequentially inoculated, along with *S. cerevisiae* and *Metschnikowia pulcherrima* co-inoculated and sequentially inoculated, continued by spontaneous malolactic consortium to perform MLF. Each experimental trial led to the different taxonomic composition of the bacterial communities of the malolactic consortia, in terms of prokaryotic phyla and genera. Among other interesting findings, they found that MLF was delayed when *M. pulcherrima* was inoculated and was even inhibited when the inoculated yeast strain was *T. delbrueckii* [89]. Thus, an antagonistic effect of *M. pulcherrima* and especially *T. delbrueckii* on lactic bacteria population has been proven, which may be due to the ability of *T. delbrueckii* to produce "toxins" or killer factors that prevent bacteria growth [68] and to the ability of *M. pulcherrima* to produce high amounts of pulcherrimin (an iron chelator) that inhibits the growth of bacteria [90].

3. Yeast Modulation of Wine Color and Pigment Formation

Anthocyanins and their derivatives, originating in the grapes, are the main pigments responsible for the red wine color, and their structural modifications result in a characteristic variation of color in red wines, from pale ruby (young red wine) to deep purple-red color (aged red wine) [91,92]. Such variations can also result in changes in wine mouthfeel and flavor.

Saccharomyces yeasts can directly or indirectly contribute to wine color, altering color parameters such as intensity and tonality by: (i) increasing the formation of stable pigment precursors (vinyl phenols, acetaldehyde, and pyruvic acid); and (ii) modifying the pH due to organic acid metabolism (production or consumption) [73].

Pyruvic acid and acetaldehyde promote the formation of vitisins of types A and B, respectively [93, 94], during must fermentation, (Figure 3A). Vitisins contribute more to wine color parameters than unmodified anthocyanins and exhibit a hypsochromic shift, i.e., a change of spectral band position (in the absorption, reflectance, transmittance, or emission spectrum) to a shorter wavelength corresponding to a higher frequency. Vitisins can change towards an orange-red hue due to the long conjugation afforded by pyran ring [95]. Moreover, their color expression remains stable against discoloration due to the presence of sulfur dioxide (bleaching capacity) or changes in pH [95] (Figure 4).

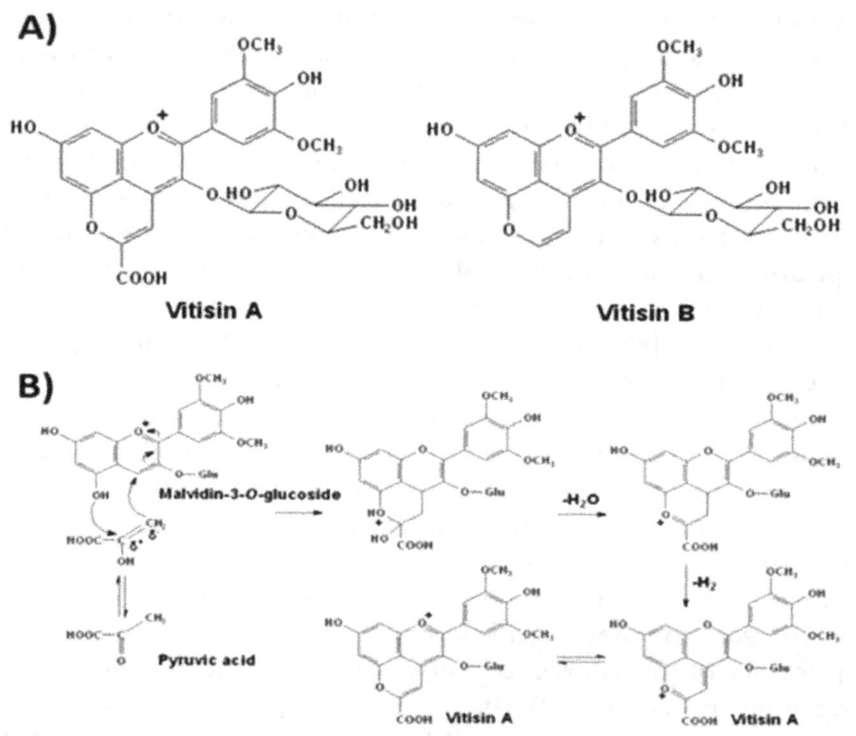

Figure 3. (**A**) Structures of vitisin A and vitisin B generated from malvidin-3-*O*-glucoside. (**B**) The formation mechanism of vitisin A produced by condensation of an anthocyanin (malvidin-3-*O*-glucoside) with pyruvic acid [92,96,97].

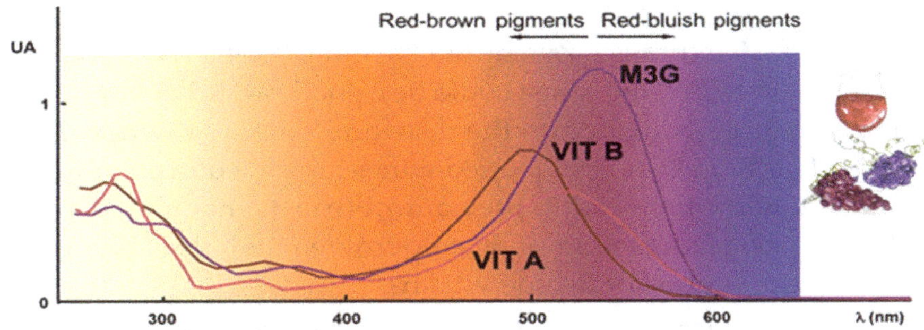

Figure 4. UV-visible spectra of malvidin-3-*O*-glucoside and vitisins A and B. Adapted from [98].

During the fermentation process, practices like pellicular maceration that increase the extraction of anthocyanins from the skins of grape berries will promote the formation of vitisins [14]; also, acetaldehyde and pyruvic acid, as mentioned before, are important intermediate compounds in yeast metabolism and influence wine vitisin content [93,99,100].

During sugars' fermentation by yeasts, pyruvate is metabolized into acetaldehyde, with the latter being the terminal electron acceptor in the formation of ethanol. Acetaldehyde and pyruvate, formed in yeast cytoplasm, are rapidly metabolized (the first is reduced to ethanol, and the second is either decarboxylated to acetaldehyde or used in the formation of acetyl-CoA). However, some of the acetaldehyde and pyruvate molecules, through cell lysis, pass to the wine medium and are sufficiently reactive to attack other molecules, enabling the transformation of anthocyanins into compounds such as pyranoanthocyanins and their secondary generated pigments (anthocyanin oligomers and polymeric anthocyanin) [92,97]. Pyroanthocyanins are the most important group of anthocyanin derivatives present in fermented beverages, including wine, and the A-type vitisins or carboxy-pyroanthocyanins are produced, as mentioned before, by condensation of anthocyanin with pyruvic acid [97] (Figure 3B).

In terms of vitisin kinetic formation, during *S. cerevisiae* fermentation, type-A vitisins are produced in the first six days of fermentation (when pyruvic acid is available). At the end of fermentation, when nutrients are limited, the amount of acetaldehyde is high enough to lead to the formation of type-B vitisins [93,99]. So, to generate a more pleasant red wine color, before fermentation the winemaker must select the wine yeast strains that will be able to increase anthocyanin extraction and/or can produce more pyruvic acid and acetaldehyde. Postponing or starting MLF early can prevent the consumption of pyruvic acid and acetaldehyde by lactic acid bacteria [101], increasing the possibility for vitisin synthesis.

Oxidative fermentation (fermentation in barrels or with micro-oxygenation) and wine aging in wood give rise to pyruvic acid and acetaldehyde, increasing the vitisin levels and, consequently, color intensity and stability [97]. Yellowish α-pyranone-anthocyanins called oxovitisins were also described by He et al. [102] in aged red wines derived from the direct oxidation of A-type vitisins. Moreover, A-type vitisins are the pyranoanthocyanins detected in higher concentrations in port wines. Port wine is made by stopping the fermentation process with the addition of wine spirit "aguardente", leaving the wine with a high concentration of sugars. So, when fermentation is stopped, the pyruvic acid concentration is relatively high and increases after wine fortification [97].

Vinylphenolic pyroanthocyanin adducts result from the condensation between vinyl phenols and anthocyanins. These color compounds also show high color stability [94]. Yeast strains are also able to affect the concentration and the composition of wine tannins as well as the degree of tannin polymerization [103], thus, indirectly throughout yeast actions affecting stabilization of anthocyanins, and consequently, stabilization of color can occur due to reaction between anthocyanins and tannins forming pigmented tannins and through copigmentation of anthocyanins [104].

Additionally, what about non-*Saccharomyces* wine yeasts? Well, we have already mentioned that an improvement in fermentation quality, efficiency, and wine pleasantness is obtained when sequential or co-inoculation of non-*Saccharomyces* and *Saccharomyces* yeast is performed.

Torulaspora delbrueckii is one non-*Saccharomyces* species available commercially (Viniflora® Harmony.nsac and Viniflora® Melody.nsac, Zymaflore® Alpha, BIODIVA®, and Viniferm NS-TD® are some commercial examples), therefore, this yeast could be a good candidate for wine color improvement as it is reported to have a positive influence on the taste and aroma of wines. For instance, Pinotage grape must inoculate with *T. delbrueckii* originated red wines improved in color intensity (anthocyanins) and mouthfeel (flavanols) when compared to the control (musts inoculated with *S. cerevisiae*) [105]. However, *T. delbrueckii* has poor production of acetaldehyde [106], thus being a poor candidate for wine color improvement in the context of B-type vitisins.

Other non-*Saccharomyces* yeasts, not yet available as commercial products, but studied in the academic community, could be good candidates for wine color improvement. Medina et al. [107] reported that in the case of co-fermentation of *Metschnikowia* and *Hanseniaspora* with *S. cerevisiae*, only an increase in the content of B-type vitisins occurred, probably due to the enhanced acetaldehyde formation.

Further experiments performed by sequential inoculation of *Schizosaccharomyces pombe* and *Lachancea thermotolerans* exposed an increase of type-A vitisins when compared with the control *S. cerevisiae* [108]. Also, several authors detected interesting features in non-*Saccharomyces* yeasts. *P. guilliermondii* strains presenting a high hydroxycinnamate decarboxylase activity may improve the formation of vinylphenolic pyranoanthocyanins; non-*Saccharomyces* yeasts, such as *Candida valida*, *Metschnikowia pulcherrima*, *Kloeckera apiculata* and *Starmerella bombicola*, which synthesize and release pectolytic enzymes, can improve wine color due to the extraction of a greater amount of polyphenolic compounds during fermentation and by facilitating clarification and filtration processes [73,98,109,110].

4. The Role of *Saccharomyces* and non-*Saccharomyces* Mannoproteins in Aroma and Color of Wines

Mannoproteins are highly glycosylated glycoproteins located on the external layer of the yeast cell wall, representing 35% to 40% of the *S. cerevisiae* cell wall (Figure 5) [111]. Mannoproteins are composed of 10% to 20% protein and 80% to 90% D-mannose associated with residues of D-glucose and N-acetylglucosamine [112].

Figure 5. Schematic representation of the yeast cell wall. The yeast cell wall is composed of mannan–oligosaccharide (mannoproteins), complex polymers of β-(1,3)/(1,6) glucan, and chitin. As shown in Figure 5, mannoproteins are located on the surface of the cell wall.

In wine, we can find two groups of mannoproteins: one made up of those secreted into wine by yeast during alcoholic fermentation (100–150 mg/L) with molecular weights from 5000 to more than 800,000 Da [113], the other one composed by those that are released into the wine due to the autolysis of yeasts during aging on lees, probably through the cleavage of linkages between mannoproteins, glucans, and chitin [113,114].

The presence of these mannoproteins in wines has many positive consequences, from the reduction of the protein haze in white wines [115] to decreasing astringency of red wines, by increased inhibition of tannin aggregation [116,117]. Among other positive factors, mannoproteins also interact with wine volatile compounds [118]. So, these yeast-derived glycoprotein complexes can have positive effects on the technological and sensorial properties of wines [119].

In terms of improving wine palatability and mouth feel, yeast mannoproteins promote the increase of wine sweetness [120] and improve the aroma persistence and complexity [114,121]. However, the number of mannoproteins released by yeast into wine can vary concerning the strain and the chemical–physical and compositional conditions of the wine system [121,122].

Their presence can also affect the release of volatile compounds, affecting the final perception of the wine [121]. The physicochemical interactions between aroma compounds and mannoproteins depends on the nature of the volatile, since a greater degree of interaction is often observed with hydrophobic compounds [123], as well as the conformational structure of the mannoproteins [114]; moreover, Chalier et al. [114] demonstrated that both the glycosidic and peptidic parts of the mannoproteins may interact well with the aroma compounds.

It has been shown that the use of mannoproteins (in low amounts) or the contact of the wine with fine lees increases the levels of esters (ethyl hexanoate, methyl, and ethyl hexadecanoate, which present fruity aromas), due to the esterification of fatty acids released during yeast fermentation or yeast autolysis [121,124], while higher amounts increase, in excess, fatty acid content, producing yeasty, herbaceous, and cheese-like smells [125]. However, it has been suggested that mannoproteins can be used to remove or reduce the incidence of wine off-flavors—4-ethylguaiacol and 4-ethylphenol. The sorption of these compounds to the yeast walls could be due to their interactions with the functional groups of the mannoproteins and the free amino acids on the surface of the cell walls [126].

The release of mannoproteins into the wine is not a physiological characteristic of just *S. cerevisiae* yeast strains. In 2014, Domizio and collaborators [127], studied eight non-*Saccharomyces* wine strains (*H. osmophila*, *L. thermotolerans*, *M. pulcherrima*, *P. fermentans*, *S. ludwigii*, *S. bacillaris*, *T. delbrueckii*, and *Z. florentinus*) in mixed inocula fermentations of a synthetic polysaccharide-free grape juice for their ability to release mannoproteins. The eight non-*Saccharomyces* yeasts confirmed a higher capacity to release polysaccharides when compared to *S. cerevisiae*. Moreover, Pérez-Través et al. [128] also studied the ability of natural hybrids of *S. cerevisiae* × *S. krudriavzevii* to release mannoproteins. Interestingly, they found that this strain, in the fermentation conditions studied, was able to produce a higher quantity of mannoproteins when compared with the sample in which only *S. cerevisiae* was used. Furthermore, the authors also found that the genome interaction in hybrids creates a biological ecosystem that boosts the release of mannoproteins.

As mentioned before, the presence of mannoproteins in wines, namely red wines, promotes tannin stability and reduction of astringency [116,117]. The interaction between mannoproteins and wine phenolic compounds is a matter of interest, as studies show that they may have an impact on color stability [129,130]. However, results are contradictory, as some authors state that there was no positive interaction between mannoproteins and color compounds and that the interaction between mannoproteins and tannins results in a decrease of wine tannin content due to the precipitation of tannin and mannoprotein [112,120,131,132], thus being responsible for a decrease in wine color intensity or lower filterability [133], whereas, others state that mannoproteins appear to stabilize anthocyanin-derived pigments, from a colloidal point of view, avoiding their aggregation and further precipitation [134]. The study of the exhaustive pigment composition of wines has shown that the addition of mannoproteins can stabilize type-A vitisins and other derivative pigments [134].

5. Final Remarks

Wines are complex and evolve physiochemically and sensorially through time. Most of the wine aroma and flavor compounds are produced or released during wine fermentation due to microbial activity of *Saccharomyces*, non-*Saccharomyces* yeast genera, and lactic acid bacteria. A true challenge for winemakers is the selection of a "fit-for-purpose" microbial starter culture or culture strains that can have a crucial role in optimizing flavor, aroma, and color of wines, among other sensory properties.

Co-inoculation involving *S. cerevisiae* and non-*Saccharomyces* yeasts species may result in the death or loss of variability of non-*Saccharomyces*, once *S. cerevisiae* dominates the fermentation and is stress-resistant to the inhibitory ethanol effect. So, several authors suggest that the sequential inoculation (non-*Saccharomyces* followed by *S. cerevisiae*) is a better technique than the mixed culture, allowing a higher expression of the metabolism of non-*Saccharomyces*. Nevertheless, the ratio of inoculation (non-*Saccharomyces* vs. *Saccharomyces*) must be taken into account, especially if the wine should present a special and desirable characteristic such as the expression of a peculiar aroma-flavor, or even the inhibition of the production of a specific family of compounds like, for instance, higher alcohols.

Concerning the MLF, the co-inoculation process has been adopted by some winemakers, in warm climates, where high concentrations of ethanol can inhibit lactic acid bacteria (LAB) growth. Co-inoculation of musts at the beginning of vinification can also accelerate the process without an excessive increase in volatile acidity. However, winemakers must be aware of the possible interactions between yeasts (*Saccharomyces* and non-*Saccharomyces*) and LAB during wine processing. LAB feed on dead and lysed yeast cells but some non-*Saccharomyces* may delay (*M. pulcherrima*) or inhibit (*T. delbrueckii*) bacterial growth, thus inhibiting the occurrence of MLF.

Regarding wine color characteristics, especially red wine, *Saccharomyces* and non-*Saccharomyces* yeasts can directly or indirectly contribute to wine intensity and tonality, by increasing the formation of stable pigment precursors (vinyl phenols, acetaldehyde, and pyruvic acid) and by modifying the pH due to organic acid metabolism. Pyruvic acid is necessary for vitisin synthesis; these important pigments contribute more to wine color parameters than unmodified anthocyanins. Concerning acetaldehyde, this compound has several negative impacts (one on health, the other as a potent binder of SO_2), so its formation, although beneficial to wine color, should be avoided. Also, it is important to choose the right time for promoting MLF, either spontaneously or by inoculation of a commercial LAB strain, to prevent consumption of pyruvic acid by lactic acid bacteria and to promote vitisin synthesis.

Finally, mannoproteins may have a positive effect on sensory perception of red wine, reducing astringency and bitterness and encouraging aroma revelation and odor complexity, but further studies are necessary in order to unravel the possible stabilization mechanism and the relationship between *Saccharomyces* and non-*Saccharomyces* mannoprotein characteristics and their ability to stabilize wine color.

So, in conclusion, knowledge and control of yeast and bacteria can help winemakers enhance the sensory quality of their wines for flavor and color.

Acknowledgments: The author wants to acknowledge Interreg Program for the financial support of the Project IBERPHENOL, Project Number 0377_IBERPHENOL_6_E, co-financed by European Regional Development Fund (ERDF) through POCTEP 2014-2020.

References

1. Spence, C. On the psychological impact of food colour. *Flavour* **2015**, *4*, 2044–7248. [CrossRef]
2. Mattes, R.D. Fat Taste in Humans: Is It a Primary. In *Fat Detection: Taste, Texture, and Post Ingestive Effects*; Montmayeur, J.P., le Coutre, J., Eds.; CRC Press: Boca Raton, FL, USA, 2010; pp. 167–193.
3. Melis, M.; Tomassini Barbarossa, I. Taste Perception of Sweet, Sour, Salty, Bitter, and Umami and Changes Due to l-Arginine Supplementation, as a Function of Genetic Ability to Taste 6-n-Propylthiouracil. *Nutrients* **2017**, *9*, 541. [CrossRef] [PubMed]
4. Chaudhari, N.; Roper, S.D. The cell biology of taste. *J. Cell Biol.* **2010**, *190*, 285–296. [CrossRef] [PubMed]
5. Raghu, M.A. Study to Explore the Effects of Sound Vibrations on Consciousness. *Int. J. Soc. Work Hum. Serv. Pract.* **2018**, *6*, 75–88. Available online: http://www.hrpub.org/download/20180730/IJRH2-19290514.pdf (accessed on 15 September 2019).

6. Barham, P.; Skibsted, L.H.; Bredie, W.L.P.; Frøst, M.B.; Møller, P.; Risbo, J.; Snitkjær, P.; Mortensen, L.M. Molecular Gastronomy: A New Emerging Scientific Discipline. *Chem. Rev.* **2010**, *110*, 2313–2365. [CrossRef]
7. Feher, J. The Chemical senses. In *Quantitative Human Physiology, an Introduction*, 2nd ed.; Feher, J., Ed.; Academic Press: London, UK, 2017; pp. 427–439.
8. Swiegers, J.; Bartowsky, E.; Henschke, P.; Pretorius, I. Yeast and bacterial modulation of wine aroma and flavour. *Aust. J. Grape Wine Res.* **2005**, *11*, 139–173. [CrossRef]
9. Bloem, A.; Bertrand, A.; Lonvaud-Funel, A.; de Revel, G. Vanillin production from simple phenols by wine-associated lactic acid bacteria. *Lett. Appl. Microbiol.* **2007**, *44*, 62–67. [CrossRef]
10. Bloem, A.; Lonvaud-Funel, A.; de Revel, G. Hydrolysis of glycosidically bound flavour compounds from oak wood by *Oenococcus Oeni*. *Food Microbiol.* **2008**, *25*, 99–104. [CrossRef]
11. Regodón-Mateos, J.; Pérez-Nevado, F.; Ramírez-Fernández, M. Influence of *Saccharomyces cerevisiae* yeast strain on the major volatile compounds of wine. *Enzyme Microb. Technol.* **2006**, *40*, 151–157. [CrossRef]
12. Vilela, A.; Inês, A.; Cosme, F. Is wine savory? Umami taste in wine. *SDRP J. Food Sci. Technol.* **2016**, *1*, 100–105. [CrossRef]
13. Moreno-Arribas, M.; Polo, M. Winemaking biochemistry and microbiology: Current knowledge and future trends. *Crit. Rev. Food Sci. Nutr.* **2005**, *45*, 265–286. [CrossRef] [PubMed]
14. Victor-Freitas, A.P.; Fernandes, A.; Oliveira, J.; Teixeira, N.; Mateus, N. A review of the current knowledge of red wine colour. *OENO One* **2017**, *51*. [CrossRef]
15. Jordão, A.M.; Vilela, A.; Cosme, F. From Sugar of Grape to Alcohol of Wine: Sensorial Impact of Alcohol in Wine. *Beverages* **2015**, *1*, 292–310. [CrossRef]
16. Belda, I.; Ruiz, J.; Esteban-Fernández, A.; Navascués, E.; Marquina, D.; Santos, A.; Moreno-Arribas, M. Microbial Contribution to Wine Aroma and Its Intended Use for Wine Quality Improvement. *Molecules* **2017**, *22*, 189. [CrossRef]
17. García, V.; Vásquez, H.; Fonseca, F.; Manzanares, P.; Viana, F.; Martínez, C.; Ganga, M. Effects of using mixed wine yeast cultures in the production of chardonnay wines. *Rev. Argent. Microbiol.* **2010**, *42*, 226–229.
18. Jolly, N.P.; Varela, C.; Pretorius, I.S. Not your ordinary yeast: Non-*Saccharomyces* yeasts in wine production uncovered. *FEMS Yeast Res.* **2014**, *14*, 215–237. [CrossRef]
19. Polásková, P.; Herszage, J.; Ebeler, S. Wine flavor: Chemistry in a glass. *Chem. Soc. Rev.* **2008**, *37*, 2478–2489. [CrossRef]
20. Ruiz, J.; Kiene, F.; Belda, I.; Fracassetti, D.; Marquina, D.; Navascués, E.; Calderón, F.; Benito, A.; Rauhut, D.; Santos, A.; et al. Effects on varietal aromas during wine making: A review of the impact of varietal aromas on the flavor of wine. *Appl. Microbiol. Biotechnol.* **2019**, *103*, 7425–7450. [CrossRef]
21. Lanaridis, P.; Salaha, M.J.; Tzourou, I.; Tsoutsouras, E.; Karagiannis, S. Volatile Compounds in Grapes and Wines From Two Muscat Varieties Cultivated In Greek Islands. *J. Int. Sci. Vigne Vin* **2002**, *36*, 39–47. [CrossRef]
22. Ong, P.K.C.; Acree, T.E. Similarities in the Aroma Chemistry of Gewürztraminer Variety Wines and Lychee (*Litchi chinesis* Sonn.) Fruit. *J. Agric. Food Chem.* **1999**, *47*, 665–670. [CrossRef]
23. Arévalo Villena, M.; Úbeda Iranzo, J.; Cordero Otero, R.; Briones Pérez, A. Optimization of a rapid method for studying the cellular location of β-glucosidase activity in wine yeasts. *J. Appl. Microbiol.* **2005**, *99*, 558–564. [CrossRef] [PubMed]
24. Marais, J. Sauvignon blanc Cultivar Aroma - A Review. *S. Afr. J. Enol. Vitic.* **1994**, *15*, 41–45. [CrossRef]
25. Carien, C.; Wessel, J. A comprehensive review on Sauvignon Blanc aroma with a focus on certain positive volatile thiols. *Food Res. Int.* **2012**, *45*, 287–298. [CrossRef]
26. Sacks, G.L.; Gates, M.J.; Ferry, F.X.; Lavin, E.H.; Kurtz, A.J.; Acree, T.E. Sensory Threshold of 1,1,6-Trimethyl-1,2-dihydronaphthalene (TDN) and Concentrations in Young Riesling and Non-Riesling Wines. *J. Agric. Food Chem.* **2012**, *60*, 2998–3004. [CrossRef] [PubMed]
27. Darriet, P.; Thibon, C.; Dubourdieu, D. Aroma and Aroma Precursors in Grape Berry. In *Aroma and Aroma Precursors in Grape Berry*; Hernâni Gerós, M., Manuela, C., Serge, D., Eds.; Bentham Science Publishers: Sharjah, UAE, 2012; pp. 111–136. [CrossRef]
28. Herderich, M.J.; Siebert, T.E.; Parker, M.; Capone, D.L.; Mayr, C.; Zhang, P.; Geffroy, O.; Williamson, P.; Francis, I.L. Synthesis of The Ongoing Works on Rotundone, an Aromatic Compound Responsible of the Peppery Notes in Wines. *Internet J. Enol. Vitic.* **2013**, *6*, 1–6.

29. Claus, H.; Mojsov, K. Enzymes for Wine Fermentation: Current and Perspective Applications. *Fermentation* **2018**, *4*, 52. [CrossRef]
30. Swiegers, J.H.; Capone, D.L.; Pardon, K.H.; Elsey, G.M.; Sefton, M.A.; Francis, I.L.; Pretorius, I.S. Engineering volatile thiol release in *Saccharomyces cerevisiae* for improved wine aroma. *Yeast* **2007**, *24*, 561–574. [CrossRef]
31. Gamero, A.; Belloch, C.; Querol, A. Genomic and transcriptomic analysis of aroma synthesis in two hybrids between *Saccharomyces cerevisiae* and *S. kudriavzevii* in winemaking conditions. *Microb. Cell Fact.* **2015**, *14*, 128. [CrossRef]
32. Ehrlich, F. Über die bedingungen der fuselölbildung und über ihren zusammenhang mit dem eiweissaufbau der hefe. *Ber. Dtsch. Chem. Ges.* **1907**, *40*, 1027–1047. [CrossRef]
33. Neubauer, O.; Fromherz, K. Über den Abbau der Aminosäuren bei der Hefegärung. *Hoppe-Seyler's Z Physiol. Chem.* **1911**, *70*, 326–350. [CrossRef]
34. Hazelwood, L.A.; Daran, J.M.; van Maris, A.J.A.; Pronk, J.T.; Dickinson, J.R. The Ehrlich Pathway for Fusel Alcohol Production: A Century of Research on *Saccharomyces cerevisiae* Metabolism. *Appl. Environ. Microbiol.* **2008**, *74*, 2259–2266. [CrossRef] [PubMed]
35. Parapouli, M.; Sfakianaki, A.; Monokrousos, N.; Perisynakis, A.; Hatziloukas, E. Comparative transcriptional analysis of flavour-biosynthetic genes of a native *Saccharomyces cerevisiae* strain fermenting in its natural must environment, vs. a commercial strain and correlation of the genes' activities with the produced flavour compounds. *J. Biol. Res. Thessalon.* **2019**, *26*. [CrossRef] [PubMed]
36. Gamero, A.; Quintilla, R.; Groenewald, M.; Alkema, W.; Boekhout, T.; Hazelwood, L. High-throughput screening of a large collection of non-conventional yeasts reveals their potential for aroma formation in food fermentation. *Food Microbiol.* **2016**, *60*, 147–159. [CrossRef] [PubMed]
37. Lee, S.-J.; Rathbone, D.; Asimont, S.; Adden, R.; Ebeler, S.E. Dynamic changes in ester formation during chardonnay juice fermentations with different yeast inoculation and initial brix conditions. *Am. J. Enol. Vitic.* **2004**, *55*, 346–354.
38. Swiegers, J.H.; Saerens, S.M.G.; Pretorius, I.S. Novel yeast strains as tools for adjusting the flavour of fermented beverages to market specifications. In *Biotechnology in Flavour Production*, 2nd ed.; Havkin-Frenkel, D., Dudai, N., Eds.; Wiley Online Library: Oxford, UK, 2016; pp. 62–132. [CrossRef]
39. Kruis, A.J.; Levisson, M.; Mars, A.E.; van der Ploeg, M.; Garcés Daza, F.; Ellena, V.; Kengen, S.W.M.; van der Oost, J.; Weusthuis, R.A. Ethyl acetate production by the elusive alcohol acetyltransferase from yeast. *Metab. Eng.* **2017**, *41*, 92–101. [CrossRef] [PubMed]
40. Kruis, A.J.; Brigida, G.; Jonker, T.; Mars, A.E.; van Rijswijck, I.M.H.; Wolkers-Rooijackers Judith, C.M.; Smid, E.J.; Jan, S.; Verstrepen, K.J.; Kengen, S.W.M.; et al. Contribution of Eat1 and other alcohol acyltransferases to ester production in *Saccharomyces cerevisiae*. *Front Microbiol.* **2018**, *9*, 3202. [CrossRef] [PubMed]
41. Querol, A.; Perez-Torrado, R.; Alonso-del-Real, J.; Minebois, R.; Stribny, J.; Oliveira, B.M.; Barrio, E. New trends in the uses of yeasts in oenology. In *Advances in Food and Nutrition Research*; Toldrá, F., Ed.; Elsevier: Cambridge, UK, 2018; pp. 177–210.
42. Mato, I.; Suarez-Luque, S.; Huidobro, J.F. Simple determination of main organic acids in grape juice and wine by using capillary zone electrophoresis with direct UV detection. *Food Chem.* **2007**, *102*, 104–112. [CrossRef]
43. Duan, L.L.; Shi, Y.; Jiang, R.; Yang, Q.; Wang, Y.Q.; Liu, P.T.; Duan, C.Q.; Yan, G.L. Effects of adding unsaturated fatty acids on fatty acid composition of *Saccharomyces cerevisiae* and major volatile compounds in wine. *S. Afr. J. Enol. Vitic.* **2015**, *36*, 285–295. Available online: http://www.scielo.org.za/scielo.php?script=sci_arttext&pid=S2224-79042015000200001&lng=en&tlng=en (accessed on 15 September 2019). [CrossRef]
44. Zhao, P.; Gao, J.; Qian, M.; Li, H. Characterization of the Key Aroma Compounds in Chinese Syrah Wine by Gas Chromatography-Olfactometry-Mass Spectrometry and Aroma Reconstitution Studies. *Molecules* **2017**, *22*, 1045. [CrossRef]
45. Landaud, S.; Helinck, S.; Bonnarme, P. Formation of volatile sulfur compounds and metabolism of methionine and other sulfur compounds in fermented food. *Appl. Microbiol. Biotechnol.* **2008**, *7*, 1191–1205. [CrossRef]
46. Bartowsky, E.J.; Pretorius, I.S. Microbial formation and modification of flavour and off-flavour compounds in wine. In *Biology of Microorganisms on Grapes, in Must and Wine*; König, H., Unden, G., Fröhlich, J., Eds.; Springer: Heidelberg, Germany, 2008; pp. 211–233. [CrossRef]
47. Vermeulen, C.; Gijs, L.; Collin, S. Sensorial contribution and formation pathways of thiols in foods: A review. *Food Rev. Int.* **2005**, *21*, 69–137. [CrossRef]

48. Tominaga, T.; Murat, M.L.; Dubourdieu, D. Development of a method analyzing the volatile thiols involved in the characteristic aroma of wines made from *Vitis vinifera* L. cv. Sauvignon blanc. *J. Agric. Food Chem.* **1998**, *46*, 1044–1048. [CrossRef]
49. Ferreira, D.C.; Hernandes, K.C.; Nicolli, K.P.; Souza-Silva, E.A.; Manfroi, V.; Alcaraz Zini, C.; Elisa Welke, J. Development of a method for determination of target toxic carbonyl compounds in must and wine using HS-SPME-GC/MS-SIM after preliminary GC×GC/TOFMS analyses. *Food Anal. Methods* **2019**, *12*, 108–120. [CrossRef]
50. Lago, L.O.; Nicolli, K.P.; Marques, A.B.; Zini, C.A.; Welke, J.E. Influence of ripeness and maceration of the grapes on levels of furan and carbonyl compounds in wine – Simultaneous quantitative determination and assessment of the exposure risk to these compounds. *Food Chem.* **2017**, *230*, 594–603. [CrossRef]
51. Semchyshyn, H.M. Reactive Carbonyl Species In Vivo: Generation and Dual Biological Effects. *Sci. World J.* **2014**, *2014*. [CrossRef]
52. Bozoudi, D.; Tsaltas, D. Grape Microbiome: Potential and Opportunities as a Source of Starter Cultures. In *Grape and Wine Biotechnology*; InTech: Rijeka, Croatia, 2016.
53. Jemec, P.K.; Cadez, N.; Zagorc, T.; Bubic, V.; Zupec, A.; Raspor, P. Yeast population dynamics in five spontaneous fermentations of Malvasia must. *Food Microbiol.* **2001**, *18*, 247–259. [CrossRef]
54. Combina, M.; Elía Mercado, L.; Catania, C.; Ganga, A.; Martinez, C. Dynamics of indigenous yeast populations during spontaneous fermentation of wines from Mendoza, Argentina. *Int. J. Food Microbiol.* **2005**, *99*, 237–243. [CrossRef]
55. Varela, C.; Siebert, T.; Cozzolino, D.; Rose, L.; McLean, H.; Henschke, P.A. Discovering a chemical basis for differentiating wines made by fermentation with 'wild' indigenous and inoculated yeasts: Role of yeast volatile compounds. *Aust. J. Grape Wine Res.* **2009**, *15*, 238–248. [CrossRef]
56. Restuccia, D.; Loizzo, M.R.; Spizzirri, U.G. Accumulation of Biogenic Amines in Wine: Role of Alcoholic and Malolactic Fermentation. *Fermentation* **2018**, *4*, 6. [CrossRef]
57. Clemente-Jimenez, J.; Mingorance-Cazorla, L.; Martínez-Rodríguez, S.; Las HerasVázquez, F.; Rodríguez-Vico, F. Influence of sequential yeast mixtures on wine fermentation. *Int. J. Food Microbiol.* **2005**, *98*, 301–308. [CrossRef]
58. Gobbi, M.; Comitini, F.; Domizio, P.; Romani, C.; Lencioni, L.; Mannazzu, I.; Ciani, M. *Lachancea thermotolerans* and *Saccharomyces cerevisiae* in simultaneous and sequential co-fermentation: A strategy to enhance acidity and improve the overall quality of wine. *Food Microbiol.* **2013**, *33*, 271–281. [CrossRef] [PubMed]
59. Mateo, J.J.; di Stefano, R. Description of the β-glucosidase activity of wine yeasts. *Food Microbiol.* **1997**, *14*, 583–591. [CrossRef]
60. Van Rensburg, P.; Pretorius, I. Enzymes in winemaking: Harnessing natural catalysts for efficient bio-transformations-a review. *S. Afr. J. Enol. Vitic. Spec. Issue* **2000**, *21*, 52–73.
61. Budić-Leto, I.; Zdunic, G.; Banovic, M.; Kovacevic-Ganic, K.; Tomic-Potrebujes, I.; Lovric, T. Fermentative aroma compounds and sensory descriptors of traditional Croatian dessert wine Prošek from Plavac mali cv. *Food Technol. Biotechnol.* **2010**, *48*, 530–537. [CrossRef]
62. Puertas, B.; Jiménez, M.J.; Cantos-Villar, E.; Cantoral, J.M.; Rodriguez, M. Use of *Torulaspora delbrueckii* and *Saccharomyces cerevisiae* in semi-industrial sequential inoculation to improve quality of Palomino and Chardonnay wines in warm climates. *J. Appl. Microbiol.* **2016**, *122*, 733–746. [CrossRef]
63. García, M.; Arroyo, T.; Crespo, J.; Cabellos, J.M.; Esteve-Zarzoso, B. Use of native non-Saccharomyces strain: A new strategy in D.O. "Vinos de Madrid" (Spain) wines elaboration. *Eur. J. Food Sci. Technol.* **2017**, *5*, 1–31. Available online: https://www.eajournals.org/journals/european-journal-of-food-science-and-technology-ejfst/vol-5-issue-2-april-2017/use-native-non-saccharomyces-strain-new-strategy-d-o-vinos-de-madrid-spain-wines-elaboration/ (accessed on 30 September 2019).
64. Rossouw, D.; Bauer, F.F. Exploring the phenotypic space of non-*Saccharomyces* wine yeast biodiversity. *Food Microbiol.* **2016**, *55*, 32–46. [CrossRef]
65. Wang, C.; Mas, A.; Esteve-Zarzoso, B. Interaction between *Hanseniaspora uvarum* and *Saccharomyces cerevisiae* during alcoholic fermentation. *Int. J. Food Microbiol.* **2015**, *206*, 67–74. [CrossRef]
66. Wang, C.; Mas, A.; Esteve-Zarzoso, B. The interaction between *Saccharomyces cerevisiae* and non-*Saccharomyces* yeast during alcoholic fermentation is species and strain specific. *Front. Microbiol.* **2016**, *7*, 502. [CrossRef]

67. Schaffrath, R.; Meinhard, F.; Klassen, R. Yeast Killer Toxins: Fundamentals and Applications. In *Physiology and Genetics*, 2nd ed.; Anke Schüffler, A., Ed.; Springer International Publishing AG: Cham, Switzerland, 2017. [CrossRef]
68. Velázquez, R.; Zamora, E.; Álvarez, M.L.; Hernández, L.M.; Ramírez, M. Effects of new *Torulaspora delbrueckii* killer yeasts on the must fermentation kinetics and aroma compounds of white table wine. *Front. Microbiol.* **2015**, *6*, 1222. [CrossRef]
69. El-Banna, A.A.; El-Sahn, M.A.; Shehata, M.G. Yeasts Producing Killer Toxins: An Overview. *Alex. J. Food Sci. Technol.* **2011**, *8*, 41–53. [CrossRef]
70. Branco, P.; Francisco, D.; Chambon, C.; Hébraud, M.; Arneborg, N.; Almeida, M.G.; Caldeira, J.; Albergaria, H. Identification of novel GAPDH-derived antimicrobial peptides secreted by *Saccharomyces cerevisiae* and involved in wine microbial interactions. *Appl. Microbiol. Biotechnol.* **2014**, *98*, 843–853. [CrossRef] [PubMed]
71. Loira, I.; Morata, A.; Comuzzo, P.; Callejo, M.J.; González, C.; Calderón, F.; Suárez-Lepe, J.A. Use of *Schizosaccharomyces pombe* and *Torulaspora delbrueckii* strains in mixed and sequential fermentations to improve red wine sensory quality. *Food Res. Int.* **2015**, *76*, 325–333. [CrossRef] [PubMed]
72. Curiel, J.A.; Morales, P.; Gonzalez, R.; Tronchoni, J. Different Non-Saccharomyces Yeast Species Stimulate Nutrient Consumption in *S. cerevisiae* Mixed Cultures. *Front Microbiol.* **2017**, *31*, 2121. [CrossRef]
73. Loira, I.; Morata, A.; Bañuelos, M.A.; Suárez-Lepe, J.A. Isolation, selection, and identification techniques for non-*Saccharomyces* yeasts of oenological interest. In *Biotechnological Progress and Beverage Consumption*; Grumezescu Alexandru, M., Holban Alina, M., Eds.; Elsevier Inc.: Amsterdam, The Netherlands, 2020; pp. 467–521. [CrossRef]
74. Renault, P.; Coulon, J.; de Revel, G.; Barbe, J.C.; Bely, M. Increase of fruity aroma during mixed *T. delbrueckii/S. cerevisiae* wine fermentation is linked to specific esters enhancement. *Int. J. Food Microbiol.* **2015**, *207*, 40–48. [CrossRef]
75. Renault, P.; Coulon, J.; Moine, V.; Thibon, C.; Bely, M. Enhanced 3-sulfanylhexan-1-ol production in sequential mixed fermentation with *Torulaspora delbrueckii/Saccharomyces cerevisiae* reveals a situation of synergistic interaction between two industrial strains. *Front. Microbiol.* **2016**, *7*, 293. [CrossRef]
76. Mosciano, G. Successful flavors: From formulation to QC to applications and beyond. In *Successful Flavors*; Mosciano, G., Ed.; Allured Publishing Corp.: Carol Stream, IL, USA, 2006; p. 240. ISBN 10 1932633197.
77. García, M.; Esteve-Zarzoso, B.; Crespo, J.; Cabellos, J.M.; Arroyo, T. Yeast Monitoring of Wine Mixed or Sequential Fermentations Made by Native Strains from D.O. "Vinos de Madrid" Using Real-Time Quantitative PCR. *Front. Microbiol.* **2017**, *8*. [CrossRef]
78. Loira, I.; Vejarano, R.; Bañuelos, M.A.; Morata, A.; Tesfaye, W.; Uthurry, C.; Villa, A.; Cintora, I.; Suárez-Lepe, J.A. Influence of sequential fermentation with *Torulaspora delbrueckii* and *Saccharomyces cerevisiae* on wine quality. *LWT Food Sci. Technol.* **2014**, *59*, 915–922. [CrossRef]
79. Peinado, R.; Moreno, J.; Medina, M.; Mauricio, J. Changes in volatile compounds and aromatic series in sherry wine with high gluconic acid levels subjected to aging by submerged flor yeast cultures. *Biotechnol. Lett.* **2004**, *26*, 757–762. [CrossRef]
80. Viana, F.; Gil, J.V.; Vallés, S.; Manzanares, P. Increasing the levels of 2-phenylethyl acetate in wine through the use of a mixed culture of *Hanseniaspora osmophila* and *Saccharomyces cerevisiae*. *Int. J. Food Microbiol.* **2009**, *135*, 68–74. [CrossRef]
81. Berbegal, C.; Spano, G.; Tristezza, M.; Grieco, F.; Capozzi, V. Microbial Resources and Innovation in the Wine Production Sector. *S. Afr. J. Enol. Vitic.* **2017**, *38*, 156–166. [CrossRef]
82. Alexandre, H.; Costello, P.J.; Remize, F.; Guzzo, J.; Guilloux-Benatier, M. *Saccharomyces cerevisiae-Oenococcus oeni* interactions in wine: Current knowledge and perspectives. *Int. J. Food Microbiol.* **2004**, *93*, 141–154. [CrossRef] [PubMed]
83. Nehme, N.; Mathieu, F.; Taillandier, P. Impact of the co-culture of *Saccharomyces cerevisiae-Oenococcus oeni* on malolactic fermentation performance and partial characterization of a yeast-derived inhibitory peptidic fraction. *Food Microbiol.* **2010**, *27*, 150–157. [CrossRef] [PubMed]
84. Zapparoli, G.; Tosi, E.; Azzolini, M.; Vagnoli, P.; Krieger, S. Bacterial inoculation strategies for the achievement of malolactic fermentation in high-alcohol wines. *S. Afr. J. Enol. Vitic.* **2009**, *30*, 49–55. [CrossRef]
85. Abrahamse, C.E.; Bartowsky, E.J. Timing of malolactic fermentation inoculation in Shiraz grape must and wine: Influence on chemical composition. *World J. Microbiol. Biotechnol.* **2012**, *28*, 255–265. [CrossRef]

86. Knoll, C.; Fritsch, S.; Schnell, S.; Grossmann, M.; Krieger-Weber, S.; Du Toit, M.; Rauhut, D. Impact of different malolactic fermentation inoculation scenarios on Riesling wine aroma. *World J. Microbiol. Biotechnol.* **2012**, *28*, 1143–1153. [CrossRef]
87. Antalick, G.; Perello, M.; de Revel, G. Co-inoculation with Yeast and LAB Under Winery Conditions: Modification of the Aromatic Profile of Merlot Wines. *S. Afr. J. Enol. Vitic.* **2013**, *34*, 223–232. [CrossRef]
88. Cañas, P.M.; Romero, E.G.; Pérez-Martín, F.; Seseña, S.; Palop, M.L. Sequential inoculation *versus* co-inoculation in Cabernet Franc wine fermentation. *Food Sci Technol Int.* **2015**, *21*, 203–212. [CrossRef]
89. Berbegal, C.; Borruso, L.; Fragasso, M.; Tufariello, M.; Russo, P.; Brusetti, L.; Spano, G.; Capozzi, V. A Metagenomic-Based Approach for the Characterization of Bacterial Diversity Associated with Spontaneous Malolactic Fermentations in Wine. *Int. J. Mol. Sci.* **2019**, *120*, 3980. [CrossRef]
90. Kántor, A.; Hutková, J.; Petrová, J.; Hleba, L.; Kačániová, M. Antimicrobial activity of pulcherrimin pigment produced by *Metschnikowia pulcherrima* against various yeast species. *J. Microbiol. Biotech. Food Sci.* **2015**, *5*, 282–285. [CrossRef]
91. Delgado-Vargas, F.; Jiménez-Aparicio, A.R.; Paredes-Lopez, O. Natural Pigments: Carotenoids, Anthocyanins, and Betalains — Characteristics, Biosynthesis, Processing, and Stability. *Crit. Rev. Food Sci. Nut.* **2000**, *40*, 173–289. [CrossRef] [PubMed]
92. Fei, H.; Na-Na, L.; Lin, M.; Qiu-Hong, P.; Jun, W.; Malcolm, J.R.; Chang-Qing, D. Anthocyanins and their variation in red wines. II. Anthocyanin derived pigments and their color evolution. *Molecules* **2012**, *17*, 1483–1519. [CrossRef]
93. Morata, A.; Gómez-Cordovés, M.C.; Colomo, B.; Suáez, J.A. Pyruvic Ácid and acetaldehyde production by different strains of *Saccharomyces cerevisiae*: Relationship with vitisin A and B formation in red wines. *J. Agric. Food Chem.* **2003**, *51*, 7402–7409. [CrossRef] [PubMed]
94. Morata, A.; Calderón, F.; Gonzalez, C.; Gómez-Cordovés, M.C.; Suarez, C.J.A. Formation of the highly stable pyranoanthocyanins (vitisins A and B) in red wines by the addition of pyruvic acid and acetaldehyde. *Food Chem.* **2007**, *100*, 1144–1152. [CrossRef]
95. Schwarz, M.; Wabnitz, T.C.; Winterhalter, P. Pathway Leading to the Formation of Anthocyanin−Vinylphenol Adducts and Related Pigments in Red Wines. *J. Agric. Food Chem.* **2003**, *51*, 3682–3687. [CrossRef]
96. Fulcrand, H.; Benabdeljalil, C.; Rigaud, J.; Chenyier, V.; Moutounet, M. A new class of wine pigments generated by reaction between pyruvic acid and grape anthocyanins. *Phytochemistry* **1998**, *47*, 1401–1407. [CrossRef]
97. Victor-Freitas, A.P.; Mateus, N. Formation of pyranoanthocyanins in red wines: A new and diverse class of anthocyanin derivatives. *Anal. Bioanal. Chem.* **2011**, *401*, 1463–1473. [CrossRef]
98. Morata, A.; Loira, I.; Suárez-Lepe, J.A. Influence of Yeasts in Wine Colour. In *Grape and Wine Biotechnology*; Morata, A., Loira, I., Eds.; IntechOpen: London, UK, 2016; pp. 285–305. [CrossRef]
99. Ruta, L.L.; Farcasanu, I.C. Anthocyanins and Anthocyanin-Derived Products in Yeast-Fermented Beverages. *Antioxidants (Basel)* **2019**, *8*, 182. [CrossRef]
100. Asenstorfer, R.E.; Markides, A.J.; Iland, P.G.; Jones, G.P. Formation of vitisin A during red wine fermentation and maturation. *Aust. J. Grape Wine Res.* **2003**, *9*, 40–46. [CrossRef]
101. Osborne, J.P.; Orduña, R.M.; Pilone, G.J.; Liu, S.Q. Acetaldehyde metabolism by wine lactic acid bacteria. *FEMS Microbiol. Lett.* **2000**, *191*, 51–55. [CrossRef]
102. He, J.R.; Oliveira, J.; Silva, A.M.-S.; Mateus, N.; Victor-Freitas, A.P. Oxovitisins: A New Class of Neutral Pyranone-anthocyanin Derivatives in Red Wines. *J. Agric. Food Chem.* **2010**, *58*, 8814–8819. [CrossRef] [PubMed]
103. Carew, A.L.; Smith, P.; Close, D.C.; Curtin, C.; Dambergs, R.G. Yeast Effects on Pinot noir Wine Phenolics, Color, and Tannin Composition. *J. Agric. Food Chem.* **2013**, *61*, 9892–9898. [CrossRef] [PubMed]
104. Kennedy, J.A. Wine colour. In *Woodhead Publishing Series in Food Science, Technology and Nutrition*; Reynolds, A.G., Ed.; Woodhead Publishing: Sawston, UK, 2010; pp. 73–104. [CrossRef]
105. Minnaar, P.P.; Ntushelo, N.; Ngqumba, Z.; van Breda, V.; Jolly, N.P. Effect of *Torulaspora delbrueckii* yeast on the anthocyanin and flavanol concentrations of Cabernet franc and Pinotage wines. *S. Afr. J. Enol. Vitic.* **2015**, *36*, 50–58. Available online: http://www.scielo.org.za/scielo.php?script=sci_arttext&pid=S2224-79042015000100013&lng=en&tlng=en (accessed on 25 September 2019). [CrossRef]

106. Pacheco, A.; Santos, J.; Chaves, S.; Almeida, J.; Leão, C.; Sousa, M.J. The Emerging Role of the Yeast Torulaspora delbrueckii. In *Bread and Wine Production: Using Genetic Manipulation to Study Molecular Basis of Physiological Responses, Structure and Function of Food Engineering*; Eissa, A.A., Ed.; IntechOpen: London, UK, 2012; pp. 339–370. [CrossRef]
107. Medina, K.; Boido, E.; Fariña, L.; Dellacassa, E.; Carrau, F. Non-*Saccharomyces* and *Saccharomyces* strains co-fermentation increases acetaldehyde accumulation: Effect on anthocyanin-derived pigments in Tannat red wines. *Yeast* **2016**, *33*, 339–343. [CrossRef] [PubMed]
108. Benito, Á.; Calderón, F.; Benito, S. The Combined Use of *Schizosaccharomyces pombe* and *Lachancea thermotolerans*-Effect on the Anthocyanin Wine Composition. *Molecules* **2017**, *22*. [CrossRef]
109. Benito, S.; Morata, A.; Palomero, F.; Gonzalez, M.; Suárez-Lepe, J. Formation of vinylphenolic pyranoanthocyanins by *Saccharomyces cerevisiae* and *Pichia guillermondii* in red wines produced following different fermentation strategies. *Food Chem.* **2011**, *124*, 15–23. [CrossRef]
110. Strauss, M.; Jolly, N.; Lambrechts, M.; Van Rensburg, P. Screening for the production of extracellular hydrolytic enzymes by non-*Saccharomyces* wine yeasts. *J. Appl. Microbiol.* **2001**, *91*, 182–190. [CrossRef]
111. Klis, F.M.; Mol, P.; Hellingwerf, K.; Brul, S. Dynamics of cell wall structure in *Saccharomyces cerevisiae*. *FEMS Microbiol. Rev.* **2002**, *26*, 239–256. [CrossRef]
112. Rodrigues, A.; Ricardo-Da-Silva, J.M.; Lucas, C.; Laureano, O. Effect of commercial mannoproteins on wine colour and tannins stability. *Food Chem.* **2012**, *131*, 907–914. [CrossRef]
113. Doco, T.; Quellec, N.; Moutounet, M. Polysaccharide patterns during the ageing of Carignan noir red wines. *Am. J. Enol. Vitic.* **1999**, *50*, 25–32.
114. Chalier, P.; Angot, B.; Delteil, D.; Doco, T.; Gunata, Z. Interactions between aroma compounds and whole mannoprotein isolated from *Saccharomyces cerevisiae* strains. *Food Chem.* **2007**, *100*, 22–30. [CrossRef]
115. Dupin, I.; Mc Kinnon, B.M.; Ryan, C.; Boulay, M.; Markides, A.J.; Jones, G.P.; Williams, P.J.; Waters, E.J. *Saccharomyces cerevisiae*, mannoproteins that protect wine from protein haze: Their release during fermentation and lees contact and a proposal mechanism of action. *J. Agric. Food Chem.* **2000**, *48*, 3098–3105. [CrossRef] [PubMed]
116. Vidal, S.; Francis, L.; Williams, P.; Kwitkowski, M.; Gawel, R.; Cheynier, V.; Waters, E. The mouth-feel properties of polysaccharides and anthocyanins in a wine like medium. *Food Chem.* **2004**, *85*, 519–525. [CrossRef]
117. Riou, V.; Vernhet, A.; Doco, T.; Moutonnet, M. Aggregation of grape seed tannins in model wine – effect of wine polysaccharides. *Food Hydrocol.* **2002**, *16*, 17–23. [CrossRef]
118. Juega, M.; Nunez, Y.P.; Carrascosa, A.V.; Martinez-Rodriguez, A.J. Influence of yeast mannoproteins in the aroma improvement of white wines. *J Food Sci.* **2012**, *77*, M499–M504. [CrossRef] [PubMed]
119. Ribéreau-Gayon, P.; Dubourdieu, D.; Donèche, B.; Lonvaud, A. *Handbook of Enology, Volume 1: The Microbiology of Wine and Vinifications*; John Wiley and Sons: Hoboken, NJ, USA, 2006; Volume 1.
120. Guadalupe, Z.; Palacios, A.; Ayestarán, B. Maceration enzymes and mannoproteins: A possible strategy to increase colloidal stability and colour extraction in red wines. *J. Agric. Food Chem.* **2007**, *55*, 4854–4862. [CrossRef]
121. Costa, G.P.; Nicolli, K.P.; Welke, J.E.; Manfroi, V.; Zini, C.A. Volatile Profile of Sparkling Wines Produced with the Addition of Mannoproteins or Lees before Second Fermentation Performed with Free and Immobilized Yeasts. *J. Braz. Chem. Soc.* **2018**, *29*, 1866–1875. [CrossRef]
122. Doco, T.; Vuchot, P.; Cheynier, V.; Moutounet, M. Structural modification of wine arabinogalactans during aging on lees. *Am. J. Enol. Vitic.* **2003**, *54*, 150–157.
123. Lubbers, S.; Voilley, A.; Feuillat, M.; Charpentier, C. Influence of mannoproteins from yeast on the aroma intensity of a model wine. *LWT-Food Sci. Technol.* **1994**, *27*, 108–114. [CrossRef]
124. Braschi, G.; Ricci, A.; Grazia, L.; Versari, A.; Patrignani, F.; Lanciotti, R. Mannoprotein Content and Volatile Molecule Profiles of Trebbiano Wines Obtained by *Saccharomyces cerevisiae* and *Saccharomyces bayanus* Strains. *Fermentation* **2019**, *5*, 66. [CrossRef]
125. Comuzzo, P.; Tat, L.; Tonizzo, A.; Battistutta, F. Yeast derivatives (extracts and autolysates) in winemaking: Release of volatile compounds and effects on wine aroma volatility. *Food Chem.* **2006**, *99*, 217–230. [CrossRef]
126. Nieto-Rojo, R.; Ancín-Azpilicueta, C.; Garrido, J.J. Sorption of 4-ethylguaiacol and 4-ethylphenol on yeast cell walls, using a synthetic wine. *Food Chem.* **2014**, *152*, 399–406. [CrossRef] [PubMed]

127. Domizio, P.; Liu, Y.; Bisson, L.F.; Barile, D. Use of non-Saccharomyces wine yeasts as novel sources of mannoproteins in wine. *Food Microbiol.* **2014**, *43*, 5–15. [CrossRef] [PubMed]
128. Pérez-Través, L.; Querol, A.; Pérez-Torrado, R. Increased mannoprotein content in wines produced by *Saccharomyces kudriavzevii* × *Saccharomyces cerevisiae* hybrids. *Int. J. Food Microbiol.* **2016**, *237*, 35–38. [CrossRef]
129. Escot, S.; Feulliat, M.; Dulau, L.; Charpentier, C. Release of polysaccharides by yeasts and the influence of released polysaccharides on colour stability and wine astringency. *Aust. J. Grape Wine Res.* **2001**, *7*, 153–159. [CrossRef]
130. Poncet-Legrand, C.; Doco, T.; Williams, P.; Vernhet, A. Inhibition of grape seed tannin aggregation by wine mannoproteins: Effect of polysaccharide molecular weight. *Am. J. Enol. Vitic.* **2007**, *58*, 87–91.
131. Guadalupe, Z.; Ayestarán, B. Effect of commercial mannoprotein addition on polysaccharide, polyphenolic, and colour composition in red wines. *J. Agric. Food Chem.* **2008**, *56*, 9022–9029. [CrossRef]
132. Guadalupe, Z.; Martínez, L.; Ayestarán, B. Yeast mannoproteins in red winemaking. Effect on polysaccharide, polyphenolic and colour composition. *Am. J. Enol. Vitic.* **2010**, *61*, 191–200.
133. Morata, A.; Gomez-Cordoves, M.C.; Suberviola, J.; Bartolome, B.; Colomo, B.; Suarez, J.A. Adsorption of anthocyanins by yeast cell walls during the fermentation of red wines. *J. Agric. Food Chem.* **2003**, *51*, 4084–4088. [CrossRef]
134. Alcalde-Eon, C.; García-Estévez, I.; Puente, V.; Rivas-Gonzalo, J.C.; Escribano-Bailón, M.T. Color Stabilization of Red Wines. A Chemical and Colloidal Approach. *J. Agric. Food Chem.* **2014**, *62*, 6984–6994. [CrossRef]

5

Two Novel Strains of *Torulaspora delbrueckii* Isolated from the Honey Bee Microbiome and their use in Honey Fermentation

Joseph P. Barry [1,†], Mindy S. Metz [1,†], Justin Hughey [2], Adam Quirk [2] and Matthew L. Bochman [1,3,*]

1 Molecular and Cellular Biochemistry Department, 212 South Hawthorne Drive, Simon Hall MSB1, Room 405B, Indiana University, Bloomington, IN 47405, USA; jpbarry@indiana.edu (J.P.B.); msmetz@indiana.edu (M.S.M.)
2 Cardinal Spirits, 922 S Morton St, Bloomington, IN 47403, USA; justin@cardinalspirits.com (J.H.); quirk@cardinalspirits.com (A.Q.)
3 Wild Pitch Yeast, Bloomington, IN 47405, USA
* Correspondence: bochman@indiana.edu
† These authors contributed equally to this work.

Abstract: Yeasts are ubiquitous microbes found in virtually all environments. Many yeast species can ferment sugar into ethanol and CO_2, and humans have taken advantage of these characteristics to produce fermented beverages for thousands of years. As a naturally abundant source of fermentable sugar, honey has had a central role in such fermentations since Neolithic times. However, as beverage fermentation has become industrialized, the processes have been streamlined, including the narrow and almost exclusive usage of yeasts in the genus *Saccharomyces* for fermentation. We set out to identify wild honey- or honey-bee-related yeasts that can be used in honey fermentation. Here, we isolated two strains of *Torulaspora delbrueckii* from the gut of a locally collected honey bee. Both strains were able to ferment honey sugar into mead but failed to metabolize more than a modest amount of wort sugar in trial beer fermentations. Further, the meads fermented by the *T. delbrueckii* strains displayed better sensory characteristics than mead fermented by a champagne yeast. The combination of *T. delbrueckii* and champagne yeast strains was also able to rapidly ferment honey at an industrial scale. Thus, wild yeasts represent a largely untapped reservoir for the introduction of desirable sensory characteristics in fermented beverages such as mead.

Keywords: yeast; mead; fermentation

---

1. Introduction

Mankind has been fermenting foods and beverages for thousands of years [1–4], which certainly predates our knowledge that yeasts were the microbial agents responsible for metabolizing sugar into alcohol. Honey, which is produced by honey bees such as *Apis mellifera*, is a natural source of abundant, readily fermentable sugar and has been found as an ingredient in some of the earliest known fermented beverages [2]. When a dilute honey solution is fermented on its own without the addition of fruit/fruit juice or other additives, this creates a traditionally strong (8–18% alcohol by volume, ABV) beverage called mead (reviewed in [5]).

Much like craft brewing, mead making is currently experiencing a renewed interest at the amateur and professional levels worldwide. Indeed, one average, a new meadery opens in the United States every 3 days, and one opens every 7 days in the rest of the world [6]. With this proliferation of mead making has come experimentation with the traditional methods, including aging mead in oak or previously used spirit barrels and fermentation with non-*Saccharomyces* microbes to generate sour and "funky" meads [6]. In many ways, this again mirrors the craft brewing scene and current popularity of sour beer [7–9].

Another factor driving innovation in fermented beverage production is the "local movement" [10,11], i.e., the push to use locally sourced ingredients. In the case of mead, it is often relatively simple to utilize locally produced honey, but few meaderies use indigenous microbes unless they rely on those naturally found in the honey itself. Such microbes are naturally resistant to high osmotic stress from concentrated sugar in honey and typically come from pollen/flower, bee gut microbiome, and dirt/dust contamination of the honey (reviewed in [12,13]). These organisms include multiple species from dozens of different genera of bacteria (e.g., *Bacillus*, *Citrobacter*, and *Lactobacillus*), yeast (e.g., *Saccharomyces*, *Schizosaccharomyces*, and *Pichia*), and molds.

Here, we describe the bio-prospecting for novel yeasts to use in honey fermentation. We concentrated on ultra-local strains by attempting to enrich for ethanol-tolerant yeasts in honey, the hive, and honey bees themselves. Ultimately, we recovered two strains of the yeast *Torulaspora delbrueckii* from the honey bee microbiome and characterized them for their ability to ferment honey. Many *T. delbrueckii* strains have previously been examined for their fermentative abilities and were found to improve wine quality (see [14] and references therein) and create distinct beer flavors in beer [15–17]. Compared to *S. cerevisiae*, *T. delbrueckii* is generally found to generate lower levels of ethanol [16] but a wider range of fruity aromas [18,19]. However, in many of these cases, *T. delbreuckii* is used in combination with *S. cerevisiae* for sequential fermentations [14] or mixed culture fermentations [15].

Currently, there are no reports of *T. delbrueckii* being used to make mead and how this species may impact the sensory qualities of the final product. Here, we found that, compared to a traditional *Saccharomyces cerevisiae* control strain (WLP715), the mead produced by the *T. delbrueckii* strains YH178 and YH179 scored more favorably during sensory analysis. Further, *T. delbrueckii* YH178 and YH179 were successfully used in mixed honey fermentations with *S. cerevisiae* WLP715 on a production scale. These data suggest that *T. delbrueckii* YH178 and YH179 have beneficial honey fermentation profiles and are suitable for mead production by homebrewers and professional mead makers.

2. Results and Discussion

2.1. Isolation of T. delbrueckii Strains YH178 and YH179

We initially set out to isolate ethanol-tolerant yeasts from multiple bee-related sources. Samples of honey, honeycomb, propolis, and fragments of a wooden beehive were surveyed for yeasts, but no microbes were recovered using our wild yeast enrichment protocol [7]. This failure was likely not due to the stringency (e.g., high ethanol and osmotic stress tolerance) of the method used because we previously recovered hundreds of wild ethanol-tolerant yeasts from a variety of environmental sources. However, as only small volumes of samples were surveyed from a single hive, a more extensive search may have yielded suitable yeasts. Alternatively, the lack of ethanol-tolerant yeasts may be attributed to the well-characterized hygienic behavior of honey bees, which are known to keep their hives clean in order to avoid disease [20–23].

We next focused our bio-prospecting efforts on the bees themselves. We again failed to isolate ethanol-tolerant yeast from the outer surface of several bees via a quick rinse with yeast enrichment medium or prolonged incubation of a bee submerged in the same medium at 30 °C with aeration. However, when a bee was submerged in yeast enrichment medium and its abdomen was ruptured, microbial growth was visible in the culture after 24 h. Plating for single colonies on Wallerstein Laboratories nutrient (WLN) agar revealed that two distinct strains of yeast were present, which differed in color (Figure 1A). These strains were designated YH178 (darker green on WLN agar) and YH179 (pale green) as part of our larger yeast hunting efforts and were identified as *T. delbrueckii* isolates by rDNA sequencing (Figure 1B). For comparison, the *S. cerevisiae* strain WLP715 is shown, which attains an even lighter shade of green than YH179 on WLN agar (Figure 1A).

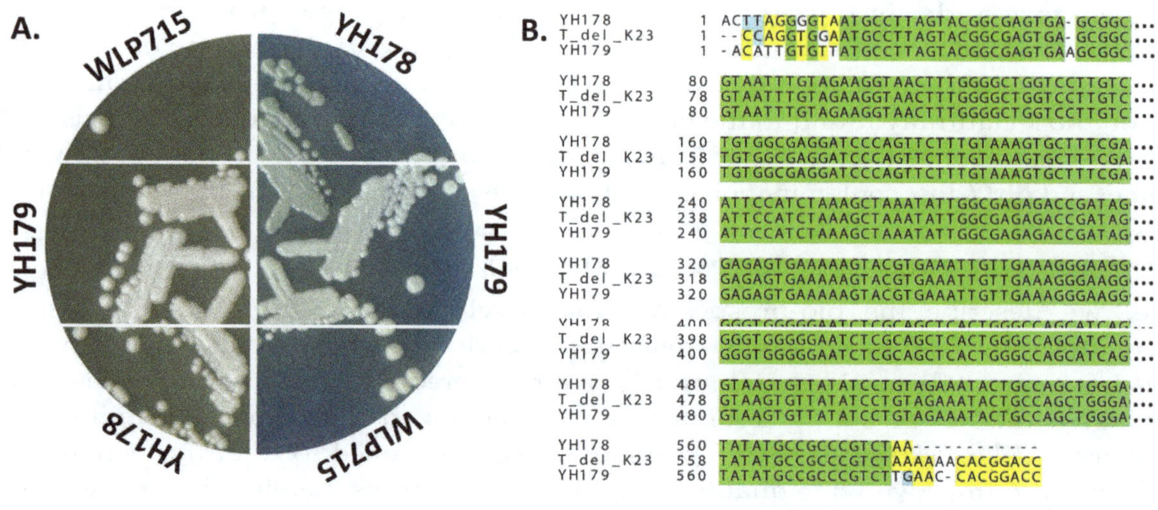

Figure 1. YH178 and YH179 are both strains of *T. delbrueckii*. (**A**) Colony morphology of YH178, YH179, and *S. cerevisiae* WLP715 on YPD (left) and WLN (right) agar plates. The indicated strains were streaked for single colonies and grown at 30 °C for 2 days prior to capturing images of the plates with a flatbed scanner. (**B**) Sequence alignments revealed that YH178 and YH179 are isolates of *T. delbrueckii*. A portion of the 26S rDNA of YH178 and YH179 was PCR-amplified and sequenced, and the sequences are aligned to the closest related strain in the NCBI nucleotide database. T_del_K23 denotes the 26S rDNA gene of *T. delbrueckii* strain K23 (GenBank accession KP852445.1). The ellipsis at the end of each line represents the missing sequence that was truncated due to spatial constraints. The full alignment is shown in Figure S1. Green: fully conserved; yellow: conserved among two sequences; blue: similarity (pyrimidines); dashes: sequence gaps due to indels.

Although no colony size variation was noted on either YPD or WLN plates at the end of a 2-day incubation (Figure 1A), YH178 attained this colony size faster than YH179 when observed prior to the end point of growth (data not shown). This suggested that YH178 has a faster doubling time than YH179. However, plotting growth curves for these strains grown in YPD medium revealed that YH178 had a shorter lag phase (<4 h; Figure 2A) to exponential growth than YH179 (>6 h; Figure 2B). Otherwise, these strains displayed similar doubling times and final cell densities.

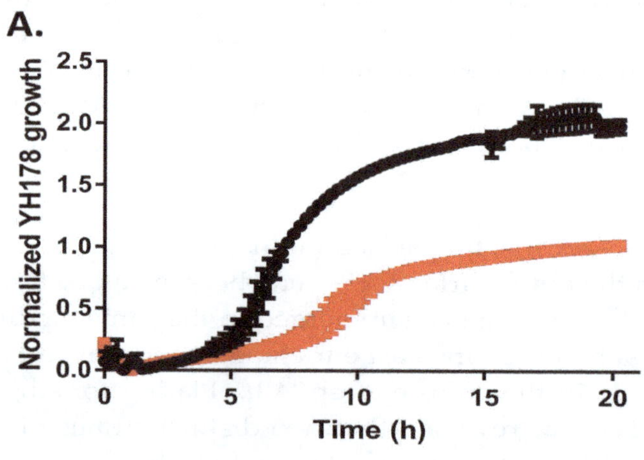

Figure 2. *Cont.*

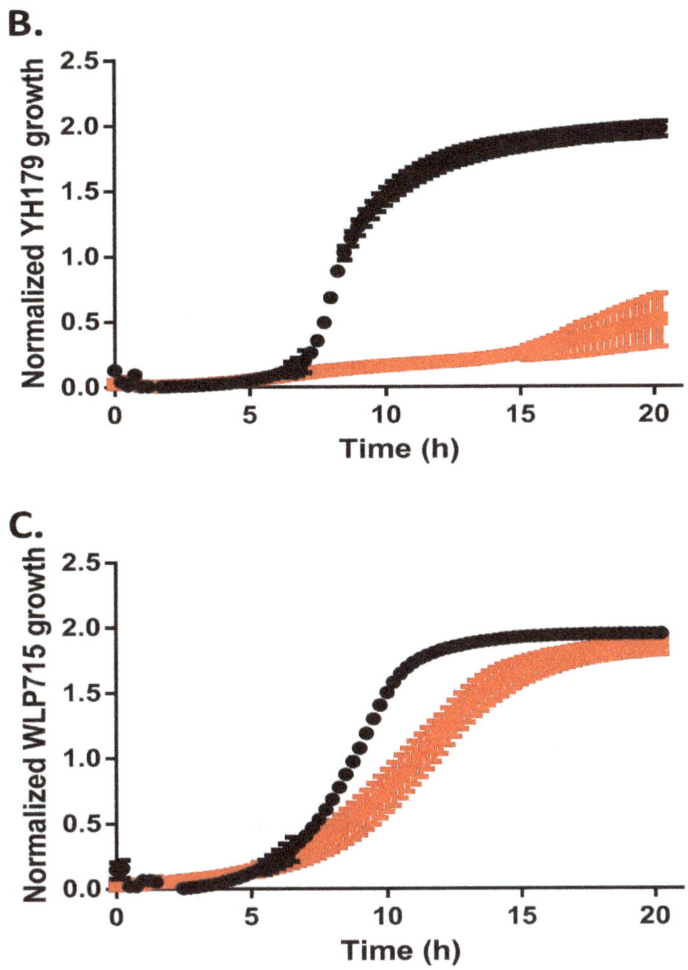

Figure 2. YH178 and YH179 grow poorly in the presence of maltose. YH178 (**A**), YH179 (**B**), and *S. cerevisiae* WLP715 (**C**) were grown to saturation overnight in YPD medium, back-diluted into YPD (black) or YPM (red) medium, and grown overnight in a 96-well plate. Growth was monitored by recording the OD_{660} of each culture and is plotted vs. time. The plotted points are the averages of three independent experiments, and the error bars are the standard deviations.

2.2. YH178 and YH179 Can Use Honey as a Carbon Source during Fermentation

To determine if YH178 and/or YH179 could ferment honey, we inoculated a diluted honey solution with an equivalent number of cells from the *T. delbrueckii* strains, singly and in combination. As a positive control for fermentation, we also used *S. cerevisiae* WLP715, which is marketed as an appropriate yeast for wine, mead, and cider fermentations [24]. The fermentations were monitored for 20 days, and the results are shown in Figure 3A. At approximately 12 h after inoculation, all cultures displayed signs of fermentation, and WLP715 quickly attenuated the honey solution to a final alcohol by volume (ABV) of 11.5% in 2 weeks (Table 1). The $t_{1/2}$ (time necessary for 50% attenuation) was <2 days.

In contrast, YH178, YH179, and the combination thereof fermented the honey solution at a much more modest pace. The $t_{1/2}$ for these fermentations was ~4–4.5 days, and while the final ABV for YH178 was identical to that of WLP715 (11.5%), the YH179 and YH178/YH179 combination fermentations only reached 11% and 11.2%, respectively (Table 1). Subsequent experiments did, however, reveal that additional incubation time of the YH179 and YH178/YH179 cultures led to the same terminal ABV of 11.5% by 35 and 28 days, respectively.

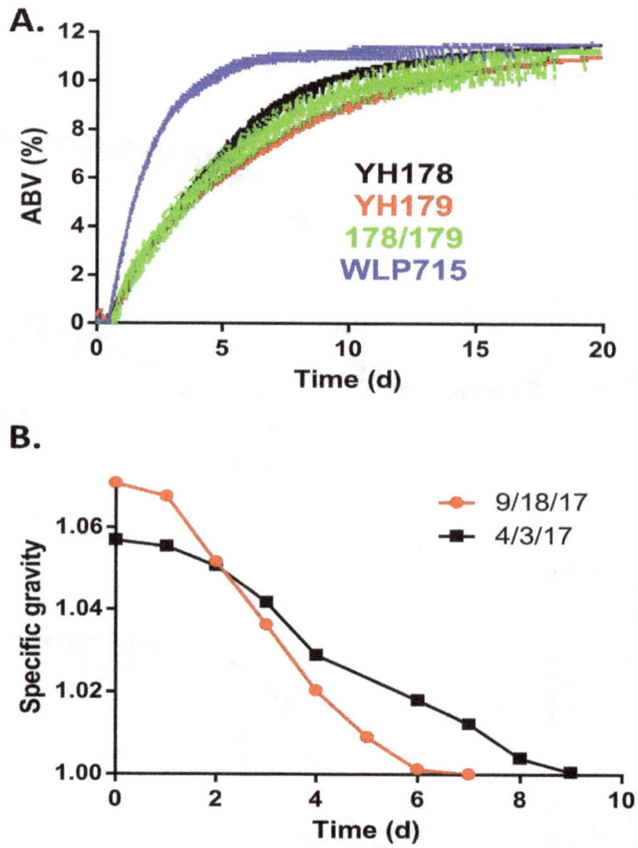

Figure 3. The *T. delbrueckii* isolates can ferment honey into mead. (**A**) The indicated strains were used to inoculate a honey solution, and fermentation of the honey was monitored with a digital hydrometer. The amount of alcohol produced is plotted vs. time. These results are representative of three independent fermentations for each strain or combination of strains (178/179 = a 1:1 mixture of YH178 and YH179 cultures). (**B**) Production-scale honey fermentations using a combination of YH178, YH179, and WLP715 yeasts. These fermentations were conducted in April (3 April 2017) and September (18 September 2017) of 2017 under similar conditions but with slightly different starting gravities of the honey solution.

Table 1. Experimental fermentations with *Saccharomyces cerevisiae*, *Torulaspora delbrueckii*, and *Hanseniaspora vineae* strains.

Strain	Honey Fermentation [1]			Beer Fermentation		
	Final Gravity	ABV [2]	Final pH [3]	Final Gravity	ABV	Final pH
S. cerevisiae WLP001	N.D.	N.D.	N.D.	1.011	4.33%	4.40
S. cerevisiae WLP715	0.992	11.5%	4.25	N.D.	N.D.	N.D.
T. delbrueckii YH178	0.992	11.5%	4.15	1.041	0.39%	5.40
T. delbrueckii YH179	0.996	11.0%	4.23	1.042	0.26%	5.40
YH178 + YH179	0.994	11.2%	4.20	N.D.	N.D.	N.D.
H. vineae YH72	1.018	8.2%	3.31	1.016	3.68%	3.26

[1] The original gravity of the honey solution was 1.080, and that of the wort was 1.044. [2] Abbreviations: ABV: alcohol by volume; N.D.: not determined. [3] The initial pH of the honey solution was 4.62, and that of the wort was 5.43.

It is perhaps unsurprising that gut microbiota from the honey bee can metabolize honey, as this is a food source for *A. mellifera* itself. Honey is mainly composed of the hexose monosaccharides glucose and fructose, and most yeast species can utilize one or both as carbon sources for cellular metabolism [25]. Conversely, unadulterated honey contains no maltose [26], which is a glucose disaccharide and the primary sugar in beer wort [27]. We tested YH178 and YH179 for their ability to ferment wort and found that attenuation of the beer was extremely poor after two weeks of fermentation (Table 1). In contrast, the *S. cerevisiae* WLP001 ale yeast attenuated to expected levels. The basal level of fermentation displayed by YH178 and YH179 is likely attributable to the *T. delbrueckii* cells consuming the limited quantities of glucose, fructose, and/or sucrose found in typical worts [27]. Indeed, growth curves in YP medium supplemented with 2% maltose (YPM medium) rather than glucose (YPD medium) demonstrate that YH178 grows much more poorly in the YPM than YPD medium (Figure 2A). Similarly, YH179 displayed only minimal growth in YPM medium (Figure 2B). The WLP715 strain is able to utilize maltose though, showing only a modest decrease in growth rate in YPM medium compared to YPD medium and no reduction in final cell density (Figure 2C). While others have successfully isolated yeasts from wasps [28] that can ferment wort into beer, not all yeasts from hymenoptera gut microbiomes are capable of this feat. Thus, individuals interested in isolating yeasts for a particular fermentation milieu are cautioned to select for crucial phenotypes (e.g., maltose utilization) early in the bio-prospecting process.

2.3. *YH178 and YH179 Produced Superior Meads Compared to WLP715*

We next compared the sensory characteristics of the meads produced via the above fermentations. This analysis was performed by 10 panelists using a method developed to quantify eight organoleptic qualities of mead [29] on a 20-point scale. Table 2 shows the results of our sensory analysis of mead fermented with YH178, YH179, the combination of YH178 and YH179, and *S. cerevisiae* WLP715. Overall, the meads fermented with YH178 or YH179 were scored more highly than that produced by WLP715 (though not significantly preferred, $p \geq 0.052$). The WLP715 mead was notable for its very forward ethanol heat, which was more pleasantly masked by YH178 and YH179 despite having comparable ABVs. However, the sensory panel also detected obvious differences in the meads made with YH178 and YH179. For instance, the aroma of the YH178 mead was more reminiscent of honey than YH179 mead, which some panelists reported as "funky" or simply lacking in honey character. This again corroborates our findings that YH178 and YH179 are distinct strains of *T. delbrueckii*. The mead fermented with the combination of YH178 and YH179 scored highly on aroma but had a grassy flavor that some panelists did not enjoy. Regardless, all of these meads received a "standard" score (12–16 points) [29], indicating that YH178 and YH179 are suitable for home and commercial brewing of mead.

Table 2. Sensory evaluation of mead samples.

Yeast [1]	Appearance	Aroma	Acidity	Balance	Body	Flavor	Finish	Overall Quality	Total
YH178	2 ± 0 [2]	3.7 ± 0.9	1 ± 0.5	1 ± 0.5	0.9 ± 0.7	1.5 ± 0.3	1.3 ± 0.5	1.4 ± 0.2	12.8 ± 2.4
YH179	2 ± 0	3.7 ± 1.2	1.1 ± 0.7	1.7 ± 0.4	1.4 ± 0.5	1.8 ± 0.4	1.8 ± 0.4	1.8 ± 0.4	15.3 ± 2.4
178/179	2 ± 0	4.5 ± 1.2	0.8 ± 0.6	0.9 ± 0.2	0.8 ± 0.6	1 ± 0.5	1.1 ± 0.2	1.2 ± 0.2	12.3 ± 1.9
WLP715	2 ± 0	4.3 ± 0.8	0.8 ± 0.6	1.3 ± 0.4	0.6 ± 0.2	1.2 ± 0.6	1.1 ± 0.4	1.4 ± 0.3	12.7 ± 2.2
YH72	2 ± 0	5.2 ± 0.7	1.5 ± 0	2 ± 0	0.7 ± 0.2	2 ± 0	2 ± 0	2 ± 0	17.4 ± 0.8

[1] YH178 and YH179 are *Torulaspora delbrueckii* isolates, 178/179 is a mixture of both *T. delbrueckii* strains, WLP715 is a *Saccharomyces cerevisiae* strain, and YH72 is a lactic acid-producing isolate of *Hanseniaspora vineae* [8]. [2] The values presented are the averages ± standard deviations.

Knowing that acidity would be ranked, we also fermented mead with *Hanseniaspora vineae* strain YH72 (Table 2), which produces lactic acid during fermentation [8], as an additional control. As in beer fermentation, the final pH of the *H. vineae* YH72 mead dropped to pH = 3.31 due to the lactate created by this strain. Surprisingly, during our sensory analysis, the YH72 mead was ranked highest in all categories except one (Body) and achieved a quality score defining it as a "superior" (16–20 points) mead [29] ($p \leq 0.017$ vs. all other meads). This was in part due to its assertive peach aroma, pleasant acidic finish, and good balance. It should, however, be noted that the final gravity of this mead was only 1.018, yielding an ABV of 8.2%. Thus, comparing it to dryer meads across all sensory categories can be misleading. Indeed, previous large-scale sensory analysis of sweet and dry meads indicates that consumers prefer sweet meads [30].

2.4. *YH178 and YH179 Were Used in the Creation of Honey Schnapps*

Based on the performance of YH178 and YH179 in mead fermentation, we sought to determine how well these strains functioned in production-scale (>20 hL) fermentations. Two large, independent fermentations of diluted honey were conducted as described in the Materials and Methods. A mixed yeast culture containing approximately equal cell counts of YH178, YH179, and WLP715 was used in both cases. As shown in Figure 3B, this yeast blend was able to ferment the solution to a specific gravity of ~1.000 in <10 days during both trials. This is consistent with the kinetics of fermentation displayed by these strains individually in laboratory-scale fermentations (Figure 3A), indicating that YH178 and YH179 can be used for industrial honey fermentations without any additional "domestication" or adaptation, at least in the presence of WLP715. Future experiments with YH178 and YH179 alone are planned, as well as trials of serial re-pitching of these strains into fermentations to select for an increased fermentation rate. Indeed, the desirable rapidity with which WLP715 ferments honey is one of the reasons that it was added to the production-scale fermentations performed in this work. Regardless, the fermented honey solutions were distilled into a schnapps that was back-sweetened with honey and were well received by consumers [31].

3. Materials and Methods

3.1. *Strains and Culture Conditions*

T. delbrueckii strains YH178 and YH179 were supplied by Wild Pitch Yeast, LLC (Bloomington, IN, USA), and *S. cerevisiae* strains WLP001 and WLP715 were purchased from White Labs (San Diego, CA, USA). All strains were routinely grown on yeast extract, peptone, and dextrose (YPD; 1% (w/v) yeast extract, 2% (w/v) peptone, and 2% (w/v) glucose) plates containing 2% (w/v) agar at 30 °C and in YPD liquid culture at 30 °C with aeration unless otherwise noted. WLN agar contained 4 g/L yeast extract, 5 g/L tryptone, 50 g/L glucose, 0.55 g/L KH_2PO_4, 0.425 g/L KCl, 0.125 g/L $CaCl_2$, 0.125 g/L $MgSO_4$, 2.5 mg/L $FeCl_3$, 2.5 mg/L $MnSO_4$, 22 mg/L bromocresol green, and 15 g/L agar. All strains were stored as 15% (v/v) glycerol stocks at −80 °C. Media components were from Fisher Scientific (Pittsburgh, PA, USA) and DOT Scientific (Burnton, MI, USA). All other reagents were of the highest grade commercially available.

3.2. *Yeast Isolation*

Yeasts YH178 and YH179 were isolated from a honey bee sample as described [7]. Briefly, honey bees were collected into sterile glass jars and stored at 4 °C before being processed in the laboratory. The bee containing YH178 and YH179 was submerged in 5 mL of YPD8E5 (1% yeast extract, 2% peptone, 8% glucose, and 5% (v/v) ethanol), crushed with a sterile pipet, and the culture was incubated at 30 °C with aeration for 24 h. A small volume of the culture was then streaked onto a WLN agar plate and for 36–48 h at 30 °C. To isolate pure strains, single colonies were picked and restreaked onto fresh WLN agar plates until a uniform colony morphology was obtained.

3.3. *Strain Identification*

YH178 and YH179 were identified as *T. delbrueckii* using the procedure described in [8]. Briefly, 100 µL of saturated overnight culture was mixed with an equal volume of lysis solution (0.2 M LiOAc and 1% SDS) and incubated in a 65 °C water bath for ≥15 min to lyse the cells. The genomic (gDNA) was precipitated with 300 µL of 100% isopropanol and vortexing, and it was pelleted with cell debris for 5 min at maximum speed in a microcentrifuge. The supernatant was removed, and the gDNA was resuspended in 50 µL TE buffer (10 mM Tris-HCl, pH 8, and 1 mM EDTA). The variable D1/D2 portion of the eukaryotic 26S rDNA was then amplified by PCR from the gDNA templates using oligos

NL1 (GCATATCAATAAGCGGAGGAAAAG) and NL4 (GGTCCGTGTTTCAAGACGG) [32] and the following cycling conditions: 98 °C for 5 min; 35 cycles of 98 °C for 30 s, 55 °C for 30 s, and 72 °C for 30 s; and 72 °C for 10 min. The amplified DNA was purified using a PCR Purification Kit (Thermo Scientific, Waltham, MA, USA) and sequenced by ACGT, Inc. (Wheeling, IL, USA) using primer NL1. The sequence was used to query the National Center for Biotechnology Information (NCBI) nucleotide database with the Basic Local Alignment Search Tool (BLAST; http://blast.ncbi.nlm.nih.gov/Blast.cgi?CMD=Web&PAGE_TYPE=BlastHome) to identify the most closely related species.

3.4. Growth Curves

The yeast strains were grown by inoculating 5 mL of YPD liquid medium with single colonies from YPD plates and incubation overnight at 30 °C with aeration. The optical density at 660 nm (OD_{660}) of each culture was determined using a Beckman Coulter DU730 UV/Vis Spectrophotometer. Then, the cells were diluted to an $OD_{660} \approx 0.1$ in 200 µL of YPD medium or YPM medium (1% (*w*/*v*) yeast extract, 2% (*w*/*v*) peptone, and 2% (*w*/*v*) maltose) in round-bottom 96-well plates, overlaid with 50 µL of mineral oil to prevent evaporation and incubated at 30 °C with shaking in a BioTek Synergy H1 plate reader. The OD_{660} of every well was measured and recorded every 15 min for ~20 h, and the normalized values (i.e., OD_{660} reads minus the initial OD_{660} value) were plotted vs. time to generate growth curves. All growth experiments were repeated three times, and the plotted points represent the average OD_{660} values (error bars represent the standard deviation).

3.5. Laboratory-Scale Honey Fermentation

Wild flower honey from Fort Wayne, IN was provided by Creek Ridge Honey, LLC and Cardinal Spirits. The honey was diluted approximately 1:5 with sterile deionized water and YP medium to a gravity of 1.080 and final concentrations of yeast extract and peptone of 0.25% and 0.5% (*w*/*v*), respectively. Fermentations of this honey solution were performed as described in [8]. Briefly, the yeast strains were grown to saturation (~200 × 10^6 cells/mL) in 50 mL of YPD liquid medium at 30 °C with aeration and used to inoculate ~800 mL of honey solution (1.080 original gravity) in 1 L glass cylinders (30 cm tall, 7.5 cm inner diameter). Control fermentations were likewise set up in parallel using the same reagents but inoculating with 50 mL (~200 × 10^6 cells/mL) WLP715. These fermentations were incubated at an average temperature of 23.4 ± 0.3 °C, and the gravity and alcohol by volume (ABV) were monitored in real time using BeerBug digital hydrometers (Sensor Share, Richmond, VA, USA) for 4 weeks.

3.6. Laboratory-Scale Wort Fermentation

Laboratory-scale wort fermentations were performed as described [8]. Briefly, blonde wort was prepared by mashing 65.9% Pilsner (2 Row) Bel and 26.9% white wheat malt at 65 °C (149 °F) for 75 min in the presence of 1 g/bbl $CaCO_3$ and 1.67 g/bbl $CaSO_4$. During the subsequent boil, 7.2% glucose and Saaz hops (to 25 international bittering units) were added. The original gravity (OG) of this wort was 1.044. The yeasts were grown to saturation (~200 × 10^6 cells/mL) overnight at 30 °C with aeration in 5 mL YPD medium, and these starter cultures were used to inoculate approximately 400 mL of blonde ale wort in glass bottles capped with rubber stoppers and standard plastic airlocks. Control fermentations were set up identically but inoculated with 5 mL (~200 × 10^6 cells/mL) WLP001 culture. The fermentation cultures were incubated at room temperature for 2 weeks, and their final gravity was measured using a MISCO digital refractometer (Solon, OH, USA).

3.7. Mead Sensory Analysis

Sensory analysis was performed using the method developed by the University of California, Davis as described by [29]. Briefly, each mead was sampled in a blinded fashion by 10 or more individuals and immediately scored for appearance (2 points), aroma (6 points), and acidity, balance, body, flavor, finish, and overall quality (2 points each). Using these criteria, a score of 17–20 is superior, 13–16 is standard, 9–12 is below standard, and 1–8 is not acceptable.

3.8. Production-Scale Honey Fermentation and Distillation

Two large honey fermentations were performed at Cardinal Spirits, LLC (Bloomington, IN). The first was performed on 3 April 2017 with a solution of 378 L honey and 23.2 hL H_2O, which had an initial specific gravity of 1.0569 (14.02 Brix) and an initial pH of 4.4. Diammonium phophate (DAP, 1.05 kg; BSG Wine, Napa, CA, USA) and Fermax (587 g; BSG HandCraft, Shakopee, MN, USA) were added to this solution as yeast nutrients, and then a 1:1:1 mixture of YH178, YH179, and WLP715 was added to inoculate the medium. The culture was maintained at a temperature of 28.5–30 °C in a stainless steel fermenter, and gravity and pH were monitored daily. After 1 day of fermentation, 1.05 kg DAP and 587 g Fermax were again added to the fermentation, and when the gravity decreased to <6 Brix, 587 g Fermax was added one last time. The fermentation proceeded for ~9 days, reaching a terminal gravity of 0.13 Brix (7.35% ABV) prior to distillation to approximately 150 proof.

The second large-scale fermentation was performed on 18 September 2017 with a solution of 549 L honey and 29.2 hL H_2O, which had an initial specific gravity of 1.0708 (17.24 Brix) and an initial pH of 5.04. The same nutrient addition schedule and yeast blend as above was used. The culture was maintained at a temperature of 28.5–31 °C in a stainless steel fermenter, and gravity and pH were again monitored daily. The fermentation proceeded for ~7 days, reaching a terminal gravity of 0 Brix (9.14% ABV) prior to distillation to approximately 150 proof. In both this and the previous cases, the distillate was diluted to ~80 proof and back-sweetened with 0.45 kg honey/3.78 L distillate. After resting this liquid for 1 month on a fining agent to precipitate out proteins and pollen, the final product was a yellow-tinted clear liquid (visually reminiscent of white wine) and marketed as a 74.2 proof honey schnapps. Further details are available upon request.

3.9. Statistical Analysis

All statistical analyses were performed using GraphPad Prism 6 software. Differences between values were compared using Student's *t*-test, and *p*-values <0.05 were considered statistically significant.

Acknowledgments: We thank the members of our sensory analysis panels for providing valuable flavor and aroma data, as well as members of the Bochman laboratory for critically reading this manuscript and providing feedback. This work was supported by startup funds from Indiana University, a Translational Research Pilot Grant from the Johnson Center for Entrepreneurship in Biotechnology, and funds from Wild Pitch Yeast, LLC (to MLB).

Author Contributions: Matthew L. Bochman, Justin Hughey and Adam Quirk conceived the study and designed the experiments. Joseph P. Barry, Mindy S. Metz, Justin Hughey and Matthew L. Bochman performed the experiments, and all authors analyzed and interpreted the results. Matthew L. Bochman wrote the manuscript. All authors read and approved the final version of the manuscript.

References

1. McGovern, P.E.; Mirzoian, A.; Hall, G.R. Ancient Egyptian herbal wines. *Proc. Natl. Acad. Sci. USA* **2009**, *106*, 7361–7366. [CrossRef] [PubMed]
2. McGovern, P.E.; Zhang, J.; Tang, J.; Zhang, Z.; Hall, G.R.; Moreau, R.A.; Nunez, A.; Butrym, E.D.; Richards, M.P.; Wang, C.S.; et al. Fermented beverages of pre- and proto-historic China. *Proc. Natl. Acad. Sci. USA* **2004**, *101*, 17593–17598. [CrossRef] [PubMed]
3. Crewe, L.; Hill, I. Finding Beer in the Archaeological Record: A Case Study from Kissonerga-Skalia on Bronze Age Cyprus. *Levant* **2012**, *44*, 205–237. [CrossRef]
4. Rogers, A. *Proof: The Science of Booze*; Mariner Books: Wilmington, DE, USA, 2015.

5. Iglesias, A.; Pascoal, A.; Choupina, A.B.; Carvalho, C.A.; Feas, X.; Estevinho, L.M. Developments in the fermentation process and quality improvement strategies for mead production. *Molecules* **2014**, *19*, 12577–12590. [CrossRef] [PubMed]
6. American Mead Makers Association. *What's the Buzz? 2017 Mead Industry Report*; American Mead Makers Association: Seattle, WA, USA, 2017.
7. Osburn, K.; Ahmad, N.N.; Bochman, M.L. Bio-prospecting, selection, and analysis of wild yeasts for ethanol fermentation. *Zymurgy* **2016**, *39*, 81–88.
8. Osburn, K.; Amaral, J.; Metcalf, S.R.; Nickens, D.M.; Rogers, C.M.; Sausen, C.; Caputo, R.; Miller, J.; Li, H.; Tennessen, J.M.; et al. Primary souring: A novel bacteria-free method for sour beer production. *Food Microbiol.* **2017**, *70*, 76–84. [CrossRef] [PubMed]
9. Rogers, C.M.; Veatch, D.; Covey, A.; Staton, C.; Bochman, M.L. Terminal acidic shock inhibits sour beer bottle conditioning by Saccharomyces cerevisiae. *Food Microbiol.* **2016**, *57*, 151–158. [CrossRef] [PubMed]
10. Independent We Stand. Craft Beer Goes Even More Local with the Farm-to-Glass Movement. 2017. Available online: https://www.independentwestand.org/craft-beer-and-the-farm-to-glass-movement/ (accessed on 28 February 2018).
11. Kallenberger, M. Local Sourcing from the Craft Beer Drinker's Perspective. *New Brew.* **2016**, *34*, 84–88.
12. Olaitan, P.B.; Adeleke, O.E.; Ola, I.O. Honey: A reservoir for microorganisms and an inhibitory agent for microbes. *Afr. Health Sci.* **2007**, *7*, 159–165. [CrossRef] [PubMed]
13. Silva, M.S.; Rabadzhiev, Y.; Eller, M.R.; Iliev, I.; Ivanova, I.; Santana, W.C. Microorganisms in Honey. In *Honey Analysis*; de Toledo, V.A.A., Ed.; InTech: Rijeka, Croatia, 2017.
14. Loira, I.; Vejarano, R.; Banuelos, M.A.; Morata, A.; Tesfye, W.; Uthurry, C.; Villa, A.; Cintora, I.; Suarez-Lepe, J.A. Influence of sequential fermentation with Torulaspora delbrueckii and Saccharomyces cerevisiae on wine quality. *LWT Food Sci. Technol.* **2014**, *59*, 915–922. [CrossRef]
15. Canonico, L.; Comitini, F.; Ciani, M. Torulaspora delbrueckii contribution in mixed brewing fermentations with different Saccharomyces cerevisiae strains. *Int. J. Food Microbiol.* **2017**, *259*, 7–13. [CrossRef] [PubMed]
16. Canonico, L.; Agarbati, A.; Comitini, F.; Ciani, M. Torulaspora delbrueckii in the brewing process: A new approach to enhance bioflavour and to reduce ethanol content. *Food Microbiol.* **2016**, *56*, 45–51. [CrossRef] [PubMed]
17. Michel, M.; Kopecka, J.; Meier-Dornberg, T.; Zarnkow, M.; Jacob, F.; Hutzler, M. Screening for new brewing yeasts in the non-Saccharomyces sector with Torulaspora delbrueckii as model. *Yeast* **2016**, *33*, 129–144. [CrossRef] [PubMed]
18. Pires, E.J.; Teixeira, J.A.; Branyik, T.; Vicente, A.A. Yeast: The soul of beer's aroma—A review of flavour-active esters and higher alcohols produced by the brewing yeast. *Appl. Microbiol. Biotechnol.* **2014**, *98*, 1937–1949. [CrossRef] [PubMed]
19. Etschmann, M.M.; Huth, I.; Walisko, R.; Schuster, J.; Krull, R.; Holtmann, D.; Wittmann, C.; Schrader, J. Improving 2-phenylethanol and 6-pentyl-α-pyrone production with fungi by microparticle-enhanced cultivation (MPEC). *Yeast* **2015**, *32*, 145–157. [CrossRef] [PubMed]
20. Al Toufailia, H.M.; Amiri, E.; Scandian, L.; Kryger, P.; Ratnieks, F.L.W. Towards integrated control of varroa: Effect of variation in hygienic behaviour among honey bee colonies on mite population increase and deformed wing virus incidence. *J. Apic. Res.* **2015**, *53*, 555–562. [CrossRef]
21. Bigio, G.; Al Toufailia, H.M.; Hughes, W.O.H.; Ratnieks, F.L.W. The effect of one generation of controlled mating on the expression of hygienic behaviour in honey bees. *J. Apic. Res.* **2014**, *53*, 563–568. [CrossRef]
22. Wilson-Rich, N.; Spivak, M.; Fefferman, N.H.; Starks, P.T. Genetic, individual, and group facilitation of disease resistance in insect societies. *Annu. Rev. Entomol.* **2009**, *54*, 405–423. [CrossRef] [PubMed]
23. Schoning, C.; Gisder, S.; Geiselhardt, S.; Kretschmann, I.; Bienefeld, K.; Hilker, M.; Genersch, E. Evidence for damage-dependent hygienic behaviour towards Varroa destructor-parasitised brood in the western honey bee, *Apis mellifera*. *J. Exp. Biol.* **2012**, *215*, 264–271. [CrossRef] [PubMed]
24. *WLP715 Champagne Yeast*; White Labs: San Diego, CA, USA, 2016. Available online: https://www.whitelabs.com/yeast-bank/wlp715-champagne-yeast (accessed on 7 February 2018).
25. Zimmermann, F.K.; Entian, K.-D. *Yeast Sugar Metabolism: Biochemistry, Genetics, Biotechnology, and Applications*; Technomic Publishing Company, Inc.: Lancaster, PA, USA, 1997.
26. Fujita, I. Determination of Maltose in Honey. *Int. J. Food Sci. Nutr. Diet.* **2012**, *1*, 1–2.

27. Briggs, D.E.; Hough, J.S.; Stevens, R.; Young, T.W. *Malting and Brewing Science*, 2nd ed.; St. Edmundsbury Press: Great Britain, UK, 1981.
28. Smith, L. Walk on the Wild Side With Beer Made From Wasp Yeast. In *The Plate*; National Geographic: Washington, DC, USA, 2015. Available online: http://theplate.nationalgeographic.com/2015/09/29/walk-on-the-wild-side-with-beer-made-from-wasp-yeast/ (accessed on 28 February 2018).
29. Hernandez, C.Y.; Serrato, J.C.; Quicazan, M.C. Evaluation of Physicochemical and Sensory Aspects of Mead, Produced by Different Nitrogen Sources and Commercial Yeast. *Chem. Eng. Trans.* **2015**, *43*, 1–6. [CrossRef]
30. Gomes, T.; Dias, T.; Cadavez, V.; Verdial, J.; Sa Morais, J.; Ramalhosa, E.; Estevinho, L.M. Influence of Sweetness and Ethanol Content on Mead Acceptability. *Pol. J. Food Nutr. Sci.* **2015**, *65*, 137–142. [CrossRef]
31. Sagon, E. *Honey Schnapps: How We Made It*; Cardinal Spirits: Bloomington, IN, USA, 2017. Available online: http://cardinalspirits.com/thedrop/honey-schnapps-how-we-made-it (accessed on 2 March 2018).
32. Lee, Y.J.; Choi, Y.R.; Lee, S.Y.; Park, J.T.; Shim, J.H.; Park, K.H.; Kim, J.W. Screening wild yeast strains for alcohol fermentation from various fruits. *Mycobiology* **2011**, *39*, 33–39. [CrossRef] [PubMed]

6

Development of Novel Textile Bioreactor for Anaerobic Utilization of Flocculating Yeast for Ethanol Production

Osagie A. Osadolor *, Patrik R. Lennartsson and Mohammad J. Taherzadeh

Swedish Centre for Resource Recovery, University of Borås, SE 50190 Borås, Sweden; Patrik.Lennartsson@hb.se (P.R.L.); Mohammad.Taherzadeh@hb.se (M.J.T.)

* Correspondence: Alex.Osagie@hb.se

Academic Editor: Ronnie G. Willaert

Abstract: Process development, cheaper bioreactor cost, and faster fermentation rate can aid in reducing the cost of fermentation. In this article, these ideas were combined in developing a previously introduced textile bioreactor for ethanol production. The bioreactor was developed to utilize flocculating yeast for ethanol production under anaerobic conditions. A mixing system, which works without aerators, spargers, or impellers, but utilizes the liquid content in the bioreactor for suspending the flocculating yeast to form a fluidized bed, was developed and examined. It could be used with dilution rates greater than 1.0 h^{-1} with less possibility of washout. The flow conditions required to begin and maintain a fluidized bed were determined. Fermentation experiments with flow rate and utilization of the mixing system as process variables were carried out. The results showed enhanced mass transfer as evidenced by faster fermentation rates on experiments with complete sucrose utilization after 36 h, even at 30 times lesser flow rate.

Keywords: flocculating yeast; textile bioreactor; ethanol; mass transfer; mixing; fluidization

---

1. Introduction

Increasing energy demand and environmental awareness have influenced the progressive rise in the production and utilization of bioethanol as a transportation fuel [1]. To boost the competitiveness of bioethanol to fossil fuel, particularly with the current low prices of fossil fuels, there is the need to continue increasing the productivity of the ethanol production process while reducing the production cost. Flocculation has proven to be advantageous in improving the productivity of the bioethanol production process [2]. Some of its benefits include production at high dilution rate, improved inhibitor tolerance, longer reuse of cells, reduced contamination tendencies at high dilution rate, reduced bioreactor cost because of smaller reactor volume [3], and ease of separating cell flocs from liquid medium in the bioreactor [4].

To effectively utilize flocculation, the size of the flocs and their settling characteristics need to be well understood. It is important that the floc size is large enough to prevent washout while being small enough to allow effective passage and mass transfer of the substrate into the cells and product out of the cells. To avoid washout, mechanical stirrers that break down the flocs are not usually used in bioreactors utilizing flocculating organisms [3]. Depending on the settling rate of the flocs, the contacting pattern that they would make in a bioreactor could result in fixed or fluidized bed. Fixed or fluidized bed systems have their benefits, but for optimal mass transfer and faster production rate, fluidized bed systems are more advantageous because of the larger contacting area of the flocs [5]. To create a stable fluidized bed, the flow rate has to be between the minimum to initiate fluidization and the maximum to prevent the flocs from being carried away from the bioreactor [5]. For these reasons, the design and operation of the bioreactor to be used for propagating flocculating microorganisms is quite important. Fluidization in bioreactors can be achieved either by aeration or by utilizing high flow liquid streams [6]. Currently, airlift bioreactors are the main type of bioreactor being used for utilizing flocculating yeast for bioethanol production [3].

Aeration is required when flocculating yeast is used in an airlift bioreactor for ethanol production [3]. This reduces the ethanol yield as ethanol is optimally produced anaerobically. Besides this, bioreactor cost is high, including aeration, generating more operation cost. Bioreactors for ethanol production have to be designed in a way that they do not hinder the activity of the microorganism within them, withstand the corrosive nature of fermentation media, and provide suitable environment and control needed to optimally produce the desired product(s) [7]. The overall goal in their design is to deliver the required functions and to be economical [8]. Conventional bioreactors for ethanol production are made using stainless steel as the major material of construction and constitute 32% of the fermentation investment cost in a typical 100,000 m^3/year plant [9]. Reducing the cost of ethanol bioreactors will reduce the cost of ethanol production. Some polymeric materials (*e.g.*, polyaniline) have good corrosion-resistance properties [10] and are cheaper than stainless steel, so they could be options for making ethanol bioreactors. However, some challenges regarding their use include their tensile strength and the effectiveness of mixing in the bioreactors made with them. A novel bioreactor with textile as its material of construction was recently introduced as a cheaper alternative to bioreactors made with stainless steel [11]. The textile bioreactor has textile as its backbone material of construction which improves its strength [11]. In this paper, the conditions needed to maintain optimal contact of the flocculating yeast in the bioreactor were determined, and the textile bioreactor was developed accordingly. This enabled anaerobic production of ethanol with the flocculating yeast, while also maintaining good mixing in the bioreactor, thus creating optimal production conditions.

2. Methods

2.1. Microorganism

Naturally flocculating yeast strain *S. cerevisiae* CCUG 53310 (Culture Collection University of Gothenburg, Gothenburg, Sweden) was used for the experiments. The flocculating yeast cells were maintained on a yeast extract peptone dextrose (YPD) agar plate containing 20 g/L agar, 20 g/L D-glucose, 20 g/L peptone, and 10 g/L yeast extract at 4 °C. Before being used for fermentation, the flocculating cells were added into 800 mL YPD media containing 20 g/L D-glucose, 10 g/L yeast extract, and 20 g/L peptone in a 2 L cotton-plugged flask. Three flasks were incubated in a shaking water bath (Grant OLS 200, Grant instrument Ltd., Cambridge, UK) at 125 rpm and 30 °C for 48 h. The supernatant liquid from two of the flasks were discarded, and the sedimented flocculating yeast were rinsed with distilled water into the textile bioreactor for fermentation. The content of the third flask was used to determine the yeast concentration in the bioreactor. A starting concentration of 2 g/L dry weight of the flocs was used for all fermentation experiments.

The average particle diameter for this yeast strain was measured using the optical density and sedimentation technique [12] and it was found to be between 190 to 320 μm. This was done by transferring samples of the cell culture to a tube in which the optical density was measured using a spectrophotometer at a wavelength of 660 nm for the various samples at different times. The optical density readings from the spectrophotometer were proportional to the flocculating yeast concentration. When all the flocculating yeast had settled (settling time), the optical density read from the spectrophotometer gave constant readings, which corresponded to the concentration of the free cells. Dividing the distance from which the flocculating yeast cells fell by the settling time gave the settling velocity. The average diameter of the flocs was calculated from Stokes' law (see equation 4), using flocculating yeast density of 1140 kg/m^3, viscosity of 0.798×10^{-3} Ns/m^2 (*i.e.*, viscosity of water at 30 °C) and acceleration due to gravity of 9.81 m/s^2.

2.2. Textile Bioreactor and Its Development

The bioreactor used for this work (Ethanolic textile lab reactor ETLRII, FOV Fabrics AB, Borås, Sweden), was a 30 L laboratory scale reactor with a working volume of 25 L. Its dimensions were 110 cm length, 8 cm depth, and 34 cm width (Figure 1a). The inlet and outlet diameter were 5 cm. It had a 4 cm opening at the middle that served as a sample collection point, a gas outlet, and a thermometer stand. Tubing was connected to the inlet and outlet of the bioreactor for recirculation.

A 12 m silicone peroxide tubing with 5 mm internal diameter and 8 mm external diameter (VWR International, Leuven, Belgium) was used as the internal mixing tubing in the bioreactor. Holes of 0.42 mm at 1 cm intervals were made in the tubing. The tubing was wound around the perimeter of the bioreactor twice and then joined into five elliptical ribbons by plastic fasteners (Figure 1b). The tubing was kept in the liquid phase by means of ten stainless steel bolts, each having internal and external diameters of 2.4 and 3.5 cm respectively, and weighing 94.6 g. The two ends of the tubing used for mixing were connected to the inlet tubing from the pump. A peristaltic pump (405U/L2 Watson-Marlow, Stockholm, Sweden) was used for recirculation, feeding, and discharging the content of the bioreactor.

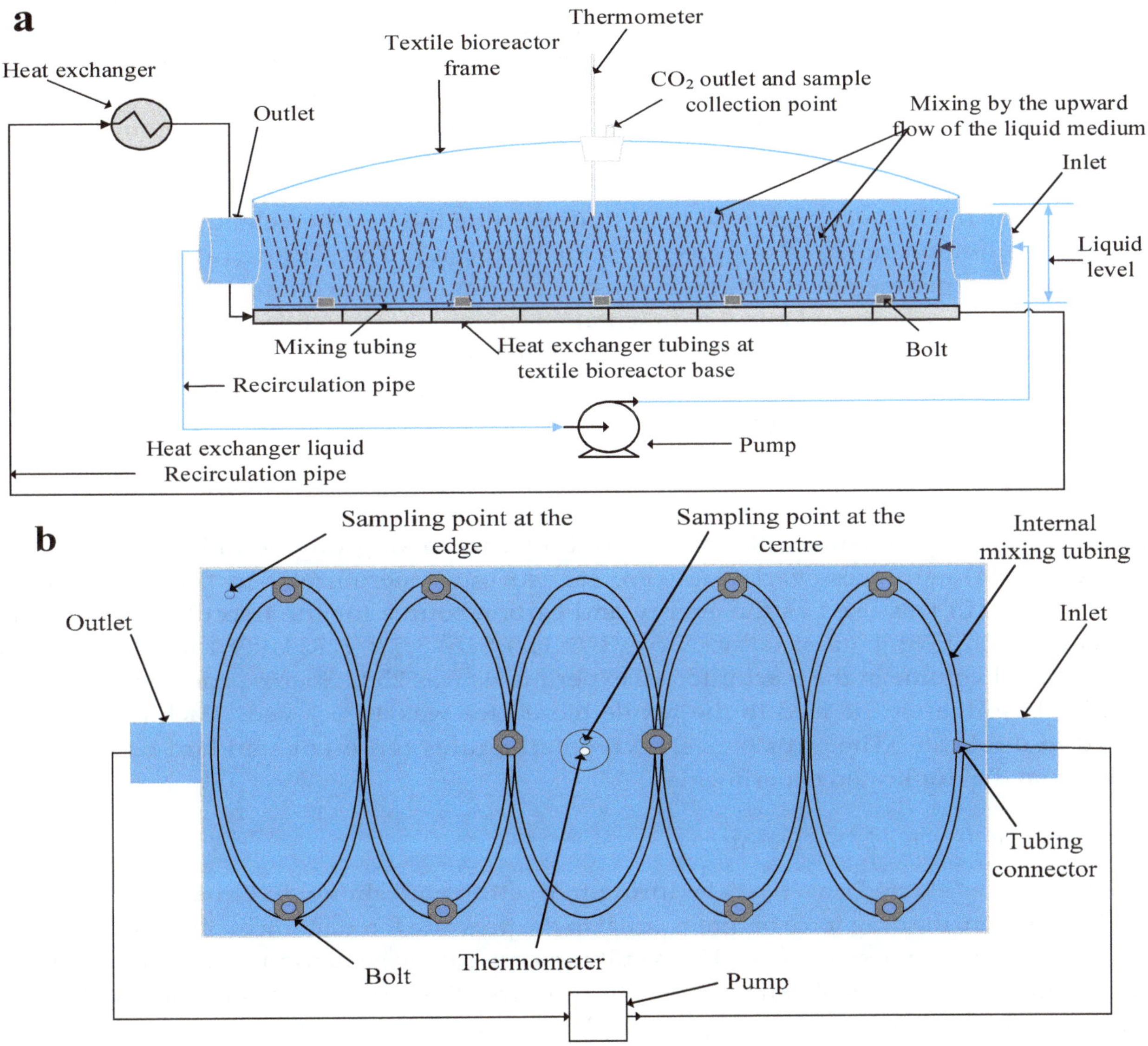

Figure 1. Developed textile bioreactor showing internal mixing inside the bioreactor from (**a**) a side view, and (**b**) a top view of the internal mixing system inside the bioreactor.

2.3. Mixing in the Bioreactor

First 20 L distilled water was fed to the bioreactor. Then, 90 ± 5 mL of 100 ppm bromophenol blue solution was added to the bioreactor at one of its ends (hereafter referred to as the injection point). The content of the bioreactor was recirculated through the internal mixing tube at flow rates of 0.002 and 0.015 volume per volume per minute (VVM). Samples were taken from the two opposite rear edges of the bioreactor. The absorbance of the samples was measured by a spectrophotometer (Biochrom Libra S60, Cambridge, UK) at 592 nm [13]. Mixing during fermentation was carried out by recirculation. To determine the effectiveness of mixing in experiments performed with internal mixing tubing, samples were taken from two sampling points, one at the center of the bioreactor and another at its edge. The samples were taken at a depth of 8 cm and from the surface of the liquid medium in the bioreactor.

2.4. Analytical Methods

Biomass was rinsed with distilled water and dried in an oven at 70 °C for 24 h, and its concentration was reported in g/L. Samples from the bioreactor were analyzed using a hydrogen-based ion exchange column (Aminex HPX-87H, Bio-road, Hercules, CA, USA) in a high performance liquid chromatography (HPLC) at 60 °C and 5 mM 0.6 mL/min H_2SO_4 eluent. A refractive index detector (Waters 2414, Waters Corporation, Milford, MA, USA) was used with the HPLC. Before being used for HPLC analysis, the samples were centrifuged for 5 min at 10,000× *g* and the liquid portion stored at −20 °C. Concentrations reported for all fermentation experiments were the amount determined by HPLC.

Scanning for the highest peak was used to determine the best wavelength for measuring the absorbance of bromophenol blue samples from the bioreactor. The settling rate of the flocculating yeast used for the experiment was determined using a sedimentation column. Biomass concentration of 2 g/L dry weight was released into a sedimentation column that was filled to 10 cm with the media for fermentation, and the content was mixed. The time taken for the flocs to settle to the base of the column was recorded as the settling time. The distance that the flocs settled from was divided by the settling time to give the settling velocity (V_s).

2.5. Experimental Setup for Fermentation

Fermentation experiments with and without internal mixing tubing and different flow rates (0.0016–0.06 VVM) as process variables were carried out anaerobically in the textile bioreactor. Sucrose (50–55 g/L) was used as the energy and carbon source for the flocculating yeast. It was supplemented with 1.0 g/L yeast extract, 7.5 g/L $(NH_4)_2SO_4$, 3.5 g/L KH_2PO_4, and 0.75 g/L $MgSO_4$ $7H_2O$. The liquid volume in the reactor for all experiments was 25 L. Sucrose concentration dropped between 51–45 g/L after the feed to the textile bioreactor was autoclaved. Each experiment was performed in duplicate. The error bars shown on all figures represent standard deviation values generated from the duplicated experiments.

2.6. Fluidization of the Flocs in the Bioreactor

The four forces exerted on a particle during free settling are shown in Equation (1) [14]. On the verge of fluidization, the drag force becomes equal to the pressure force acting on the flocs, and the force due to acceleration becomes zero [14]. The resulting force balance equation is shown in Equation (2). The minimum flow rate to establish this condition occurs when the superficial velocity (V_O) is equal to the fluid upwards velocity (V_u). To prevent the flocs from be carried away from the bioreactor, the maximum fluid upwards velocity (V_{umax}) should be equal to the flocs' settling velocity (V_s) [15]. For Reynolds numbers less than 10, these velocities can be obtained from Equations (3) and (4) [15].

Where ρ_p is the density of the particle, ρ_f is the density of the fluid, g is acceleration due to gravity, D_p is the particle diameter, μ is the viscosity, and ε is the void fraction.

$$\text{Force due to acceleration} = \text{Force due to gravity} - \text{Buoyancy force} - \text{Drag force} \tag{1}$$

$$\text{Pressure force } (F_p) = \text{Force due to gravity } (F_g) - \text{Buoyancy force } (F_B) \tag{2}$$

$$V_0 = V_u = (\rho_p - \rho_f) g D_p^2 \varepsilon^3 / 150\, \mu\, (1 - \varepsilon) \tag{3}$$

$$V_{umax} = V_s = (\rho_p - \rho_f) g D_p^2 / 18\, \mu \tag{4}$$

For the textile bioreactor prototype used in this work, the superficial velocity is defined in Equation (5), where Q is the liquid flow rate in m^3/s and A is the surface area. The velocity of the fluid (V_i) going through the mixing tubing is defined in Equation (6), where A_i is the internal area of the tubing. Using the continuity principle, the velocity at which the fluid leaves the small holes in the tubing is 142 (*i.e.*, $2.5^2/0.21^2$) times higher than that at which it enters [15]. However, as the tubing is 12 m long and the holes are spaced 1 cm apart, there were 1200 holes. So, the upward velocity (V_h) with which the fluid emerges would split accordingly. This is shown in Equation (7). For fluidization to begin and be sustained in the developed textile bioreactor, Equation (8) shows the governing criteria:

$$V_0 = Q/A = Q/(1.1 \times 0.34) = 2.674Q \text{ (m/s)} \tag{5}$$

$$V_i = Q/A_i = Q/(\pi \times r_i^2) = 5.094 \times 10^4 Q \text{ (m/s)} \tag{6}$$

$$V_h = 142 \times V_i/1200 = 6.028 \times 10^3 Q \text{ (m/s)} \tag{7}$$

$$V_0 \leq V_h \leq V_s \tag{8}$$

2.7. Statistical Analysis

All statistical analyses were performed with the MINITAB® software package. Results were analyzed with ANOVA (analysis of variance), using general linear model, and factors were considered significant when they had p-value less than 0.05. The analysis was performed on the results obtained from samples measured from the start of the experiment up until the 32nd hour, when stationary phase was reached. Ethanol and sucrose concentrations were used as the response variables, the position from which the sample was taken was used as the main factor, while time and number of runs served as blocking factors.

3. Results and Discussion

Possible ways to reduce the fermentation cost in a bioethanol plant includes process development, utilizing new and cheaper bioreactors, and making the separation process more efficient. This work combines these ideas in developing the textile bioreactor. The newly developed textile bioreactor has the high flexibility, ease of operation and installation, good mechanical strength, high thermal tolerance, low purchase cost, resistance to corrosive fermentation media, and ease of transportation of its previous prototype [11]. It has a new mixing system that eliminates the need for mixing by using either axial flow impellers or aeration spargers. This removes the purchasing and operational cost associated with the maintenance of those devices. A highly flocculating yeast strain with a settling rate of 1 cm/s was used to examine the performance of the developed textile bioreactor for bioethanol production. The efficiency of mixing (a measure of the mass transfer efficiency) in the developed textile bioreactor and the flow rate needed to maintain optimal contact between the flocculating yeast and bioreactor content were investigated in this work.

3.1. Maintaining Optimal Flocs Contact in the Textile Bioreactor

Increasing the surface contact area of enzymes or catalyst in the form of a fluidized bed generally increases the speed at which a chemical reaction takes place [14]. Thus, flocculating yeast retained in a bioreactor in the form of fluidized particles would result in more rapid utilization of the substrate in the bioreactor. Using Equations (5) to (7), different fluid upflow velocities and superficial velocities for different flow rates were generated, and their values are shown in Table 1. As seen in this table, the required velocity needed to begin fluidization is low. This is because the textile bioreactor prototype used for this analysis has a high surface area to volume ratio of 12.5 m^2/m^3. The superficial velocity reduces with increasing surface area to volume ratio, so the upflow velocity needed on the verge of fluidization would be less for bioreactors having a higher surface area to volume ratio.

Table 1. The fluid upflow velocity generated at different flow rate in the textile bioreactor. VVM: volume per volume per minute; V_0: Superficial velocity; V_i: Fluid velocity; V_h: Fluid upflow velocity; V_s: Flocs settling velocity.

Q (VVM)	V_0 (m/s) $\times 10^6$	V_i (m/s)	V_h at Different Hole Spacing (m/s)			V_s (m/s)
			1 cm Spacing	5 mm Spacing	2 mm Spacing	
0.0016	1.78	0.03	0.004	0.002	0.001	0.01
0.0120	13.37	0.25	0.030	0.015	0.006	0.01
0.0160	17.83	0.34	0.040	0.020	0.008	0.01
0.0320	35.64	0.68	0.080	0.040	0.016	0.01
0.0600	66.84	1.27	0.151	0.075	0.030	0.01

For continuous production of bioethanol using flocculating yeast, it is desirable to carry out the production at a high dilution rate and at the same time prevent washout. From Equation (8), washout could occur when the fluid upflow velocity (V_h) exceeds the flocs settling velocity (V_s). From Table 1, it can be observed that washout would occur in this textile bioreactor prototype if operated on a dilution rate greater than 0.72 h^{-1} (*i.e.*, $Q \geq 0.012$ VVM), as V_h is greater than V_s. One way to increase the dilution rate would be to increase the number of holes by reducing the space between consecutive holes in the mixing tubing. From Table 1, reducing the space between the tubes to 2 mm would allow the bioreactor to be operated at a dilution rate of 1 h^{-1} with lesser tendency for washout, as V_h equals V_s with this condition. Another possibility of increasing the dilution rate would be to use mixing tubing of a length longer than 12 m. For example, using 18 m long tubing at 2 mm spacing, from equation 7, at $Q = 0.02$ VVM (dilution rate of 1.2), V_h becomes 0.007 m/s, which is less than the settling velocity of 0.01 m/s. Normally, at a dilution rate greater than the growth rate, washout would occur, but with this type of configuration for the mixing tubing, even with dilution rate higher than growth rate, washout would not occur.

3.2. Mixing as a Means of Reducing Mass Transfer Limitations

Efficient mass transfer is important for optimal performance in any bioreactor, as it helps to facilitate transfer of substrate into and product out of the microorganism, and prevent improper cell growth in the bioreactor [7]. Mass transfer is influenced by two factors: the diffusional flux between the cells and the liquid media, and the bulk flow of the liquid media. The first factor is influenced by the diffusivity of the product or substrate into the liquid media. The second is enhanced by the mixing system in the bioreactor, like agitation, aeration, or the use of stirrers [16]. With the theoretical understanding of how the flow rate can influence good contacting pattern in the previous section, experimental verifying the theoretical concepts are presented in this section.

For bioreactors utilizing flocculating yeast, mass transfer limitations need to be adequately considered. There needs to be a balance between having flocs of large sizes, favorable for the enhancement of cell retention in the bioreactor, and reducing the size of the flocs, which is necessary for the reduction of the mass transfer limitations. Characterizing solute transport into and out of the flocs is challenging because of the fragile nature of the flocs and the difficulty in deciding the geometry the flocs actually have inside the bioreactors [3]. The approach in this article was to relate the mass transfer due to diffusion to the coefficient of diffusivity, and the flow rate of the liquid media being recirculated as a measure of that due to the bulk flow of the liquid [17].

Figure 2 shows the result of the absorbance of bromophenol blue measured across the textile bioreactor at different flow rates. At 0.015 VVM, equilibrium was reached in the textile bioreactor in 15 min. At a flow rate of 0.002 VVM, equilibrium was reached in 50 min. At 25 °C, the diffusion coefficient of bromophenol blue in water is 3×10^{-10} m^2/s [18], that of ethanol in water is 1.24×10^{-9} m^2/s, that of sucrose in water is 5.24×10^{-10} m^2/s, and that of water in water is 2.45×10^{-9} m^2/s [19]. The lower diffusivity of bromophenol blue than that of the product and substrate serves as a benchmark for understanding the influence of diffusivity on mass transfer under limiting product or substrate conditions [20]. Achieving equilibrium in 50 min at a flow rate of 0.002 VVM is sufficient, as the sampling time for experiments on bioethanol production is usually measured in hours [21].

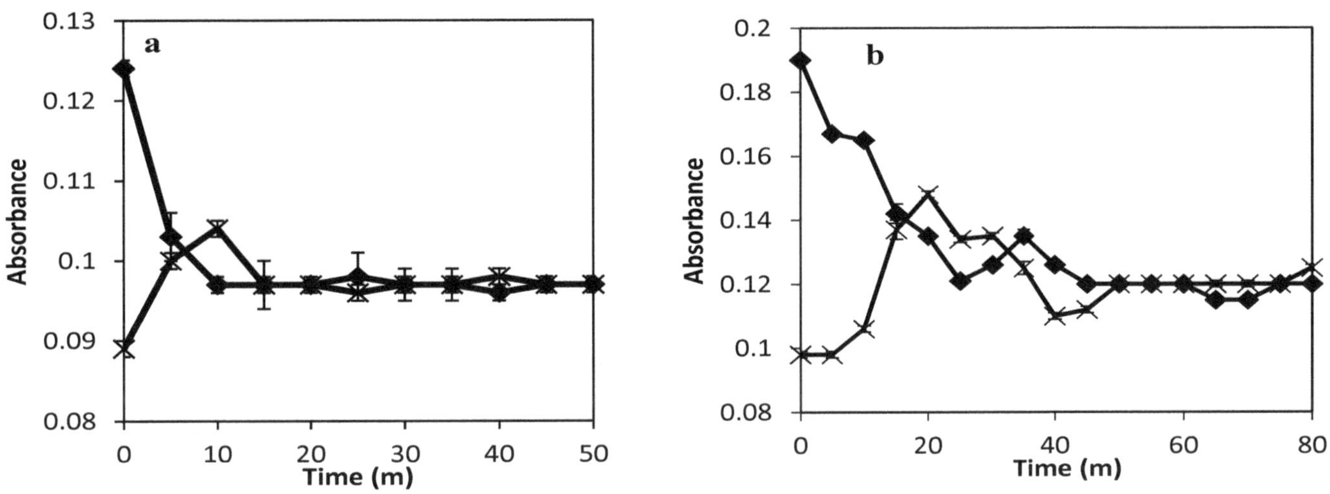

Figure 2. Bromophenol blue absorbance variation in the textile bioreactor at (**a**) a flow rate of 0.015 VVM, and (**b**) at a flow rate of 0.002 VVM at the injection point (♦) and at the opposite end (×).

3.2.1. Mass Transfer Enhancement by Internal Mixing Tubing in the Textile Bioreactor

The internal tubing had holes of 0.42 mm diameter which were 1 cm spread apart all through the textile bioreactor. Both ends of the tubing were connected to the input stream from the pump. The size of the holes and the continuous inflow of the liquid medium generate a pressure difference between the input stream and the fluid inside the tubing. A diameter of 0.42 mm was used because using a larger diameter would result in the incoming fluid leaving the tubing only from the entrance, and a smaller size would generate high back pressure on the pump.

The holes in the internal tubing create upward flow of the liquid stream, which helps in re-suspending the settling flocs, thus improving the mass transfer rate inside the textile bioreactor. To test this idea, fermentation experiments were carried out in the textile bioreactor. Figure 3a shows the result with internal tubing to aid mixing at 0.0016 VVM, while Figure 3b shows the result when mixing is not aided at a flow rate of 0.032 VVM. A lower flow rate was used for the case with aided mixing to show that the result obtained was not due to the flow rate but due to the mixing system itself—as demonstrated in Figure 2a, higher flow rate has been shown previously to improve mass transfer. The same starting concentration of 2 g/L dry weight flocculating yeast was used in both cases. For the aided mixing case (Figure 3a), the fermentation process and sucrose consumption was complete after 36 h, while the unaided mixing experiment (Figure 3b) still had more than 4 g/L of unconsumed sucrose after 48 h. The longer time in Figure 3b is attributable to the settling of the flocs at the bottom of the textile bioreactor, which causes uneven consumption of the substrate in the reactor. The superficial velocity for Figure 3b from Table 1 is 3.6×10^{-5} m/s, which can be met by the velocity of the CO_2 gas bubbles rising from the bottom of the reactor, at the onset of the fermentation experiment. This could explain the higher substrate consumption during the first 28 h of fermentation than in the latter part. However, when the sucrose at the bottom of the reactor is consumed, there will no longer be CO_2 bubbles to re-suspend the settled flocs, so the flocs at the bottom of the reactor, because of the mass transfer limitation, would go into the stationary phase faster, resulting in longer fermentation time or incomplete utilization of the substrate. In Figure 3a, the upward flow of the liquid stream helped to keep the flocs uniformly distributed in the textile bioreactor through the duration of the fermentation experiment. This shows that the developed mixing system for the textile bioreactor is effective in preventing ineffective substrate utilization and increasing the fermentation rate. From Figure 3a, the average peak ethanol concentration was 22.13 ± 0.93 g/L. Using the average fermentation time gave the specific ethanol productivity in the developed textile bioreactor with aided mixing at 0.0016 VVM as 0.29 ± 0.01 g-ethanol/g-biomass/h. The best specific productivity from a gas lift reactor with recycle was reported as 0.045 g-ethanol/g-biomass/h with no sugar loss, and 0.43 g-ethanol/g-biomass/h with significant sugar loss [22]. For an airlift reactor, optimum specific productivity of 0.4 g-ethanol/g-biomass/h with significant sugar loss has been reported [23]. Comparing the reported specific productivity of different bioreactors for ethanol production from literature with that of the developed textile bioreactor, it can be seen that for an optimal combination of high specific productivity and complete sugar utilization, the developed textile bioreactor performs better.

3.2.2. Higher Flow Rate with and without Internal Tubing

The mass transfer rate can be enhanced to an extent by increasing the bulk flow of the liquid media [16]. The effect of higher flow rate on the fermentation rate was investigated. For the textile bioreactor with internal mixing tubing, a flow rate of 0.012 VVM (7.5 times higher than the previously examined flow rate) was used as the recirculation rate in the textile bioreactor with internal mixing tubing. Figure 4a shows the result of this experiment. From Figure 4a, the fermentation rate was slightly faster in the first 32 h of fermentation in comparison to the experiment with a recirculation rate of 0.0016 VVM. However, the substrate was fully consumed after 36 h in both cases. This shows that mixing is not the rate limiting step in both cases, but rather sucrose hydrolysis and slow fructose utilization [22]. This limitation can be handled in several ways such as operating the reactor in a fed-batch mode [24], or by maintaining a high concentration of flocs in the bioreactor [22]. However, for continuous production, using a flow rate of 0.012 VVM at 1 cm hole spacing would cause washout to occur much faster than that at a flow rate of 0.0016 VVM (Table 1). This did not affect the fermentation rate in the experiment performed at a flow rate of 0.012 VVM because the cells were recycled back into the textile bioreactor.

For the case without the mixing tubing, a flow rate of 0.06 VVM (1.9 times higher than the previously examined flow rate) was used as the recirculation rate. The result of this experiment is shown in Figure 4b. Comparing the experiment performed at 0.06 VVM with that at 0.032 VVM shows that the higher recirculation flow rate increased the fermentation rate. However, the higher flow rate was not sufficient to cause complete utilization of the sucrose within 48 h because the settling rate of the flocs is higher than the turbulence created by the high flow rate. Comparing the experiment performed at 0.0016 VVM with that performed at 0.06 VVM shows that the developed mixing system is much more effective in overcoming mass transfer limitations associated with mixing, even with 30 times slower flow rate.

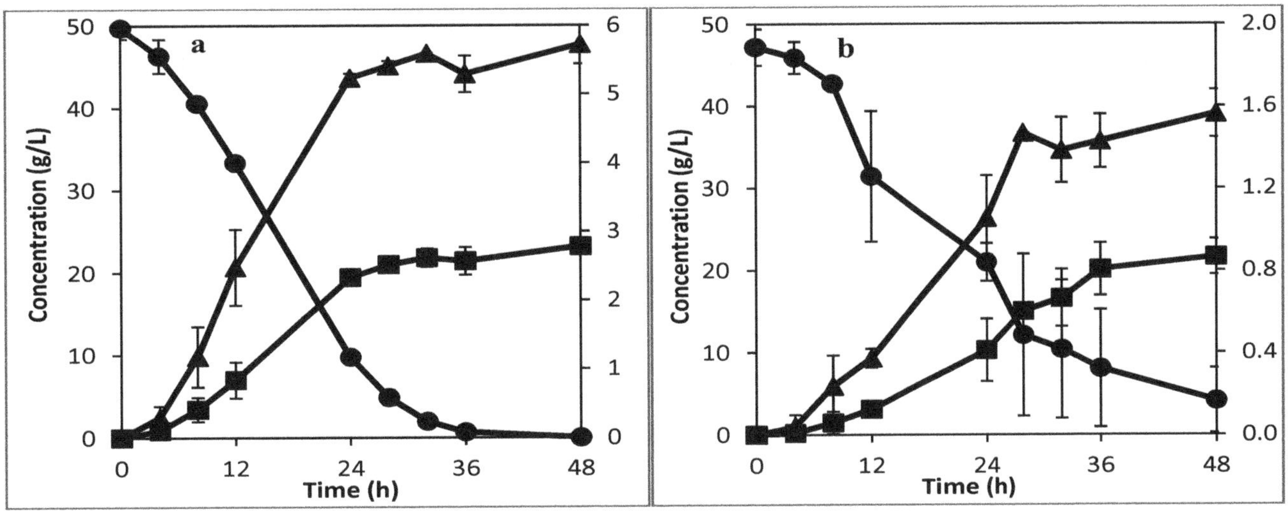

Figure 3. Experiment with enhanced mixing at (**a**) a recirculation rate of 0.0016 VVM , and (**b**) without enhanced mixing at 0.032 VVM, showing ethanol (■) and sucrose (●) concentration in the primary axis (right hand side), and glycerol concentration (▲) in the secondary axis.

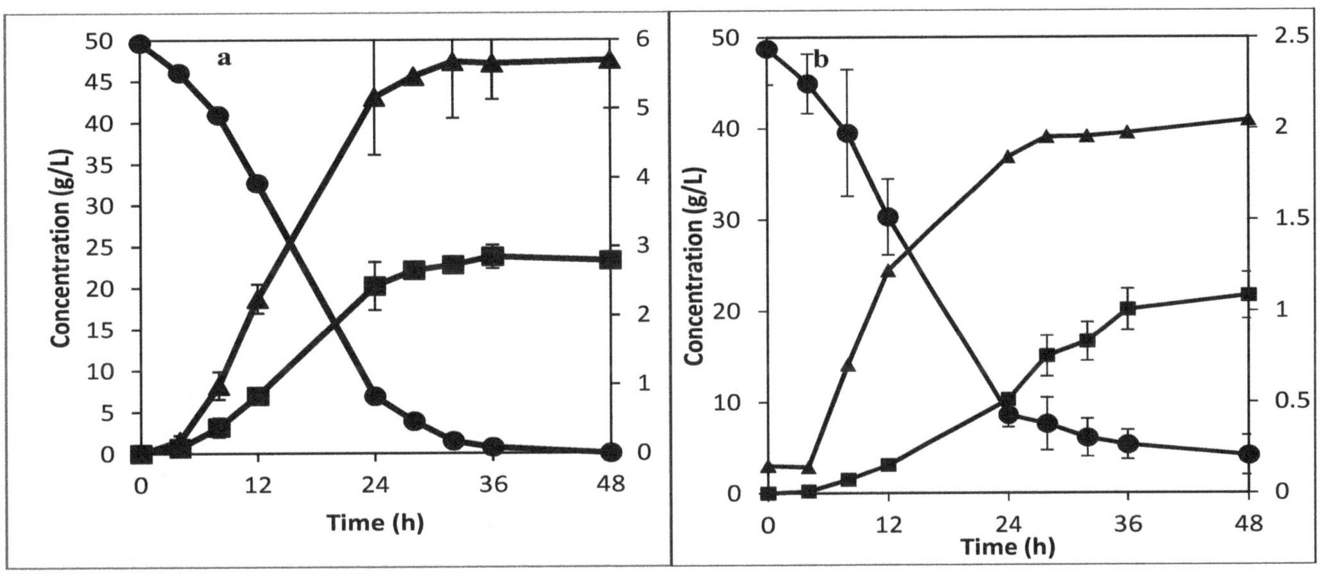

Figure 4. Experiments with (**a**) enhanced mixing at 0.012 VVM, and (**b**) without enhanced mixing at a recirculation rate of 0.06 VVM , showing ethanol (■) and sucrose (●) concentration in the primary axis (right hand side), and glycerol concentration (▲) in the secondary axis (left hand side).

3.3. *Mixing along the Edges of the Textile Bioreactor*

One of the limitations of a rectangular reactor is the possibility of it having poor mixing, especially at the edges [25]. Ethanol and sucrose concentrations across different sampling positions in the developed textile bioreactor at a recirculation rate of 0.0016 VVM are shown in Table 2. From Table 2 it can be seen that the concentrations of ethanol and sucrose from different sampling points and depths are similar. ANOVA using a general linear model of ethanol and sucrose concentration as responses and sampling position as factor gave p-values of 0.861 for ethanol and 0.733 for sucrose, so the position from which the sample is drawn is not statistically significant, meaning that the mixing is uniform at all points in the textile bioreactor. This uniformity in the textile bioreactor implies that the flocs responsible for the conversion of substrate to product is uniformly distributed across the developed textile bioreactor. From this it can be seen that the developed textile bioreactor does not have regions with significant ethanol or sucrose concentration variation, so there will be efficient and fast utilization of the substrate across it.

Table 2. Ethanol and sucrose concentrations from different sampling point in the textile bioreactor.

Time (h)	Run	Ethanol Concentration (g/L)				Sucrose Concentration (g/L)			
		Edge 8 cm Deep	Edge Surface	Centre 8 cm Deep	Centre Surface	Edge 8 cm Deep	Edge Surface	Centre 8 cm Deep	Centre Surface
0	1	0.00	0.00	0.00	0.00	48.96	48.46	48.75	48.75
4	1	0.86	0.45	0.51	0.47	46.22	44.97	44.84	44.84
8	1	2.39	2.30	2.15	2.38	43.61	41.91	40.97	40.97
12	1	5.14	5.41	5.23	5.44	36.16	38.04	36.01	33.70
24	1	17.91	18.24	18.87	18.85	11.61	5.52	12.11	9.34
28	1	20.46	19.95	20.60	20.44	3.21	2.13	3.14	5.50
32	1	21.20	21.25	21.68	21.04	1.35	1.35	1.38	1.39
0	2	0.00	0.00	0.00	0.00	48.19	50.40	50.68	50.68
4	2	1.18	1.18	1.27	1.21	46.86	46.86	47.72	47.72
8	2	4.56	4.41	4.41	4.44	40.87	39.94	40.10	40.10
12	2	8.60	8.27	8.23	8.53	32.41	30.77	31.84	32.98
24	2	20.03	19.94	20.01	19.98	10.00	10.14	10.03	10.16
28	2	21.93	21.65	21.88	21.52	4.34	4.16	4.31	4.22
32	2	22.60	22.05	22.65	22.59	2.30	2.82	2.87	2.55

3.4. Ethanol Production Process Development and Cost Reduction

Ethanol could be produced by batch, fed-batch, or continuous modes of production. Conventionally, ethanol is produced anaerobically, and the yeast cells are centrifuged and either recycled or channeled somewhere else [26]. However, for a 100,000 m^3/year ethanol production facility, the centrifuge accounts for 18% of the fermentation investment cost [9], excluding the energy cost. This can be eliminated by using flocculating yeast. Mixing in bioreactors using flocculating yeast is carried out using aeration (so the bioreactor is operated as a bubble column or an airlift reactor) as stirrers breaks the flocs [3]. Despite the Crabtree effect that favors ethanol production by *S. cerevisiae* under aerobic conditions [27], aerobic conditions cause more biomass production, which reduces the amount of ethanol that can be produced [28]. The developed textile bioreactor has been shown to both anaerobically use flocculating yeast for ethanol production and maintain good mixing, thus it can combine the benefits of optimal ethanol production and elimination of centrifugation cost.

Using the same volume of textile bioreactor in place of a stainless steel reactor has been previously shown to reduce the fermentation investment cost by 19% [11]. Combining this with the savings obtained by not using a centrifuge would result in 37% reduction in the investment cost of a 100,000 m^3/year ethanol production facility [9].

4. Conclusions

The developed textile bioreactor showed a better combination of specific ethanol productivity and complete sugar utilization than both gas lift bioreactors with recycle and airlift bioreactors when using flocculating yeast for ethanol production. The results of the anaerobic mixing efficiency with high settling flocculating yeast showed excellent mixing in comparison with the previous prototype. This makes it possible to combine optimal anaerobic ethanol production rates and the process benefits of using flocculating yeast for ethanol production. Additionally, for yeast flocs having average particle diameter between 190 to 320 µm, there is the possibility of operating the bioreactor at a dilution rate of more than 1 h^{-1} with less chances of washout by using mixing tubing configurations that makes the fluid upflow velocity less than the flocs settling velocity. Using flocculating yeast as the fermenter in the same volume of the developed textile bioreactor as the conventionally used bioreactor in a 100,000 m^3/year ethanol production facility can give a 37% reduction in fermentation investment cost.

Acknowledgments: The authors would like to appreciate Magnus Lundin, Peter Therning, and Regina Patinvoh for their technical assistance and valuable discussions. The textile bioreactor used for this article was kindly provided by FOV Fabrics AB (Borås, Sweden).

Author Contributions: Osagie A. Osadolor did the experiment design, performed all the experiment, wrote most of the manuscript and had the idea. Patrik R. Lennartsson was responsible for part of the manuscript. Mohammad J. Taherzadeh was responsible for part of the manuscript. All authors have given their approval for this final version of the manuscript.

References

1. Balat, M.; Balat, H. Recent trends in global production and utilization of bio-ethanol fuel. *Appl. Energ.* **2009**, *86*, 2273–2282. [CrossRef]
2. Mussatto, S.I.; Dragone, G.; Guimarães, P.M.R.; Silva, J.P.A.; Carneiro, L.M.; Roberto, I.C.; Vicente, A.; Domingues, L.; Teixeira, J.A. Technological trends, global market, and challenges of bio-ethanol production. *Biotechnol. Adv.* **2010**, *28*, 817–830. [CrossRef] [PubMed]
3. Domingues, L.; Vicente, A.A.; Lima, N.; Teixeira, J.A. Applications of yeast flocculation in biotechnological processes. *Biotechnol. Bioprocess. Eng.* **2000**, *5*, 288–305. [CrossRef]
4. Verstrepen, K.J.; Klis, F.M. Flocculation, adhesion and biofilm formation in yeasts. *Mol. Microbiol.* **2006**, *60*, 5–15. [CrossRef] [PubMed]
5. Ergun, S.; Orning, A.A. Fluid flow through randomly packed columns and fluidized beds. *J. Ind. Eng. Chem.* **1949**, *41*, 1179–1184. [CrossRef]
6. Muroyama, K.; Fan, L.S. Fundamentals of gas-liquid-solid fluidization. *AIChE. J.* **1985**, *31*, 1–34. [CrossRef]
7. Blakebrough, N. Fundamentals of fermenter design. *Pure Appl. Chem.* **1973**, *36*, 305–316. [CrossRef]
8. Mark, R.; Wilkins, A.H.A. Fermentation. In *Food and Industrial Bioproducts and Bioprocessing*; John Wiley & Sons: Hoboken, NJ, USA, 2012.
9. Maiorella, B.L.; Blanch, H.W.; Wilke, C.R.; Wyman, C.E. Economic evaluation of alternative ethanol fermentation processes. *Biotechnol. Bioeng.* **2009**, *104*, 419–443. [CrossRef] [PubMed]
10. Wei, Y.; Wang, J.; Jia, X.; Yeh, J.-M.; Spellane, P. Polyaniline as corrosion protection coatings on cold rolled steel. *Polymer* **1995**, *36*, 4535–4537. [CrossRef]
11. Osadolor, O.A.; Lennartsson, P.R.; Taherzadeh, M.J. Introducing textiles as material of construction of ethanol bioreactors. *Energies* **2014**, *7*, 7555–7567. [CrossRef]
12. Van Hamersveld, E.H.; van der Lans, R.; Luyben, K. Quantification of brewers' yeast flocculation in a stirred tank: Effect of physical parameters on flocculation. *Biotechnol. Bioeng.* **1997**, *56*, 190–200. [CrossRef]
13. Nagai, Y.; Unsworth, L.D.; Koutsopoulos, S.; Zhang, S. Slow release of molecules in self-assembling peptide nanofiber scaffold. *J. Control. Release* **2006**, *115*, 18–25. [CrossRef] [PubMed]
14. Kunii, D.; Levenspiel, O. *Fluidization Engineering*; Elsevier: Amsterdam, The Netherlands, 2013.
15. McCabe, W.L.; Smith, J.C.; Harriott, P. *Unit Operations of Chemical Engineering*; McGraw-Hill: New York, NY, USA, 1993; Volume 5.
16. Benitez, J. *Principles and Modern Applications of Mass Transfer Operations*; John Wiley & Sons: Hoboken, NJ, USA, 2011.
17. Bergman, T.L.; Incropera, F.P.; Lavine, A.S. *Fundamentals of Heat and Mass Transfer*; John Wiley & Sons: Hoboken, NJ, USA, 2011.
18. West, J.; Gleeson, J.P.; Alderman, J.; Collins, J.K.; Berney, H. Structuring laminar flows using annular magnetohydrodynamic actuation. *Sensor Actuator B* **2003**, *96*, 190–199. [CrossRef]
19. Hayduk, W.; Laudie, H. Prediction of diffusion coefficients for nonelectrolytes in dilute aqueous solutions. *AIChE. J.* **1974**, *20*, 611–615. [CrossRef]
20. Bosma, T.N.P.; Middeldorp, P.J.M.; Schraa, G.; Zehnder, A.J.B. Mass transfer limitation of biotransformation: Quantifying bioavailability. *Environ. Sci. Technol.* **1996**, *31*, 248–252. [CrossRef]
21. Elander, R.T.; Putsche, V.L. *Ethanol from Corn: Technology and Economics*; Taylor and Francis: Washington, DC, USA, 1996; pp. 329–349.
22. Fontana, A.; Ghommidh, C.; Guiraud, J.P.; Navarro, J.M. Continuous alcoholic fermentation of sucrose using flocculating yeast. The limits of invertase activity. *Biotechnol. Lett.* **1992**, *14*, 505–510. [CrossRef]

23. Sousa, M.L.; Teixeira, J.A.; Mota, M. Comparative analysis of ethanolic fermentation in two continuous flocculation bioreactors and effect of flocculation additive. *Bioprocess. Eng.* **1994**, *11*, 83–90. [CrossRef]
24. Echegaray, O.F.; Carvalho, J.C.M.; Fernandes, A.N.R.; Sato, S.; Aquarone, E.; Vitolo, M. Fed-batch culture of sacchoromyces cerevisiae in sugar-cane blackstrap molasses: Invertase activity of intact cells in ethanol fermentation. *Biomass Bioenergy* **2000**, *19*, 39–50. [CrossRef]
25. Oca, J.; Masaló, I. Design criteria for rotating flow cells in rectangular aquaculture tanks. *Aquacult. Eng.* **2007**, *36*, 36–44. [CrossRef]
26. Basso, L.C.; Rocha, S.N.; Basso, T.O. *Ethanol Production in Brazil: The Industrial Process and its Impact on Yeast Fermentation*; Intech: Rijeka, Croatia, 2011.
27. Verduyn, C.; Zomerdijk, T.P.L.; van Dijken, J.P.; Scheffers, W.A. Continuous measurement of ethanol production by aerobic yeast suspensions with an enzyme electrode. *Appl. Microbiol. Biotechnol.* **1984**, *19*, 181–185. [CrossRef]
28. Taherzadeh, M.J.; Lennartsson, P.R.; Teichert, O.; Nordholm, H. Bioethanol production processes. *Biofuels Prod.* **2013**, 211–253. [CrossRef]

7

PTR-ToF-MS for the Online Monitoring of Alcoholic Fermentation in Wine: Assessment of VOCs Variability Associated with Different Combinations of *Saccharomyces/Non-Saccharomyces*

Carmen Berbegal [1,2], Iuliia Khomenko [3], Pasquale Russo [1], Giuseppe Spano [1], Mariagiovanna Fragasso [1], Franco Biasioli [3,*] and Vittorio Capozzi [4,*]

1 Department of the Sciences of Agriculture, Food and Environment, University of Foggia, Via Napoli 25, 71122 Foggia, Italy; carmen.berbegal@uv.es (C.B.); pasquale.russo@unifg.it (P.R.); giuseppe.spano@unifg.it (G.S.); mariagiovanna.fragasso@gmail.com (M.F.)

2 EnolabERI BioTecMed, Universitat de València, 46100 Valencia, Spain

3 Department of Food Quality and Nutrition, Research and Innovation Centre, Fondazione Edmund Mach (FEM), via E. Mach 1, 38010 San Michele all'Adige, Italy; iuliia.khomenko@fmach.it

4 Institute of Sciences of Food Production, National Research Council (CNR), c/o CS-DAT, Via Michele Protano, 71121 Foggia, Italy

* Correspondence: franco.biasioli@fmach.it (F.B.); vittorio.capozzi@ispa.cnr.it (V.C.)

Abstract: The management of the alcoholic fermentation (AF) in wine is crucial to shaping product quality. Numerous variables (e.g., grape varieties, yeast species/strains, technological parameters) can affect the performances of this fermentative bioprocess. The fact that these variables are often interdependent, with a high degree of interaction, leads to a huge 'oenological space' associated with AF that scientists and professionals have explored to obtain the desired quality standards in wine and to promote innovation. This challenge explains the high interest in approaches tested to monitor this bioprocess including those using volatile organic compounds (VOCs) as target molecules. Among direct injection mass spectrometry approaches, no study has proposed an untargeted online investigation of the diversity of volatiles associated with the wine headspace. This communication proposed the first application of proton-transfer reaction-mass spectrometry coupled to a time-of-flight mass analyzer (PTR-ToF-MS) to follow the progress of AF and evaluate the impact of the different variables of wine quality. As a case study, the assessment of VOC variability associated with different combinations of *Saccharomyces*/non-*Saccharomyces* was selected. The different combinations of microbial resources in wine are among the main factors susceptible to influencing the content of VOCs associated with the wine headspaces. In particular, this investigation explored the effect of multiple combinations of two *Saccharomyces* strains and two non-*Saccharomyces* strains (belonging to the species *Metschnikowia pulcherrima* and *Torulaspora delbrueckii*) on the content of VOCs in wine, inoculated both in commercial grape juice and fresh grape must. The results demonstrated the possible exploitation of non-invasive PTR-ToF-MS monitoring to explore, using VOCs as biomarkers, (i) the huge number of variables influencing AF in wine, and (ii) applications of single/mixed starter cultures in wine. Reported preliminary findings underlined the presence of different behaviors on grape juice and on must, respectively, and confirmed differences among the single yeast strains 'volatomes'. It was one of the first studies to include the simultaneous inoculation on two non-*Saccharomyces* species together with a *S. cerevisiae* strain in terms of VOC contribution. Among the other outcomes, evidence suggests that the addition of *M. pulcherrima* to the coupled *S. cerevisiae*/*T. delbrueckii* can modify the global release of volatiles as a function of the characteristics of the fermented matrix.

Keywords: volatile organic compounds; proton-transfer reaction-mass spectrometry; *Saccharomyces cerevisiae*; *Metschnikowia pulcherrima*; *Torulaspora delbrueckii*; wine; flavor

1. Introduction

Wine is the result of alcoholic fermentation (AF) performed by yeasts that convert the sugars present in grape must into ethanol and carbon dioxide. During this fermentation, other chemical changes are produced as a consequence of yeast metabolic activities. Among the chemical changes, a consistent part of volatile organic compounds (VOCs) is released, influencing wine flavor [1,2]. The interest in the monitoring of this bioprocess is high due to (i) the vast number of variables that can affect AF performances, and (ii) the crucial relevant impact of AF on wine quality. Non-separative approaches based on direct injection mass spectrometry (DIMS) have recently emerged as an alternative for the high-throughput and cost-effective quantitative profiling of volatiles in food and beverages [3]. To the best of our knowledge, no study has explored the potential of DIMS techniques to assess online VOC variability in association with alcoholic fermentation in wine [4].

Saccharomyces cerevisiae has a leading role in performing AF in wine [5,6]. However, an increasing interest has been given to non-*Saccharomyces* yeasts as drivers of the differentiation of the quality of final wines [7,8]. Non-*Saccharomyces* yeasts can possess enzymatic activities different from the *S. cerevisiae* enzymatic inventory, catalyzing the synthesis and the release (from non-volatile bound precursors) of VOCs able to modulate aromatic wine complexity [9,10]. Moreover, they may influence other characteristics such as glycerol and mannoprotein content, volatile acidity, color stability, and ethanol levels of wines [11,12]. Usually, as a reason for non-optimal fermentative performances, non-*Saccharomyces* yeasts are used in combination with *S. cerevisiae* strains. Some studies have shown that the strategy of co-inoculating *S. cerevisiae* starter together with selected non-*Saccharomyces* yeasts at high cell density produces wines with distinctive characteristics [13]. The interaction between the different yeast species influences the content of VOCs associated with fermentations [14]. Among the non-*Saccharomyces* species, *Torulaspora delbrueckii*, *Metschnikowia pulcherrima*, *Candida zemplinina*, and *Hanseniaspora uvarum* are mostly cited and have been intensively investigated [9,15–21]. Strains belonging to the species *Lachancea thermotolerans*, *Metschnikowia fructicola*, *Schizosaccharomyces pombe*, *T. delbrueckii*, *Kluyveromyces thermotolerans*, *Pichia kluyveri*, and *M. pulcherrima* are commercialized or have patented applications [16,22]. Belonging to the class of direct injection mass spectrometry (DIMS) approaches, proton transfer reaction mass spectrometry (PTR-MS) is an established method for the rapid, direct, and non-invasive online monitoring of VOCs characterized by short response time and high sensitivity [23]. The coupling of proton transfer ionization with time-of-flight (ToF) mass spectrometers and automated sampling offers several advantages in terms of mass resolution, throughput, and reproducibility [24–26]. This analytical strategy has found several applications in the food field (e.g., [27–31]), with a specific interest in bioprocess monitoring associated with microbial-based processes (e.g., [32–34]). Furthermore, several studies have applied PTR-based approaches to monitor VOC release associated with yeast metabolisms [35], often in association with food matrices [36–38]. In the case of matrices containing ethanol, consistent experimental efforts have been performed to avoid the adverse effects of a high concentration of this alcohol (primary ion depletion and ethanol–ethanol/water clusters formation responsible for the loss of efficiency in the qualitative/quantitative detection) [34,39–42].

In the present work, PTR-ToF-MS was used for the online monitoring of AF in wine and to compare the performance of four (autochthonous and commercial) yeast strains, both in single cultures and in multiple inoculations, using two diverse model matrices as substrates (real grape must and commercial grape juice). This study also aimed to preliminarily explore the interest in PTR-ToF-MS analysis of flavor-related volatile compounds in the control, design, and application of single/mixed starter cultures for wine.

2. Materials and Methods

2.1. Microorganisms and the Determination of Microbial Population

The following microorganisms were used for grape juice and grape must inoculation: the commercially available *Saccharomyces cerevisiae* strain DV10 (Lallemand, Montreal, QC, Canada, autochthonous characterized *S. cerevisiae* I6 strain from the Apulian region (Southern Italy) [43], and the commercially available non-*Saccharomyces* strains *Metschnikowia pulcherrima* FLAVIA (Lallemand, USA) and *Torulaspora delbrueckii* BIODIVA (Lallemand, Montreal, QC, Canada). Yeast starters were purchased in active-dried form. Rehydration procedures were done according to the suppliers' instructions. Starter cultures were prepared by growing pure cultures of the yeast strains separately grown in liquid Yeast Peptone Dextrose (YPD) medium (2% glucose, 2% Bacto peptone, 1% yeast extract) at 28 °C.

The viable count of yeasts during the AF was enumerated on Wallerstein Laboratory (WL) agar medium (Sigma-Aldrich, St. Louis, USA). WL discriminates between the used yeast species by colony morphology and color (*S. cerevisiae* produces large white colonies, whereas non-*Saccharomyces* yeasts produce green colonies on this medium). Plates were incubated at 28 °C for 48 h.

2.2. Micro-Vinifications and Wine Analysis

Starter cultures were prepared by growing strains in YPD medium as described above and then inoculating the strains into commercial red grape juice (Vitafit, Lidl Stiftung & Co., Neckarsulm, Germany) and red grape must from Apulian autochthonous grape varieties (20° Babo; 7.2 g/L total acidity; 3.5 g/L malic acid; pH 3.5; free ammonium 163.5 mg/L). Fermentations were performed inoculating at concentrations of 1×10^6 cfu/mL (colony-forming units per milliliter) of *M. pulcherrima* FLAVIA, 1×10^6 cfu/mL of *T. delbrueckii* BIODIVA and 1×10^4 cfu/mL of *S. cerevisiae* strains (DV10 or I6). Each fermentation experiment was carried out by performing three simultaneous independent repetitions. With these four biotypes, 14 different combinations of strains were carried out (Table 1). Fermentative kinetics from grape must were monitored daily by gravimetric determinations for seven days. With this purpose, samples were weighed daily to follow the weight loss caused by CO_2 production.

Table 1. Microorganisms employed in different grape must/juice fermentations (trials 1–15).

Sample Code	Inoculated Yeast Cultures
1	*S. cerevisiae* DV10
2	*S. cerevisiae* I6
3	*M. pulcherrima* FLAVIA
4	*T. delbrueckii* BIODIVA
5	*S. cerevisiae* DV10 + *S. cerevisiae* I6
6	*S. cerevisiae* DV10 + *M. pulcherrima* FLAVIA
7	*S. cerevisiae* DV10 + *T. delbrueckii* BIODIVA
8	*S. cerevisiae* I6 + *M. pulcherrima* FLAVIA
9	*S. cerevisiae* I6 + *T. delbrueckii* BIODIVA
10	*S. cerevisiae* DV10 + *S. cerevisiae* I6 + *M. pulcherrima* FLAVIA
11	*S. cerevisiae* DV10 + *S. cerevisiae* I6 + *T. delbrueckii* BIODIVA
12	*S. cerevisiae* DV10 + *M. pulcherrima* FLAVIA + *T. delbrueckii* BIODIVA
13	*S. cerevisiae* I6 + *M. pulcherrima* FLAVIA + *T. delbrueckii* BIODIVA
14	*S. cerevisiae* DV10 + *S. cerevisiae* I6 + *M. pulcherrima* FLAVIA + *T. delbrueckii* BIODIVA
15	Uninoculated grape must/juice

2.3. Samples Preparation and PTR-ToF-MS (Proton Transfer Reaction-Time of Flight-Mass Spectrometry) Analysis

Nano-vinifications were performed in the vials using the above yeast combinations (Table 1) in commercial red grape juice and fresh red grape must. Nitrogen flux in the vial headspace assured maintaining the conditions comparable with those present in vinification. While the manufacturer sterilized the commercial grape juice, the must was not treated to reduce microbial presence. When present in the same trial, yeasts were co-inoculated in the juice/must. The resulting AF was monitored for three days. The whole experiment was performed in five replicates. For the measurements, a commercial PTR-ToF-MS 8000 apparatus from Ionicon Analytik GmbH (Innsbruck, Austria) was used in its standard configuration (V mode). The air associated with the headspace of the sample was directly injected in the PTR-MS drift tube. An argon dilution system was applied after headspace sampling. The dilution ratio was one part of headspace to three parts of argon. The argon flow rate was 120 sccm and was controlled by a multigas controller (MKS Instruments, Inc, Andover, MA, USA). Ionization conditions were as follows: 110 °C drift tube temperature, 2.30 mbar drift pressure, and 550 V drift voltage. These conditions led to an E/N ratio of about 140 Td (1 Townsend = 10–17 cm^2 V^{-1} s^{-1}). The inlet line was a PEEK capillary tube (internal diameter 0.04 in.) heated at 110 °C, with a flow set at 40 sccm. The acquisition rate of the instrument was one spectrum per second.

2.4. Data Analysis

Deadtime correction, internal calibration of mass spectral data, and peak extraction were performed according to the procedure described by Cappellin et al. [44,45]. Peak intensity in ppbv was estimated using the formula described by Lindinger et al. [46] using a constant value for the reaction rate coefficient ($k = 2.10^{-9}$ cm^3 s^{-1}). This approach introduces a systematic error for the absolute concentration for each compound that is, in most cases, below 30% and could be accounted for if the actual rate constant coefficient is available [45]. All data detected and recorded by the PTR-ToFMS were processed and analyzed using MATLAB R2017a (MathWorks Inc., Natick, MA, USA) and R (R Foundation for Statistical Computing, Vienna, Austria). Principal component analysis, analysis of variance, and Tukey's post-hoc test were performed to spot the differences in the volatile aroma compounds emitted by the 28 grape must and juice fermentations used in this study.

3. Results

3.1. Alcoholic Fermentation Kinetics and Yeast Dynamics

The kinetics of the 14 fermentations in red grape must were monitored daily for seven days, evaluating the loss of weight due to the production of CO_2 (data not shown). All fermentations were completed in four days except for sample 3 (inoculated with a single culture of *M. pulcherrima* FLAVIA), which was not able to complete the AF. The interactions between *Saccharomyces* spp. and both non-*Saccharomyces* spp. of enological interest, *M. pulcherrima* FLAVIA and *T. delbrueckii* BIODIVA, were investigated in terms of cell density. The differential morphology of the colonies on WL medium allowed us to calculate the proportion of each yeast species in different phases of AF (Figure 1). Only when both *S. cerevisiae* strains were co-inoculated the viable cell count was considered as total *S. cerevisiae* viable cells without distinguishing between DV10 and I6 strains.

Results from plate counting revealed that the maximum cell density of the single cultures was obtained after 48–72 h of the grape must inoculation both for *S. cerevisiae* (with an initial cell population of 1×10^4 cfu/mL) and non-*Saccharomyces* yeasts (with an initial cell population of 1×10^6 cfu/mL) (Figure 1, Experiments 1–4). In terms of single cultures studied, *M. pulcherrima* FLAVIA reached the lowest cell concentration (slightly more than 1×10^7 cfu/mL) after 72 h of inoculation (Figure 1, Experiment 3). Conversely, *T. delbrueckii* BIODIVA, even if with a different profile, achieved a biomass concentration comparable to those of the two *S. cerevisiae* strains. Considering the strain combinations, when two *S. cerevisiae* strains (DV10 and I6) were inoculated simultaneously, the growth behavior was the same as when they were inoculated in a single form and reached the maximum yeast population after 48 h of the inoculation (Figure 1, Experiment 5). Results from the co-inoculation of one strain of *S. cerevisiae* with *M. pulcherrima* FLAVIA (Figure 1 Experiments 6 and 8) showed that in 24 h, the *S. cerevisiae* strains were able to overtake the non-*Saccharomyces* yeast concentration, and in 48–72 h achieved the maximum yeast population. On the other hand, the *M. pulcherrima* FLAVIA population decreased drastically after 72 h of inoculation. Contrary to *M. pulcherrima* FLAVIA, yeast *T. delbrueckii* BIODIVA presented a high cell density when it was co-inoculated with one of the *S. cerevisiae* strains, and with similar cell concentration levels to *S. cerevisiae* (Figure 1, Experiments 7 and 9). In the same way, when the two *S. cerevisiae* strains, DV10 and I6, were co-inoculated with *M. pulcherrima* FLAVIA or with *T. delbrueckii* BIODIVA strains, the growth behavior of the non-*Saccharomyces* strains was the same than when they were inoculated with only one *S. cerevisiae* strain (Figure 1, Experiments 10 and 11). The simultaneous inoculation of one *S. cerevisiae* strain with both, *T. delbrueckii* BIODIVA and *M. pulcherrima* FLAVIA (Figure 1, Experiments 12 and 13), revealed that *S. cerevisiae* strains needed more time to reach the maximum cell concentration than when co-inoculated with only one non-*Saccharomyces* strain, and *T. delbrueckii* BIODIVA presented a higher population than *S. cerevisiae* strains for 48 h.

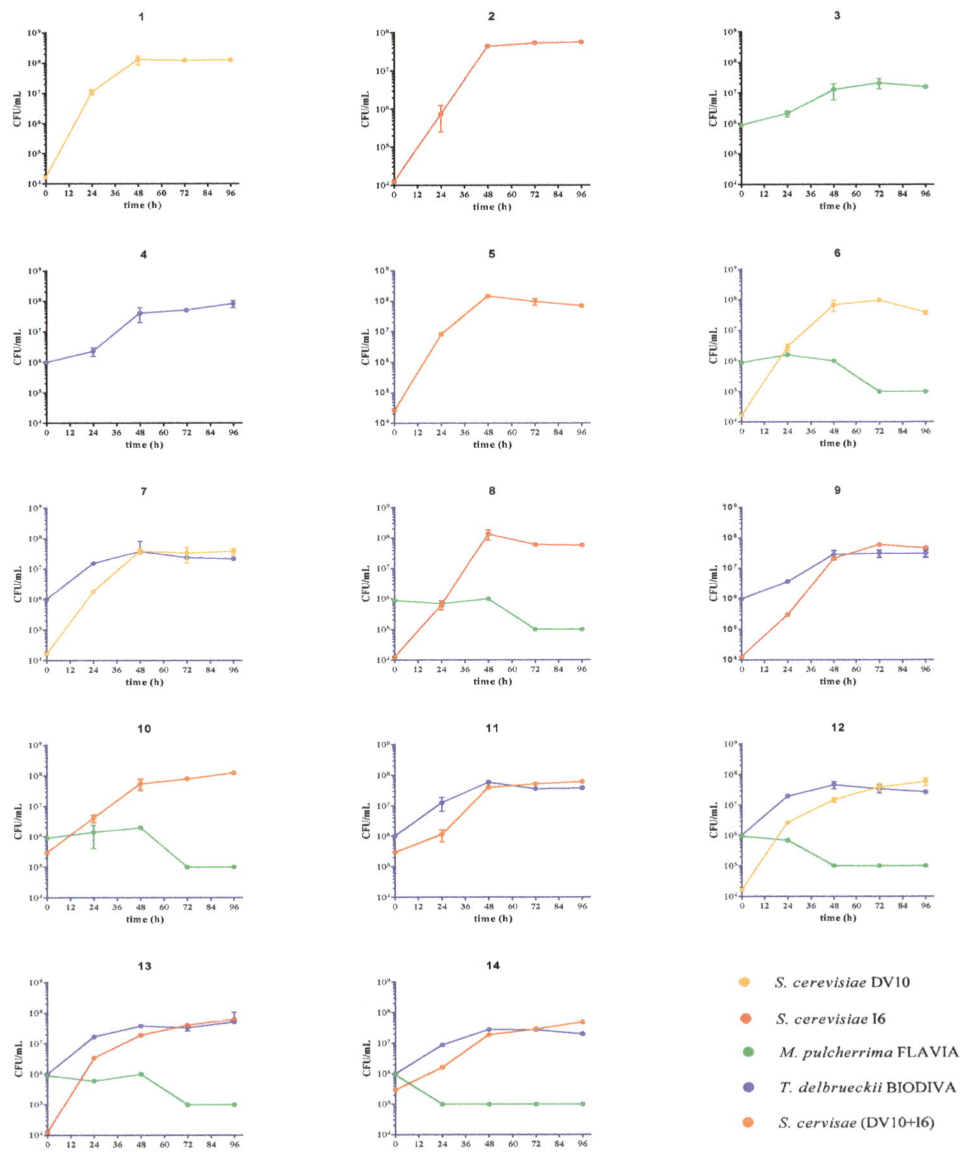

Figure 1. Viable cell count (cfu/mL) of different yeast single or mixed culture (Table 1) inoculated. The cell enumeration was performed on Wallerstein Laboratory agar medium that discriminates *S. cerevisiae* (large white colonies) from non-*Saccharomyces* yeasts (green colonies).

Inoculating the four starter cultures simultaneously (Figure 1, Experiment 14) triggered, as in the previous cases, that *S. cerevisiae* strains required more time to reach the maximum cell population and, the maximum cell concentration was lower than when they were inoculated in a single culture form. Furthermore, the population of *S. cerevisiae* and *T. delbrueckii* BIODIVA presented similar population levels after 48 h of inoculation. Otherwise, the *M. pulcherrima* FLAVIA population decreased from the inoculation time. Overall, biological interactions influenced single yeast growth behavior. Nevertheless, in all of the studied experimental modes, the most significant changes related to yeast population occurred during the first 72 h, which led us to focus on this temporal interval for the online monitoring of VOCs associated with the considered experimental modes using the PTR-ToF-MS technology.

3.2. Automated Monitoring Volatile Organic Compound (VOC) Evolution in Red Grape Must and Juice Fermentation by Different Yeast Mixed Cultures

A preliminary data exploration has been made to visualize the results of the PTR-ToF-MS grape must and juice analysis through a principal component analysis (PCA). The first and second PCA components (Figure 2) accounted for 84% of total variability and showed that the two matrices (grape juice and must) used in this study led to clear changes of VOC release. Differences in the distribution of variances were also observed concerning single yeasts or yeast combinations. It is easy to follow a time-dependent dimension of the phenomena, observing the increasing dimensions of the symbols in Figure 2.

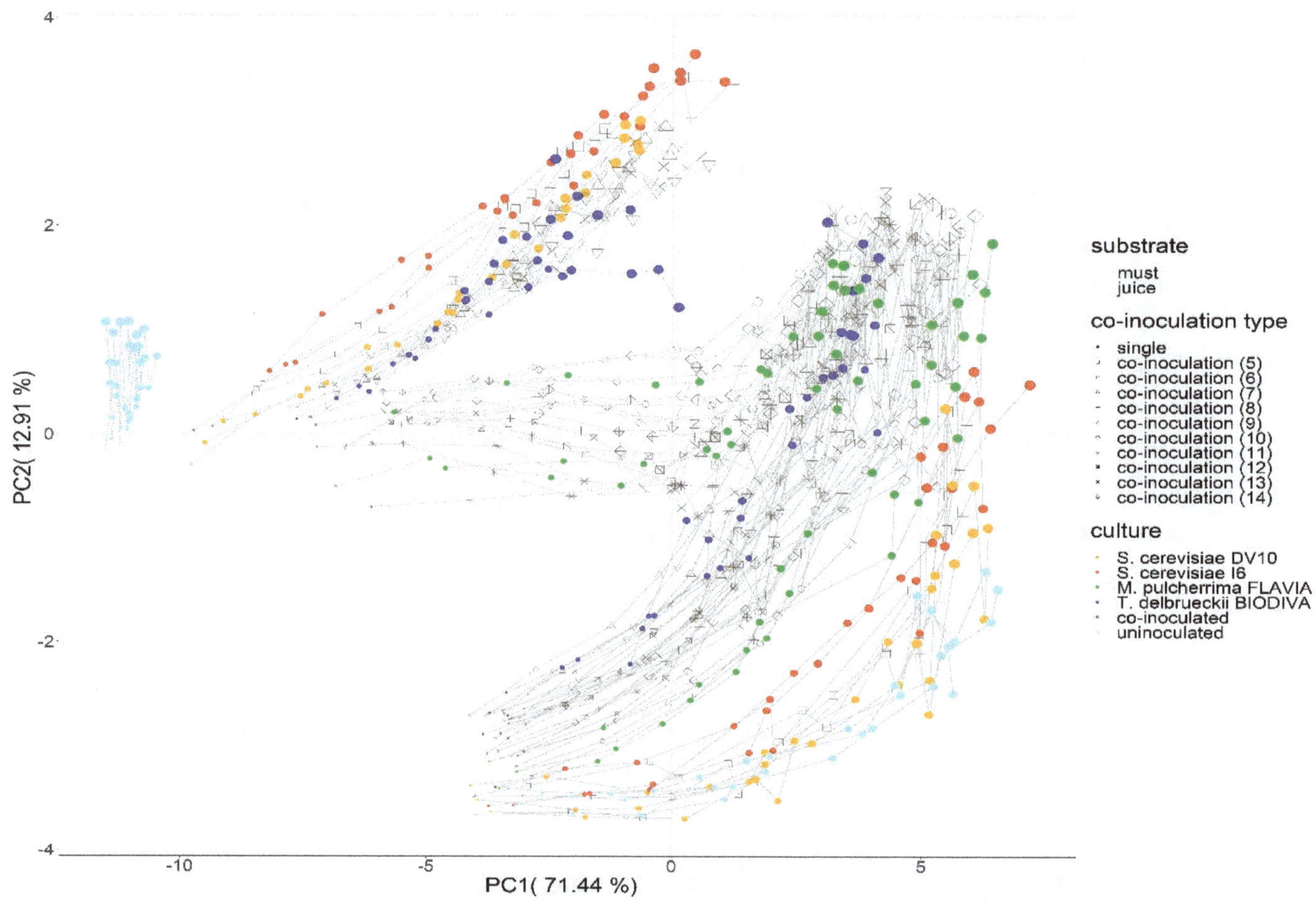

Figure 2. Score plot of the principal component analysis of volatile organic compound (VOC) emission evolution associated with the first three days of AF for each trial tested in this study. Data were logarithmically transformed and centered. Different colors indicate the different yeast managements, medium, and blank samples. The size of the points grew with the time of measurements. For a detailed view of the figure, the original image is included in the Supplementary Materials (Figure S1).

Separations and different evolutions were evident by comparing the matrices 'grape must' and 'grape juice'. In the 'must' assays, it was clear the partition between the trials with only *S. cerevisiae* strains inoculated (sample codes 1, 2, and 5; Table 1) and those that included, in the starter cultures, non-*Saccharomyces* strains (sample codes 3, and 4; Table 1) (Figure 2). Additionally, a diverse behavior was noticeable for the fermentations inoculated with pure cultures of *M. pulcherrima* FLAVIA and *T. delbrueckii* BIODIVA strains, respectively. All the experiments that included both *S. cerevisiae* and non-*Saccharomyces* strains (sample codes 6–14; Table 1) followed a trend that appeared more similar to the non-*Saccharomyces* pure cultures. Concerning the 'juice' experimental plan, an uniform trend was confirmed for the samples inoculated with the *S. cerevisiae* strains (Figure 2). In contrast, the behavior observed for the pure inoculation of *M. pulcherrima* FLAVIA strain (sample code 3; Table 1) was radically different. The pure culture of *T. delbrueckii* BIODIVA strain (sample code 4; Table 1) and all the combinations of *S. cerevisiae–T. delbrueckii* (sample codes 7, 9, and 11; Table 1) followed trajectories closer to those of *S. cerevisiae* strains than to the *M. pulcherrima* one. In contrast, all the other trials (sample codes 6, 8, 10, and 12–14; Table 1) observed patterns of evolution similar to *M. pulcherrima*. Concerning these last trials, some samples also included *T. delbrueckii* BIODIVA among the inoculated strains. In the case of PCA, the loading plot (Figure 3) indicates the mass peaks related to the observed evolution of the VOC profile in Figure 2.

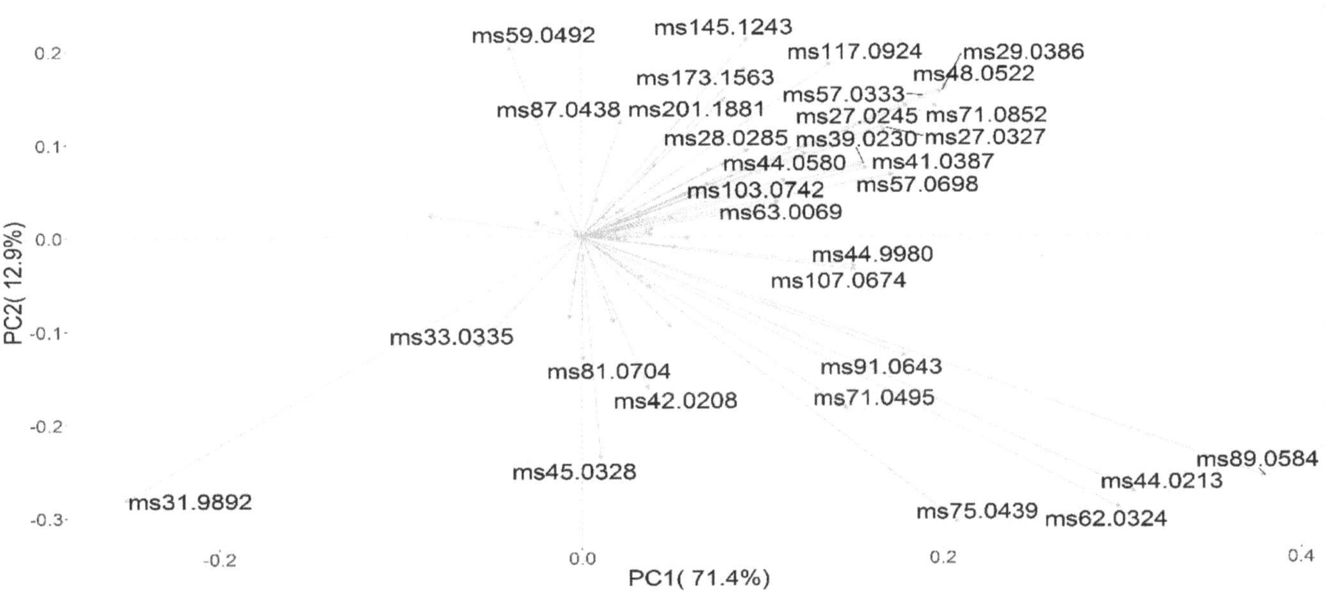

Figure 3. Loading plot of principal component analysis of the mass peaks (ms) related to the observed evolution of VOCs profile in Figure 2. For a detailed view of the figure, the original image is included in the Supplementary Materials (Figure S2).

More than 70 mass peaks were identified among the four yeast commercial starters during online monitoring throughout the three days of fermentation. For each of these mass peaks, it was possible to perform a tentative identification (allowing a possible link of the ion with a given molecule/molecular fragment) and to follow the evolution of the intensity in the time, allowing a direct analytic determination to evaluate the yeast metabolic activity during the progress of AF.

More specifically, differences between the matrices used were observed when the score plot of the PCA analysis on the distribution of variances associated with VOC emission during the first three days of AF was represented separately for each trial (Figure 4). Negligible VOC evolution was evident in uninoculated grape juice and slow evolution in grape must as revealed by the first PCA dimension, which is related to the increase of volatile concentration in the sample headspace (Figure 4, uninoculated trial, experiment 15). Regarding the inoculated yeasts, differences in the VOC emissions were also found. For example, *M. pulcherrima* and *T. delbrueckii* in single culture (Figure 4, experiments

3 and 4) tended to reach a lesser concentration of VOCs in juice, while both *S. cerevisiae* kept producing more volatile compounds with time (Figure 4, experiments 1 and 2). Moreover, this graph confirmed that there were differences between the different yeast combinations inoculated, as they were arranged in the graph according to different patterns. This effect is of particular interest, if we consider that together with the effect of different strains/species combinations we also tested the impact of the increasing microbial diversity of the starter cultures inoculated.

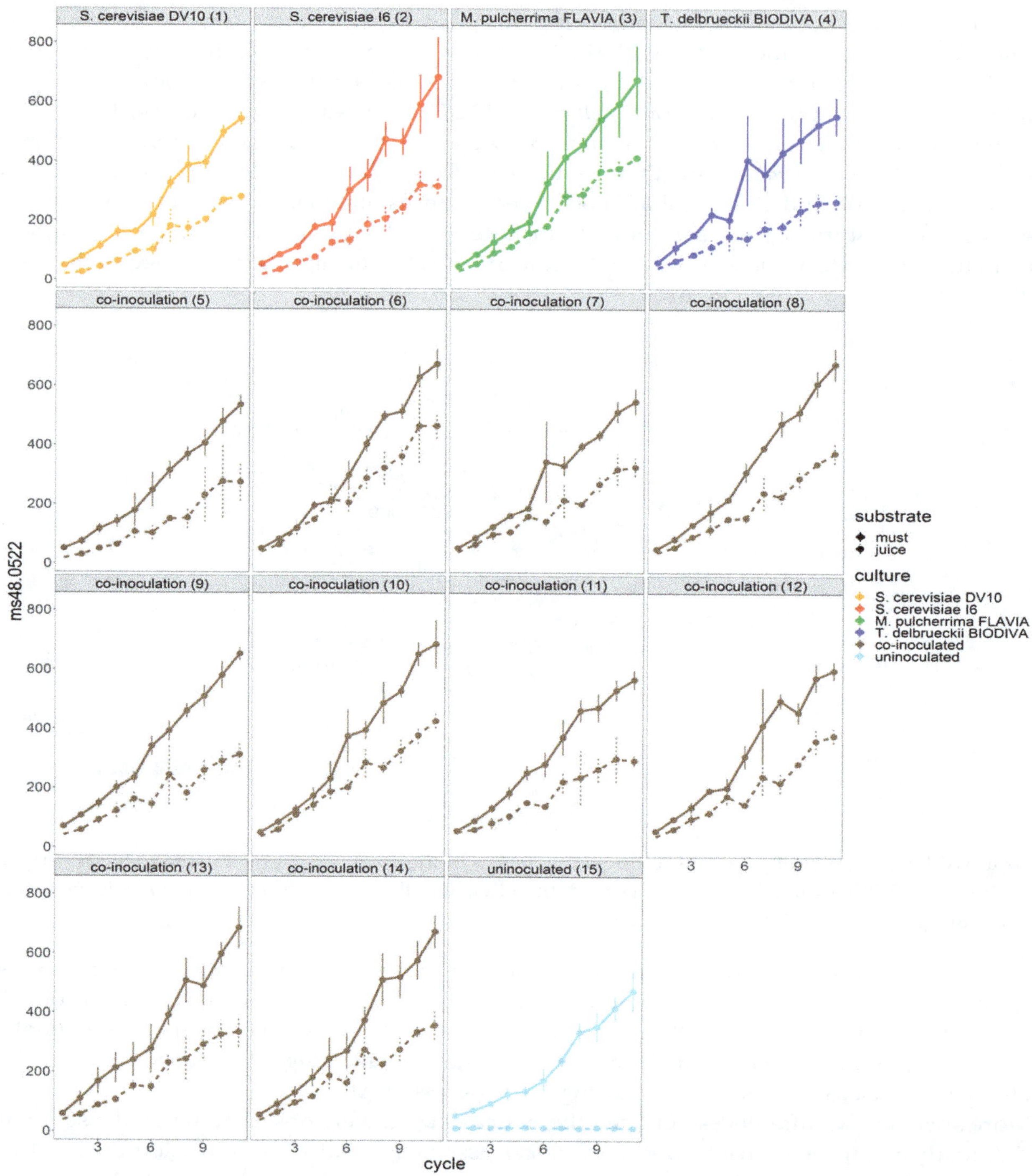

Figure 4. Score plot of principal component analysis of VOC emission evolution associated with the first three days of AF, separately represented for each trial (Table 1). Continuous lines indicate grape must and broken lines indicate grape juice. For a detailed view of the figure, the original image is included in the Supplementary Materials (Figure S3).

4. Discussion

Wine is a peculiar commodity in the agrifood sector in terms of business opportunities and innovative trends [47]. The management of AF deeply affects the optimization of the product quality and the improvement of process sustainability [48–52]. Several variables can influence the performance of AF such as grape variety, yeast species, yeast strain, nutrient availability for the yeasts, temperature of the process, addition of chemical compounds, and technological regimen [49,53–55]. The fact that these variables are often intimately connected leads to a huge 'oenological space' that needs to be explored. This observation explains the high interest in approaches tested to monitor this bioprocess [53,56] including those using VOCs as target molecules [57–61]. Furthermore, it is important to underline that the study of VOC diversity has a dual significance; on one hand, VOC variability is the effect of yeast metabolism, on the other, VOCs represent the molecular basis of the sensory perception of wine tasting [1,62,63]. Among the DIMS techniques, no study has delved into the survey of the untargeted diversity of volatiles associated with the wine headspace in order to (i) monitor online the progress of AF, and (ii) evaluate the impact of the different variables of wine quality [4].

As a case study, the assessment of VOCs variability associated with different combinations of *Saccharomyces*/non-*Saccharomyces* was selected. In fermented beverages such as wine, a relevant field of study deals with the contribution of microbiological resources to the organoleptic and sensory properties of the final product [64,65]. In the winemaking process, some of the most characteristic flavor and aroma components are synthesized by yeasts during the AF [66]. *S. cerevisiae* is the main responsible microorganism of the AF in wine, but nowadays, non-*Saccharomyces* yeasts are used in industry to improve flavor, aroma, and stability [16,22,64,65]. This heterogeneous class of eukaryotic microorganisms detains a wide enzymatic diversity [67–69]. In this light, it appears comprehensible the interest in the formation of volatile compounds by both *Saccharomyces* and non-*Saccharomyces* yeasts, which are important to maximize the sensorial quality of the final products. The present study, in particular, tested the PTR-ToF-MS-based approach recently optimized to compare the performance of different yeasts in cultural media [35]. This technology has been successfully employed in fermented foods and beverages to monitor the effect of different microorganisms responsible for the fermentative process, for instance, to discriminate wines inoculated with different malolactic starters [39], monitor lactic fermentation driven by different yoghurt commercial starter cultures [33], and characterize single commercial yeast starters in bread productions [36]. This study proposed the first application of PTR-ToF-MS for the AF monitoring in wine, demonstrating the high potential of this analytical approach to explore the huge number of variables influencing this bioprocess crucial in winemaking.

Concerning the yeast population kinetics, the respective inoculation of non-*Saccaromyces* and *Saccharomyces* strains to promote/drive alcoholic fermentation in wine were generally performed (i) by inoculating together the strains (generally with a ratio 100:1 in favor of the non-*Saccharomyces* strain) (simultaneous inoculation) or (ii) inoculating the *Saccharomyces* strain with a delay of 24–48 h compared to the non-*Saccharomyces* inoculation (sequential inoculation) [70–72]. Both approaches aimed to maximize the development of non-*Saccharomyces*, concretizing an advantage for these yeasts [73]. The oenological objective is to counteract the fermentative advantage of *S. cerevisiae*, allowing non-*Saccharomyces* yeasts to influence wine quality [74]. The findings reported in the present article suggest that simultaneous inoculation led to good growth/survival for the tested non-*Saccharomyces* in combination with the selected *S. cerevisiae* strain. Cell concentration remained particularly high for *T. delbrueckii*, confirming the ability of this non-*Saccharomyces* yeast to survive at high ethanol concentrations [75]. For *M. pulcherrima*, the evidence was only partially in accordance to that reported by Dutraive et al. [17], who observed an initial decline of this yeast between the second and the third day after the inoculation, but followed by the complete annulment of the population.

The analysis included both commercial grape juice and fresh grape must to test the efficacy of the technique both in model conditions and in the real winemaking conditions. In fact, commercial grape juice, together with synthetic grape must [76] represents a common model medium for the fermentative studies in oenology (e.g., [77,78]). We found an evolution of volatiles during the three days of the study, which was in accordance with the evolution of yeast cell counts carried out during the AF. The results reported from the analysis of 'volatomes' associated with the development of single yeast species depicted different trends that could be coherent with different claimed aromatic properties for three commercialized strains, which received a considerable interest in the scientific literature (*S. cerevisiae* DV10, e.g., [79–83]; *M. pulcherrima* FLAVIA, e.g., [17,19,84–87]; *T. delbrueckii* BIODIVA, e.g., [17,84–88]). The study highlighted a global separation of VOC variability associated with the headspaces of the two tested matrices that can be ascribable to the chemical differences and/or to variable microbiological properties of the two media. Interesting, the behavior VOCs released by *M. pulcherrima* FLAVIA radically changed, shifting from fresh must to commercial juice, meaning that chemical/microbiological determinants of these media can directly or indirectly modulate VOC production by the yeasts. Even if the effect of abiotic and biotic interaction in the wine environment have been extensively investigated [89–91], further studies are needed to understand the biology affecting this phenotype, particularly in light of the huge intraspecific variability in terms of oenological properties within the species *M. pulcherrima* [92]. Intriguingly, variable patterns in must versus juice have also been observed in the trials where *M. pulcherrima* was co-inoculated with *S. cerevisiae* and with *T. delbrueckii*. A few studies have delved into the compatibility of *S. cerevisiae* combined in the same vinification with more than one non-*Saccharomyces* species [69,93]. Except for the coupled *Lachancea thermotolerans* and *Schizosaccharomyces pombe* that (used both in combination but not with also *S. cerevisiae*) has been extensively explored [94–96], only one study has tested the sensory impact (but not the VOC analysis) of this non-*Saccharomyces* multiple inoculation in wine [69]. In fact, usually, the articles considered the impact on volatile diversity of single strains or mixed starters composed of one *S. cerevisiae* strain and one non-*Saccharomyces* species. While the effect of multiple *Saccharomyces* yeast co-inoculations on volatile wine composition has been assessed (yeast inocula differed substantially in volatile thiols and other flavor compounds) [97,98], the interactions among different non-*Saccharomyces* wine yeast species need to be further elucidated [69]. The present findings suggest that the addition of *M. pulcherrima* to the coupled (in the case of the strains we tested) *S. cerevisiae/T. delbrueckii* can modify the global release of volatiles during the AF in wine as a function of the fermented matrix.

The different behaviors of the 'volatomes' associated with the single and mixed cultures showed promising results in terms of variability of the single mass peaks. The study of individual peak mass profiles during the three first days of AF in association with the tested yeast combinations will be the natural follow-up of the present communication. The objective will be focused to elucidate the single mass peaks/molecules responsible for the strain/species-specific differences and the specific yeast interactions/combinations, but also for the selection of candidate 'volatile' markers for the rapid screening of new microbial resources for 'flavoring' starter culture design in wine fermentations. Some findings have corroborated the evidence that the complexity of the microbial starter cultures inoculated can be among the levers capable of improving sensory wine complexity, assuring the safety of the productions (also for the possible exploitation in terms of biocontrol activity) [12,99–101]. It is interesting to underline that the tested strategy could find an application also in testing the interaction of yeast with malolactic bacteria [102–105]. Furthermore, it is important to stress how the proposed exploration of the phenotypic space of yeast activity in oenology can open new research lines for fundamental research in the field of yeast biology [106–108].

5. Conclusions

PTR-ToF-MS, combining high sensitivity/accuracy without neither sample preparation nor sample destruction, allows rapid real-time determination of volatile organic compounds (VOCs). In this paper, preliminary findings on the application of this analytical approach for the online monitoring of alcoholic fermentation in wine is proposed. The study explored different single and multiple inoculation of diverse oenological yeasts both in commercial grape juice and fresh must. The experiment highlighted a variability of the global volatiles in association with (i) the different yeast species, (ii) the different yeast combinations, and (iii) the different fermenting matrices. The evidence demonstrates the potential of PTR-ToF-MS in monitoring experimental variables associated with alcoholic fermentation in wine, opening new opportunities to manage this crucial phase, thus improving the quality of the final products and optimizing the processes.

Author Contributions: Conceptualization, C.B., I.K., P.R., G.S., M.F., F.B., and V.C.; Methodology, C.B., I.K., P.R., G.S., M.F., F.B., and V.C.; Investigation, C.B., I.K., P.R., M.F., F.B., and V.C.; Resources, G.S., F.B., and V.C.; Data curation, C.B., I.K., F.B., and V.C.; Writing—original draft preparation, C.B., I.K., and V.C.; Writing—review and editing, P.R., G.S., M.F., and F.B.; Supervision, G.S. and F.B.; Project administration, F.B. and V.C.; Funding acquisition, F.B. and V.C. All authors have read and agreed to the published version of the manuscript.

Acknowledgments: We would like to thank (i) the three anonymous reviewers for their suggestions and comments, (ii) Francesco De Marzo and Massimo Franchi of the Institute of Sciences of Food Production—CNR for their skilled technical support provided during the realization of this work, and (iii) Sergio Pelosi of the Institute of Sciences of Food Production—CNR for their critical reading of the manuscript.

References

1. Belda, I.; Ruiz, J.; Esteban-Fernández, A.; Navascués, E.; Marquina, D.; Santos, A.; Moreno-Arribas, M.V. Microbial Contribution to Wine Aroma and Its Intended Use for Wine Quality Improvement. *Mol. Basel Switz.* **2017**, *22*, 189. [CrossRef] [PubMed]
2. Benito, Á.; Calderón, F.; Benito, S. The Influence of Non-*Saccharomyces* Species on Wine Fermentation Quality Parameters. *Fermentation* **2019**, *5*, 54. [CrossRef]
3. Lebrón-Aguilar, R.; Soria, A.C.; Quintanilla-López, J.E. Comprehensive evaluation of direct injection mass spectrometry for the quantitative profiling of volatiles in food samples. *Philos. Trans. R. Soc. Math. Phys. Eng. Sci.* **2016**, *374*, 20150375. [CrossRef]
4. Pinu, F.R. Grape and Wine Metabolomics to Develop New Insights Using Untargeted and Targeted Approaches. *Fermentation* **2018**, *4*, 92. [CrossRef]
5. Tufariello, M.; Maiorano, G.; Rampino, P.; Spano, G.; Grieco, F.; Perrotta, C.; Capozzi, V.; Grieco, F. Selection of an autochthonous yeast starter culture for industrial production of Primitivo "Gioia del Colle" PDO/DOC in Apulia (Southern Italy). *LWT* **2019**, *99*, 188–196. [CrossRef]
6. Garofalo, C.; Tristezza, M.; Grieco, F.; Spano, G.; Capozzi, V. From grape berries to wine: Population dynamics of cultivable yeasts associated to "Nero di Troia" autochthonous grape cultivar. *World J. Microbiol. Biotechnol.* **2016**, *32*, 59. [CrossRef]
7. Capozzi, V.; Garofalo, C.; Chiriatti, M.A.; Grieco, F.; Spano, G. Microbial terroir and food innovation: The case of yeast biodiversity in wine. *Microbiol. Res.* **2015**, *181*, 75–83. [CrossRef] [PubMed]
8. Ciani, M.; Comitini, F.; Mannazzu, I.; Domizio, P. Controlled mixed culture fermentation: A new perspective on the use of non-*Saccharomyces* yeasts in winemaking. *FEMS Yeast Res.* **2010**, *10*, 123–133. [CrossRef] [PubMed]
9. Comitini, F.; Gobbi, M.; Domizio, P.; Romani, C.; Lencioni, L.; Mannazzu, I.; Ciani, M. Selected non-*Saccharomyces* wine yeasts in controlled multistarter fermentations with *Saccharomyces cerevisiae*. *Food Microbiol.* **2011**, *28*, 873–882. [CrossRef]
10. Tronchoni, J.; Curiel, J.A.; Morales, P.; Torres-Pérez, R.; Gonzalez, R. Early transcriptional response to biotic stress in mixed starter fermentations involving *Saccharomyces cerevisiae* and *Torulaspora delbrueckii*. *Int. J. Food Microbiol.* **2017**, *241*, 60–68. [CrossRef]
11. Ciani, M.; Morales, P.; Comitini, F.; Tronchoni, J.; Canonico, L.; Curiel, J.A.; Oro, L.; Rodrigues, A.J.; Gonzalez, R. Non-conventional Yeast Species for Lowering Ethanol Content of Wines. *Front. Microbiol.* **2016**, *7*, 642. [CrossRef] [PubMed]
12. Rossouw, D.; Bauer, F.F. Exploring the phenotypic space of non-*Saccharomyces* wine yeast biodiversity. *Food Microbiol.* **2016**, *55*, 32–46. [CrossRef] [PubMed]

13. Beckner Whitener, M.E.; Stanstrup, J.; Panzeri, V.; Carlin, S.; Divol, B.; Du Toit, M.; Vrhovsek, U. Untangling the wine metabolome by combining untargeted SPME–GCxGC-TOF-MS and sensory analysis to profile Sauvignon blanc co-fermented with seven different yeasts. *Metabolomics* **2016**, *12*, 53. [CrossRef]
14. Lu, Y.; Huang, D.; Lee, P.R.; Liu, S.Q. Assessment of volatile and non-volatile compounds in durian wines fermented with four commercial non-*Saccharomyces* yeasts. *J. Sci. Food Agric.* **2016**, *96*, 1511–1521. [CrossRef]
15. Medina, K.; Boido, E.; Dellacassa, E.; Carrau, F. Growth of non-Saccharomyces yeasts affects nutrient availability for *Saccharomyces cerevisiae* during wine fermentation. *Int. J. Food Microbiol.* **2012**, *157*, 245–250. [CrossRef]
16. Petruzzi, L.; Capozzi, V.; Berbegal, C.; Corbo, M.R.; Bevilacqua, A.; Spano, G.; Sinigaglia, M. Microbial Resources and Enological Significance: Opportunities and Benefits. *Front. Microbiol.* **2017**, *8*, 995. [CrossRef]
17. Dutraive, O.; Benito, S.; Fritsch, S.; Beisert, B.; Patz, C.D.; Rauhut, D. Effect of Sequential Inoculation with Non-*Saccharomyces* and *Saccharomyces* Yeasts on Riesling Wine Chemical Composition. *Fermentation* **2019**, *5*, 79. [CrossRef]
18. Du Plessis, H.; Du Toit, M.; Nieuwoudt, H.; Van der Rijst, M.; Hoff, J.; Jolly, N. Modulation of Wine Flavor using Hanseniaspora uvarum in Combination with Different *Saccharomyces cerevisiae*, Lactic Acid Bacteria Strains and Malolactic Fermentation Strategies. *Fermentation* **2019**, *5*, 64. [CrossRef]
19. Morata, A.; Loira, I.; Escott, C.; del Fresno, J.M.; Bañuelos, M.A.; Suárez-Lepe, J.A. Applications of *Metschnikowia pulcherrima* in Wine Biotechnology. *Fermentation* **2019**, *5*, 63. [CrossRef]
20. Garofalo, C.; Russo, P.; Beneduce, L.; Massa, S.; Spano, G.; Capozzi, V. Non-*Saccharomyces* biodiversity in wine and the 'microbial terroir': A survey on Nero di Troia wine from the Apulian region, Italy. *Ann. Microbiol.* **2016**, *66*, 143–150. [CrossRef]
21. Sgouros, G.; Chalvantzi, I.; Mallouchos, A.; Paraskevopoulos, Y.; Banilas, G.; Nisiotou, A. Biodiversity and Enological Potential of Non-*Saccharomyces* Yeasts from Nemean Vineyards. *Fermentation* **2018**, *4*, 32. [CrossRef]
22. Roudil, L.; Russo, P.; Berbegal, C.; Albertin, W.; Spano, G.; Capozzi, V. Non-*Saccharomyces* Commercial Starter Cultures: Scientific Trends, Recent Patents and Innovation in the Wine Sector. *Recent Pat. Food Nutr. Agric.* **2019**. [CrossRef] [PubMed]
23. Lindinger, W.; Jordan, A. Proton-transfer-reaction mass spectrometry (PTR–MS): On-line monitoring of volatile organic compounds at pptv levels. *Chem. Soc. Rev.* **1998**, *27*, 347–375. [CrossRef]
24. Jordan, A.; Haidacher, S.; Hanel, G.; Hartungen, E.; Märk, L.; Seehauser, H.; Schottkowsky, R.; Sulzer, P.; Märk, T.D. A high resolution and high sensitivity proton-transfer-reaction time-of-flight mass spectrometer (PTR-TOF-MS). *Int. J. Mass Spectrom.* **2009**, *286*, 122–128. [CrossRef]
25. Capozzi, V.; Yener, S.; Khomenko, I.; Farneti, B.; Cappellin, L.; Gasperi, F.; Scampicchio, M.; Biasioli, F. PTR-ToF-MS Coupled with an Automated Sampling System and Tailored Data Analysis for Food Studies: Bioprocess Monitoring, Screening and Nose-space Analysis. *JoVE J. Vis. Exp.* **2017**, e54075. [CrossRef]
26. Cappellin, L.; Loreto, F.; Aprea, E.; Romano, A.; Del Pulgar, J.S.; Gasperi, F.; Biasioli, F. PTR-MS in Italy: A Multipurpose Sensor with Applications in Environmental, Agri-Food and Health Science. *Sensors* **2013**, *13*, 11923–11955. [CrossRef]
27. Pico, J.; Khomenko, I.; Capozzi, V.; Navarini, L.; Bernal, J.; Gómez, M.; Biasioli, F. Analysis of volatile organic compounds in crumb and crust of different baked and toasted gluten-free breads by direct PTR-ToF-MS and fast-GC-PTR-ToF-MS. *J. Mass Spectrom.* **2018**, *53*, 893–902. [CrossRef]
28. Deuscher, Z.; Andriot, I.; Sémon, E.; Repoux, M.; Preys, S.; Roger, J.M.; Boulanger, R.; Labouré, H.; Quéré, J.L.L. Volatile compounds profiling by using proton transfer reaction-time of flight-mass spectrometry (PTR-ToF-MS). The case study of dark chocolates organoleptic differences. *J. Mass Spectrom.* **2019**, *54*, 92–119. [CrossRef]
29. van Ruth, S.M.; Koot, A.; Akkermans, W.; Araghipour, N.; Rozijn, M.; Baltussen, M.; Wisthaler, A.; Märk, T.D.; Frankhuizen, R. Butter and butter oil classification by PTR-MS. *Eur. Food Res. Technol.* **2008**, *227*, 307–317. [CrossRef]
30. Capozzi, V.; Lonzarich, V.; Khomenko, I.; Cappellin, L.; Navarini, L.; Biasioli, F. Unveiling the Molecular Basis of Mascarpone Cheese Aroma: VOCs analysis by SPME-GC/MS and PTR-ToF-MS. *Molecules* **2020**, *25*, 1242. [CrossRef]

31. Makhoul, S.; Yener, S.; Khomenko, I.; Capozzi, V.; Cappellin, L.; Aprea, E.; Scampicchio, M.; Gasperi, F.; Biasioli, F. Rapid non-invasive quality control of semi-finished products for the food industry by direct injection mass spectrometry headspace analysis: The case of milk powder, whey powder and anhydrous milk fat. *J. Mass Spectrom.* **2016**, *51*, 782–791. [CrossRef] [PubMed]
32. Romano, A.; Capozzi, V.; Spano, G.; Biasioli, F. Proton transfer reaction–mass spectrometry: Online and rapid determination of volatile organic compounds of microbial origin. *Appl. Microbiol. Biotechnol.* **2015**, *99*, 3787–3795. [CrossRef] [PubMed]
33. Benozzi, E.; Romano, A.; Capozzi, V.; Makhoul, S.; Cappellin, L.; Khomenko, I.; Aprea, E.; Scampicchio, M.; Spano, G.; Märk, T.D.; et al. Monitoring of lactic fermentation driven by different starter cultures via direct injection mass spectrometric analysis of flavour-related volatile compounds. *Food Res. Int.* **2015**, *76*, 682–688. [CrossRef] [PubMed]
34. Richter, T.M.; Silcock, P.; Algarra, A.; Eyres, G.T.; Capozzi, V.; Bremer, P.J.; Biasioli, F. Evaluation of PTR-ToF-MS as a tool to track the behavior of hop-derived compounds during the fermentation of beer. *Food Res. Int.* **2018**, *111*, 582–589. [CrossRef] [PubMed]
35. Khomenko, I.; Stefanini, I.; Cappellin, L.; Cappelletti, V.; Franceschi, P.; Cavalieri, D.; Märk, T.D.; Biasioli, F. Non-invasive real time monitoring of yeast volatilome by PTR-ToF-MS. *Metab. Off. J. Metab. Soc.* **2017**, *13*, 118. [CrossRef]
36. Capozzi, V.; Makhoul, S.; Aprea, E.; Romano, A.; Cappellin, L.; Sanchez Jimena, A.; Spano, G.; Gasperi, F.; Scampicchio, M.; Biasioli, F. PTR-MS Characterization of VOCs Associated with Commercial Aromatic Bakery Yeasts of Wine and Beer Origin. *Molecules* **2016**, *21*, 483. [CrossRef]
37. Makhoul, S.; Romano, A.; Capozzi, V.; Spano, G.; Aprea, E.; Cappellin, L.; Benozzi, E.; Scampicchio, M.; Märk, T.D.; Gasperi, F.; et al. Volatile Compound Production During the Bread-Making Process: Effect of Flour, Yeast and Their Interaction. *Food Bioprocess Technol.* **2015**, *8*, 1925–1937. [CrossRef]
38. Makhoul, S.; Romano, A.; Cappellin, L.; Spano, G.; Capozzi, V.; Benozzi, E.; Märk, T.D.; Aprea, E.; Gasperi, F.; El-Nakat, H.; et al. Proton-transfer-reaction mass spectrometry for the study of the production of volatile compounds by bakery yeast starters. *J. Mass Spectrom. JMS* **2014**, *49*, 850–859. [CrossRef]
39. Campbell-Sills, H.; Capozzi, V.; Romano, A.; Cappellin, L.; Spano, G.; Breniaux, M.; Lucas, P.; Biasioli, F. Advances in wine analysis by PTR-ToF-MS: Optimization of the method and discrimination of wines from different geographical origins and fermented with different malolactic starters. *Int. J. Mass Spectrom.* **2016**, *397–398*, 42–51. [CrossRef]
40. Romano, A.; Fischer, L.; Herbig, J.; Campbell-Sills, H.; Coulon, J.; Lucas, P.; Cappellin, L.; Biasioli, F. Wine analysis by FastGC proton-transfer reaction-time-of-flight-mass spectrometry. *Int. J. Mass Spectrom.* **2014**, *369*, 81–86. [CrossRef]
41. Sémon, E.; Arvisenet, G.; Guichard, E.; Quéré, J.L.L. Modified proton transfer reaction mass spectrometry (PTR-MS) operating conditions for *in vitro* and *in vivo* analysis of wine aroma. *J. Mass Spectrom.* **2018**, *53*, 65–77. [CrossRef] [PubMed]
42. Muñoz-González, C.; Canon, F.; Feron, G.; Guichard, E.; Pozo-Bayón, M.A. Assessment Wine Aroma Persistence by Using an *in Vivo* PTR-ToF-MS Approach and Its Relationship with Salivary Parameters. *Mol. Basel Switz.* **2019**, *24*, 1277. [CrossRef] [PubMed]
43. Garofalo, C.; El Khoury, M.; Lucas, P.; Bely, M.; Russo, P.; Spano, G.; Capozzi, V. Autochthonous starter cultures and indigenous grape variety for regional wine production. *J. Appl. Microbiol.* **2015**, *118*, 1395–1408. [CrossRef] [PubMed]
44. Cappellin, L.; Biasioli, F.; Fabris, A.; Schuhfried, E.; Soukoulis, C.; Märk, T.D.; Gasperi, F. Improved mass accuracy in PTR-TOF-MS: Another step towards better compound identification in PTR-MS. *Int. J. Mass Spectrom.* **2010**, *290*, 60–63. [CrossRef]
45. Cappellin, L.; Biasioli, F.; Granitto, P.M.; Schuhfried, E.; Soukoulis, C.; Costa, F.; Märk, T.D.; Gasperi, F. On data analysis in PTR-TOF-MS: From raw spectra to data mining. *Sens. Actuators B Chem.* **2011**, *155*, 183–190. [CrossRef]
46. Lindinger, W.; Hansel, A.; Jordan, A. On-line monitoring of volatile organic compounds at pptv levels by means of proton-transfer-reaction mass spectrometry (PTR-MS) medical applications, food control and environmental research. *Int. J. Mass Spectrom. Ion Process.* **1998**, *173*, 191–241. [CrossRef]
47. Bisson, L.F.; Waterhouse, A.L.; Ebeler, S.E.; Walker, M.A.; Lapsley, J.T. The present and future of the international wine industry. *Nature* **2002**, *418*, 696–699. [CrossRef]

48. Berbegal, C.; Fragasso, M.; Russo, P.; Bimbo, F.; Grieco, F.; Spano, G.; Capozzi, V. Climate Changes and Food Quality: The Potential of Microbial Activities as Mitigating Strategies in the Wine Sector. *Fermentation* **2019**, *5*, 85. [CrossRef]
49. Bisson, L.F. Stuck and Sluggish Fermentations. *Am. J. Enol. Vitic.* **1999**, *50*, 107–119.
50. Fleet, G.H. Wine yeasts for the future. *FEMS Yeast Res.* **2008**, *8*, 979–995. [CrossRef]
51. Pretorius, I.S. Tailoring wine yeast for the new millennium: Novel approaches to the ancient art of winemaking. *Yeast* **2000**, *16*, 675–729. [CrossRef]
52. Nardi, T. Microbial Resources as a Tool for Enhancing Sustainability in Winemaking. *Microorganisms* **2020**, *8*, 507. [CrossRef]
53. Sablayrolles, J.M. Control of alcoholic fermentation in winemaking: Current situation and prospect. *Food Res. Int.* **2009**, *42*, 418–424. [CrossRef]
54. Molina, A.M.; Swiegers, J.H.; Varela, C.; Pretorius, I.S.; Agosin, E. Influence of wine fermentation temperature on the synthesis of yeast-derived volatile aroma compounds. *Appl. Microbiol. Biotechnol.* **2007**, *77*, 675–687. [CrossRef] [PubMed]
55. Russo, P.; Berbegal, C.; De Ceglie, C.; Grieco, F.; Spano, G.; Capozzi, V. Pesticide Residues and Stuck Fermentation in Wine: New Evidences Indicate the Urgent Need of Tailored Regulations. *Fermentation* **2019**, *5*, 23. [CrossRef]
56. Di Egidio, V.; Sinelli, N.; Giovanelli, G.; Moles, A.; Casiraghi, E. NIR and MIR spectroscopy as rapid methods to monitor red wine fermentation. *Eur. Food Res. Technol.* **2010**, *230*, 947–955. [CrossRef]
57. Mouret, J.R.; Perez, M.; Angenieux, M.; Nicolle, P.; Farines, V.; Sablayrolles, J.M. Online-Based Kinetic Analysis of Higher Alcohol and Ester Synthesis During Winemaking Fermentations. *Food Bioprocess Technol.* **2014**, *7*, 1235–1245. [CrossRef]
58. Mallouchos, A.; Komaitis, M.; Koutinas, A.; Kanellaki, M. Investigation of Volatiles Evolution during the Alcoholic Fermentation of Grape Must Using Free and Immobilized Cells with the Help of Solid Phase Microextraction (SPME) Headspace Sampling. *J. Agric. Food Chem.* **2002**, *50*, 3840–3848. [CrossRef]
59. Morales, M.L.; Fierro-Risco, J.; Callejón, R.M.; Paneque, P. Monitoring volatile compounds production throughout fermentation by *Saccharomyces* and non-*Saccharomyces* strains using headspace sorptive extraction. *J. Food Sci. Technol.* **2017**, *54*, 538–557. [CrossRef]
60. Silva Ferreira, A.C.; Monforte, A.R.; Teixeira, C.S.; Martins, R.; Fairbairn, S.; Bauer, F.F. Monitoring Alcoholic Fermentation: An Untargeted Approach. *J. Agric. Food Chem.* **2014**, *62*, 6784–6793. [CrossRef]
61. Buratti, S.; Benedetti, S. Chapter 28-Alcoholic Fermentation Using Electronic Nose and Electronic Tongue. In *Electronic Noses and Tongues in Food Science*; Rodríguez Méndez, M.L., Ed.; Academic Press: San Diego, CA, USA, 2016; pp. 291–299, ISBN 978-0-12-800243-8.
62. Aguilera, T.; Lozano, J.; Paredes, J.A.; Álvarez, F.J.; Suárez, J.I. Electronic Nose Based on Independent Component Analysis Combined with Partial Least Squares and Artificial Neural Networks for Wine Prediction. *Sensors* **2012**, *12*, 8055–8072. [CrossRef] [PubMed]
63. Jiang, B.; Zhang, Z. Volatile Compounds of Young Wines from Cabernet Sauvignon, Cabernet Gernischet and Chardonnay Varieties Grown in the Loess Plateau Region of China. *Molecules* **2010**, *15*, 9184–9196. [CrossRef] [PubMed]
64. Jolly, N.P.; Varela, C.; Pretorius, I.S. Not your ordinary yeast: Non-*Saccharomyces* yeasts in wine production uncovered. *FEMS Yeast Res.* **2014**, *14*, 215–237. [CrossRef] [PubMed]
65. Berbegal, C.; Spano, G.; Tristezza, M.; Grieco, F.; Capozzi, V. Microbial Resources and Innovation in the Wine Production Sector. *S. Afr. J. Enol. Vitic.* **2017**, *38*, 156–166. [CrossRef]
66. Lambrechts, M.G.; Pretorius, I.S. Yeast and its Importance to Wine Aroma—A Review. *S. Afr. J. Enol. Vitic.* **2000**, *21*, 97–129. [CrossRef]
67. Escribano, R.; González-Arenzana, L.; Garijo, P.; Berlanas, C.; López-Alfaro, I.; López, R.; Gutiérrez, A.R.; Santamaría, P. Screening of enzymatic activities within different enological non-*Saccharomyces* yeasts. *J. Food Sci. Technol.* **2017**, *54*, 1555–1564. [CrossRef]
68. Escribano, R.; González-Arenzana, L.; Portu, J.; Garijo, P.; López-Alfaro, I.; López, R.; Santamaría, P.; Gutiérrez, A.R. Wine aromatic compound production and fermentative behaviour within different non-*Saccharomyces* species and clones. *J. Appl. Microbiol.* **2018**, *124*, 1521–1531. [CrossRef]

69. Padilla, B.; Gil, J.V.; Manzanares, P. Past and Future of Non-*Saccharomyces* Yeasts: From Spoilage Microorganisms to Biotechnological Tools for Improving Wine Aroma Complexity. *Front. Microbiol.* **2016**, *7*, 411. [CrossRef]
70. Ciani, M.; Beco, L.; Comitini, F. Fermentation behaviour and metabolic interactions of multistarter wine yeast fermentations. *Int. J. Food Microbiol.* **2006**, *108*, 239–245. [CrossRef]
71. Taillandier, P.; Lai, Q.P.; Julien-Ortiz, A.; Brandam, C. Interactions between *Torulaspora delbrueckii* and *Saccharomyces cerevisiae* in wine fermentation: Influence of inoculation and nitrogen content. *World J. Microbiol. Biotechnol.* **2014**, *30*, 1959–1967. [CrossRef]
72. Contreras, A.; Hidalgo, C.; Henschke, P.A.; Chambers, P.J.; Curtin, C.; Varela, C. Evaluation of Non-*Saccharomyces* Yeasts for the Reduction of Alcohol Content in Wine. *Appl. Environ. Microbiol.* **2014**, *80*, 1670–1678. [CrossRef] [PubMed]
73. Tristezza, M.; Tufariello, M.; Capozzi, V.; Spano, G.; Mita, G.; Grieco, F. The Oenological Potential of *Hanseniaspora uvarum* in Simultaneous and Sequential Co-fermentation with *Saccharomyces cerevisiae* for Industrial Wine Production. *Front. Microbiol.* **2016**, *7*. [CrossRef] [PubMed]
74. Russo, P.; Tufariello, M.; Renna, R.; Tristezza, M.; Taurino, M.; Palombi, L.; Capozzi, V.; Rizzello, C.G.; Grieco, F. New Insights into the Oenological Significance of *Candida zemplinina*: Impact of Selected Autochthonous Strains on the Volatile Profile of Apulian Wines. *Microorganisms* **2020**, *8*, 628. [CrossRef]
75. González-Royo, E.; Pascual, O.; Kontoudakis, N.; Esteruelas, M.; Esteve-Zarzoso, B.; Mas, A.; Canals, J.M.; Zamora, F. Oenological consequences of sequential inoculation with non-*Saccharomyces* yeasts (*Torulaspora delbrueckii* or *Metschnikowia pulcherrima*) and *Saccharomyces cerevisiae* in base wine for sparkling wine production. *Eur. Food Res. Technol.* **2015**, *240*, 999–1012. [CrossRef]
76. Renault, P.E.; Albertin, W.; Bely, M. An innovative tool reveals interaction mechanisms among yeast populations under oenological conditions. *Appl. Microbiol. Biotechnol.* **2013**, *97*, 4105–4119. [CrossRef]
77. Gobert, A.; Tourdot-Maréchal, R.; Morge, C.; Sparrow, C.; Liu, Y.; Quintanilla-Casas, B.; Vichi, S.; Alexandre, H. Non-*Saccharomyces* Yeasts Nitrogen Source Preferences: Impact on Sequential Fermentation and Wine Volatile Compounds Profile. *Front. Microbiol.* **2017**, *8*. [CrossRef]
78. Hansen, E.H.; Nissen, P.; Sommer, P.; Nielsen, J.C.; Arneborg, N. The effect of oxygen on the survival of non-*Saccharomyces* yeasts during mixed culture fermentations of grape juice with Saccharomyces cerevisiae. *J. Appl. Microbiol.* **2001**, *91*, 541–547. [CrossRef]
79. Vasserot, Y.; Mornet, F.; Jeandet, P. Acetic acid removal by *Saccharomyces cerevisiae* during fermentation in oenological conditions. Metabolic consequences. *Food Chem.* **2010**, *119*, 1220–1223. [CrossRef]
80. Jackowetz, J.N.; Dierschke, S.; Mira de Orduña, R. Multifactorial analysis of acetaldehyde kinetics during alcoholic fermentation by *Saccharomyces cerevisiae*. *Food Res. Int.* **2011**, *44*, 310–316. [CrossRef]
81. Legras, J.L.; Karst, F. Optimisation of interdelta analysis for *Saccharomyces cerevisiae* strain characterization. *FEMS Microbiol. Lett.* **2003**, *221*, 249–255. [CrossRef]
82. Garofalo, C.; Berbegal, C.; Grieco, F.; Tufariello, M.; Spano, G.; Capozzi, V. Selection of indigenous yeast strains for the production of sparkling wines from native Apulian grape varieties. *Int. J. Food Microbiol.* **2018**, *285*, 7–17. [CrossRef] [PubMed]
83. Jin, X.; Chen, W.; Chen, H.; Chen, W.; Zhong, Q. Combination of *Lactobacillus plantarum* and *Saccharomyces cerevisiae* DV10 as Starter Culture to Produce Mango Slurry: Microbiological, Chemical Parameters and Antioxidant Activity. *Molecules* **2019**, *24*, 4349. [CrossRef] [PubMed]
84. Seguinot, P.; Bloem, A.; Brial, P.; Meudec, E.; Ortiz-Julien, A.; Camarasa, C. Analyzing the impact of the nature of the nitrogen source on the formation of volatile compounds to unravel the aroma metabolism of two non-*Saccharomyces* strains. *Int. J. Food Microbiol.* **2020**, *316*, 108441. [CrossRef] [PubMed]
85. Chua, J.Y.; Lu, Y.; Liu, S.Q. Evaluation of five commercial non-*Saccharomyces* yeasts in fermentation of soy (tofu) whey into an alcoholic beverage. *Food Microbiol.* **2018**, *76*, 533–542. [CrossRef]
86. Prior, K.J.; Bauer, F.F.; Divol, B. The utilization of nitrogenous compounds by commercial non-*Saccharomyces* yeasts associated with wine. *Food Microbiol.* **2019**, *79*, 75–84. [CrossRef]
87. Ivit, N.N.; Kemp, B. The Impact of Non-*Saccharomyces* Yeast on Traditional Method Sparkling Wine. *Fermentation* **2018**, *4*, 73. [CrossRef]

88. Nardi, T.; Panero, L.; Petrozziello, M.; Guaita, M.; Tsolakis, C.; Cassino, C.; Vagnoli, P.; Bosso, A. Managing wine quality using *Torulaspora delbrueckii* and *Oenococcus oeni* starters in mixed fermentations of a red Barbera wine. *Eur. Food Res. Technol.* **2019**, *245*, 293–307. [CrossRef]
89. Chandra, M.; Oro, I.; Ferreira-Dias, S.; Malfeito-Ferreira, M. Effect of Ethanol, Sulfur Dioxide and Glucose on the Growth of Wine Spoilage Yeasts Using Response Surface Methodology. *PLoS ONE* **2015**, *10*, e0128702. [CrossRef]
90. Ciani, M.; Capece, A.; Comitini, F.; Canonico, L.; Siesto, G.; Romano, P. Yeast Interactions in Inoculated Wine Fermentation. *Front. Microbiol.* **2016**, *7*. [CrossRef]
91. Pinu, F.R.; Edwards, P.J.B.; Gardner, R.C.; Villas-Boas, S.G. Nitrogen and carbon assimilation by *Saccharomyces cerevisiae* during Sauvignon blanc juice fermentation. *FEMS Yeast Res.* **2014**, *14*, 1206–1222. [CrossRef]
92. Barbosa, C.; Lage, P.; Esteves, M.; Chambel, L.; Mendes-Faia, A.; Mendes-Ferreira, A. Molecular and Phenotypic Characterization of *Metschnikowia pulcherrima* Strains from Douro Wine Region. *Fermentation* **2018**, *4*, 8. [CrossRef]
93. Escribano-Viana, R.; Portu, J.; Garijo, P.; López, R.; Santamaría, P.; López-Alfaro, I.; Gutiérrez, A.R.; González-Arenzana, L. Effect of the Sequential Inoculation of Non-*Saccharomyces*/*Saccharomyces* on the Anthocyans and Stilbenes Composition of Tempranillo Wines. *Front. Microbiol.* **2019**, *10*. [CrossRef]
94. Benito, Á.; Calderón, F.; Palomero, F.; Benito, S. Combine Use of Selected *Schizosaccharomyces pombe* and *Lachancea thermotolerans* Yeast Strains as an Alternative to theTraditional Malolactic Fermentation in Red Wine Production. *Molecules* **2015**, *20*, 9510–9523. [CrossRef] [PubMed]
95. Benito, Á.; Calderón, F.; Benito, S. The Combined Use of *Schizosaccharomyces pombe* and *Lachancea thermotolerans*—Effect on the Anthocyanin Wine Composition. *Molecules* **2017**, *22*, 739. [CrossRef] [PubMed]
96. Benito, Á.; Calderón, F.; Benito, S. Combined Use of *S. pombe* and *L. thermotolerans* in Winemaking. Beneficial Effects Determined Through the Study of Wines' Analytical Characteristics. *Molecules* **2016**, *21*, 1744. [CrossRef] [PubMed]
97. King, E.S.; Kievit, R.L.; Curtin, C.; Swiegers, J.H.; Pretorius, I.S.; Bastian, S.E.P.; Leigh Francis, I. The effect of multiple yeasts co-inoculations on Sauvignon Blanc wine aroma composition, sensory properties and consumer preference. *Food Chem.* **2010**, *122*, 618–626. [CrossRef]
98. Gustafsson, F.S.; Jiranek, V.; Neuner, M.; Scholl, C.M.; Morgan, S.C.; Durall, D.M. The Interaction of Two *Saccharomyces* cerevisiae Strains Affects Fermentation-Derived Compounds in Wine. *Fermentation* **2016**, *2*, 9. [CrossRef]
99. Berbegal, C.; Garofalo, C.; Russo, P.; Pati, S.; Capozzi, V.; Spano, G. Use of Autochthonous Yeasts and Bacteria in Order to Control *Brettanomyces bruxellensis* in Wine. *Fermentation* **2017**, *3*, 65. [CrossRef]
100. Berbegal, C.; Spano, G.; Fragasso, M.; Grieco, F.; Russo, P.; Capozzi, V. Starter cultures as biocontrol strategy to prevent *Brettanomyces bruxellensis* proliferation in wine. *Appl. Microbiol. Biotechnol.* **2018**, *102*, 569–576. [CrossRef]
101. Capozzi, V.; Fragasso, M.; Russo, P. Microbiological Safety and the Management of Microbial Resources in Artisanal Foods and Beverages: The Need for a Transdisciplinary Assessment to Conciliate Actual Trends and Risks Avoidance. *Microorganisms* **2020**, *8*, 306. [CrossRef]
102. Berbegal, C.; Borruso, L.; Fragasso, M.; Tufariello, M.; Russo, P.; Brusetti, L.; Spano, G.; Capozzi, V. A Metagenomic-Based Approach for the Characterization of Bacterial Diversity Associated with Spontaneous Malolactic Fermentations in Wine. *Int. J. Mol. Sci.* **2019**, *20*, 3980. [CrossRef] [PubMed]
103. Englezos, V.; Cachón, D.C.; Rantsiou, K.; Blanco, P.; Petrozziello, M.; Pollon, M.; Giacosa, S.; Río Segade, S.; Rolle, L.; Cocolin, L. Effect of mixed species alcoholic fermentation on growth and malolactic activity of lactic acid bacteria. *Appl. Microbiol. Biotechnol.* **2019**, *103*, 7687–7702. [CrossRef] [PubMed]
104. Balmaseda, A.; Bordons, A.; Reguant, C.; Bautista-Gallego, J. Non-Saccharomyces in Wine: Effect Upon *Oenococcus oeni* and Malolactic Fermentation. *Front. Microbiol.* **2018**, *9*, 534. [CrossRef] [PubMed]
105. Capozzi, V.; Berbegal, C.; Tufariello, M.; Grieco, F.; Spano, G.; Grieco, F. Impact of co-inoculation of *Saccharomyces cerevisiae*, *Hanseniaspora uvarum* and *Oenococcus oeni* autochthonous strains in controlled multi starter grape must fermentations. *LWT* **2019**, *109*, 241–249. [CrossRef]
106. Willaert, R.; Kasas, S.; Devreese, B.; Dietler, G. Yeast Nanobiotechnology. *Fermentation* **2016**, *2*, 18. [CrossRef]

107. Botstein, D.; Fink, G.R. Yeast: An Experimental Organism for 21st Century Biology. *Genetics* **2011**, *189*, 695–704. [CrossRef]
108. Walker, R.S.K.; Pretorius, I.S. Applications of Yeast Synthetic Biology Geared towards the Production of Biopharmaceuticals. *Genes* **2018**, *9*, 340. [CrossRef]

8

Citric Acid Production by *Yarrowia lipolytica* Yeast on Different Renewable Raw Materials

Igor G. Morgunov *, Svetlana V. Kamzolova and Julia N. Lunina

G.K. Skryabin Institute of Biochemistry and Physiology of Microorganisms, Russian Academy of Sciences, pr-t Nauki, 5, 142290 Pushchino, Russia; kamzolova@rambler.ru (S.V.K.); luninaj@rambler.ru (J.N.L.)

* Correspondence: morgunovs@rambler.ru

Abstract: The world market of citric acid (CA) is one of the largest and fastest growing markets in the biotechnological industry. Microbiological processes for CA production have usually used the mycelial fungi *Aspergillus niger* as a producer and molasses as a carbon source. In this paper, we propose methods for CA production from renewable carbon substrates (rapeseed oil, glucose, glycerol, ethanol, glycerol-containing waste of biodiesel industry and glucose-containing aspen waste) by the mutant strain *Yarrowia lipolytica* NG40/UV5. It was revealed that *Y. lipolytica* grew and synthesized CA using all tested raw materials. The obtained results are sufficient for industrial use of most of the raw materials studied for CA production. Using rapeseed oil, ethanol and raw glycerol (which is an important feedstock of biodiesel production), a high CA production (100–140 g L^{-1}) was achieved.

Keywords: yeast *Yarrowia lipolytica*; citric acid (CA) production; raw materials

---

1. Introduction

Citric acid (CA) and its salts are widely used as an acidulate, flavoring agent and antioxidant in the production of beverages and confectionery, in infant formula, as well as in the chemical, pharmaceutical, electronic, defense, and other industries. The volume of citric acid globally exceeds two million tons per year and its production is annually increased by 5% [1].

Modern technologies of citric acid (CA) production are based on using various mutant strains of the mycelial fungi *Aspergillus niger* as a producer and molasses as a raw material. CA production by fungi is a complicated and environmentally unsafe process; as a result of its implementation, a large number of both liquid effluents containing mineral acids, ballast organic substances, cyanides, and solid wastes, primarily gypsum, are accumulated. Moreover, *A. niger* is an opportunistic pathogenic fungi and can cause allergic diseases and aspergillosis [2,3].

Over the past 40 years, the interest of researchers has focused on yeast as a producer of CA; the yeast *Yarrowia lipolytica* has been the most used CA producer [1–3]. Initially, this kind of yeast attracted the attention of researchers due to its ability to grow and synthesize CA in media with n alkanes—an available and cheap substrate [2]. However, due to changes in the world oil market, the use of this substrate has become economically unprofitable. In this regard, it is of interest to use other types of raw materials.

The choice of raw materials for developing CA biotechnology is determined by factors such as renewability, ability of the producer to assimilate the substrate with a high conversion rate, consumption value and cost price of the target product. To carry out fermentation processes, in addition to very expensive food raw materials, such as glucose [4–7] and plant oils [8–10], much cheaper substrates which are waste products of various industries, such as glycerol-containing waste of the biodiesel industry [9–17], glucose-containing wood hydrolysates [18,19], olive mill waste-water [20], and inulin [21] are used. Ethanol, a water-soluble individual compound which ensures the formation of a pure product and facilitates the isolation process, is also of great importance [22,23]. In the works of the above-mentioned researchers it was shown that all these substrates are promising for CA production and the use of glycerol-containing waste of biodiesel industry and glucose-containing wood hydrolyzates can increase the profitability of CA production process.

The aim of this work was a comparative study of CA production by the yeast *Y. lipolytica* on different types of renewable raw materials.

2. Materials and Methods

2.1. Microorganisms and Chemicals

The mutant strain *Y. lipolytica* NG40/UV5 was obtained as described previously [9,24].

Chemicals, the manufacturer of the carbon sources, and their characteristics were presented in our published articles [7,9,15,17].

2.2. Media and Cultivation Conditions

All experiments were done using the same equipment, nutrient medium and cultivation conditions. Strain *Y. lipolytica* NG40/UV5 was cultivated in a 10-L ANKUM-2M fermenter (Pushchino, Russia) with an initial volume of 5 L. The medium contained (g L^{-1}): carbon source: $(NH_4)_2SO_4$, 6; $MgSO_4 \cdot 7H_2O$, 1.4; NaCl, 0.5; $Ca(NO_3)_2$, 0.8; KH_2PO_4, 2.0; K_2HPO_4, 0.2; Difco yeast extract (BD Diagnostic Systems, Sparks, MD, USA), 1.0; trace elements (mg L^{-1}): $FeSO_4 \times 7H_2O$, 14.9; $MnSO_4 \times 4H_2O$, 0.2; $ZnSO_4 \times 7H_2O$, 8.1; $CuSO_4 \times 5H_2O$, 3.9. The fermentation conditions were maintained automatically at a constant level: Temperature 28—0.5 °C; pH 4.5.0—0.1; pO_2 50% (of air saturation); agitation rate of 800 rpm. Pulsed addition of carbon source (by 2–20 g L^{-1}) depending on the carbon source used was performed as the pO_2 value changed by 10%. Cultivation was continued for 6 days.

2.3. Assays

Biomass, concentration of CA, isocitric acid (ICA) and other organic acids were determined as described previously [7].

2.4. Calculations

Earlier, it was found that the mass yield of CA production (Y_{CA}), expressed in g of CA per g of carbon source, and the fermenter productivity, expressed in g $(L{\cdot}h)^{-1}$ were influenced by the medium dilution due to the addition of NaOH solution for maintaining a constant pH value [9,17]. In this regard, the total amount of CA in the culture broth at the end of the fermentation was used to calculate Y_{CA} and fermenter productivity. Formulas for calculation of Y_{CA} value and fermenter productivity were described earlier [9,17].

All the data presented are the mean values of three experiments and two measurements for each experiment; standard deviations were calculated (S.D. < 10%).

3. Results and Discussion

The dynamics of nitrogen consumption and the accumulation of biomass and CA by *Y. lipolytica* NG40/UV5 grown on rapeseed oil are shown in Figure 1a, while the logarithmic growth curve (μ) and the specific rate of biosynthesis of CA (q_p) are shown in Figure 1b. As it can be seen in the latter figure, the growth curve had an exponential phase (phase I) lasting for 12 h, growth retardation phase (phase II) lasting from 12 to 36 h of cultivation, and stationary phase (phase III) lasting from 36 h to the end of cultivation. The retardation of growth coincided with the exhaustion of nitrogen from the medium. The specific growth rate attained a maximum ($\mu_{max} = 0.360\ h^{-1}$) in the exponential growth phase (12 h of cultivation). This value of μmax was more than two times higher than that of the other CA-producing strain *Y. lipolytica* (0.17–0.22 h^{-1}) [4,8]. After 12 h of cultivation, μ gradually decreased to zero after 48 h of cultivation. The excretion of CA did not occur in the exponential growth phase but became active in the growth retardation and stationary phases. Within this cultivation period, the specific rate of CA production (q_p) was between 0.065 and 0.104 g CA/g·h. At the end of cultivation (144 h), the strain produced 140 g L^{-1} CA and 5.3 g L^{-1} ICA (data not shown) with CA:ICA ratio of 26.4:1. The CA production yield Y_{CA} was 1.5 g g^{-1}; the fermenter productivity was calculated to be 1.46 g $(L{\cdot}h)^{-1}$ with account for the dilution factor.

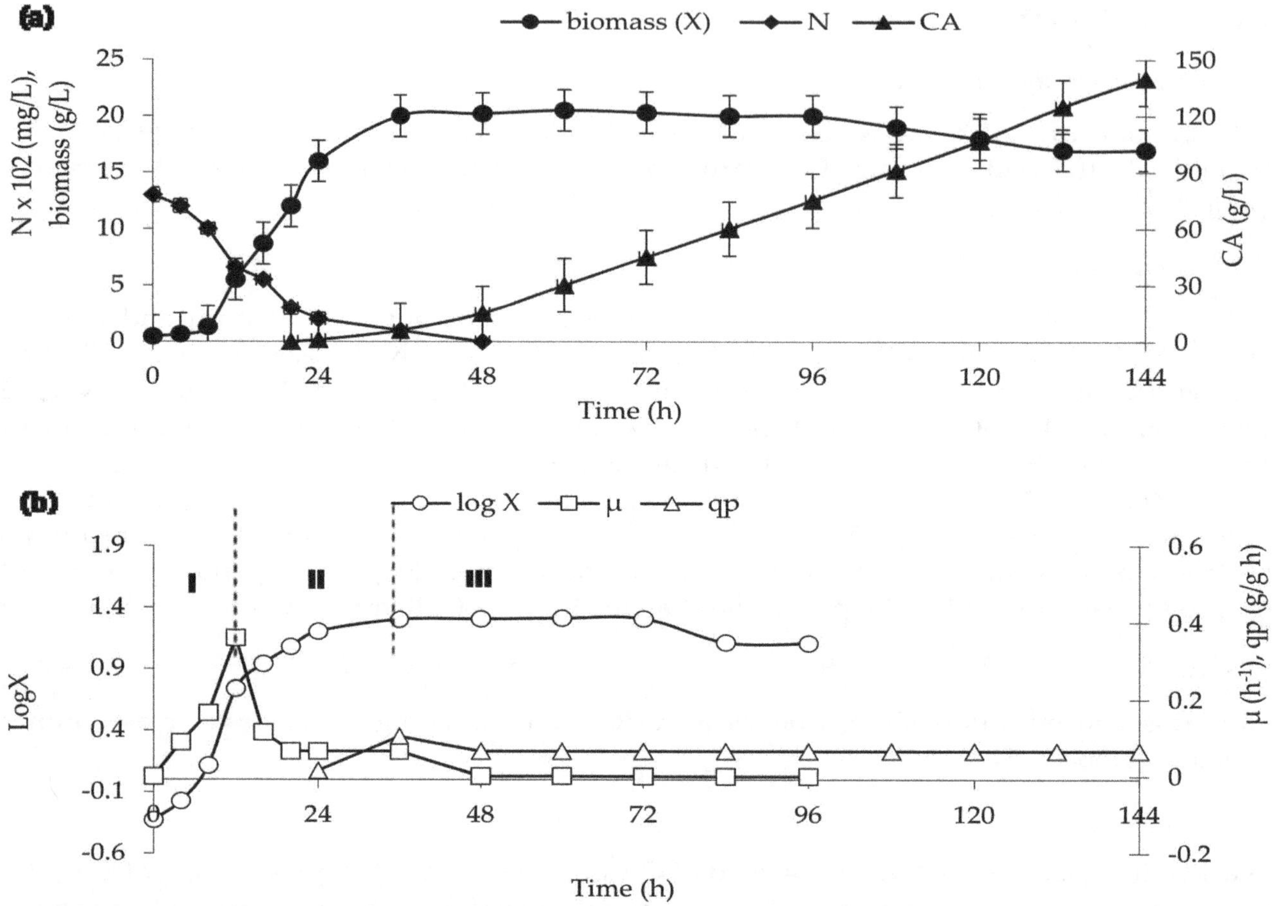

Figure 1. Time courses of nitrogen consumption, biomass accumulation, and citric acid production in *Y. lipolytica* grown on rapeseed oil (**a**) and calculated parameters of the process (**b**): I—the exponential cell growth; II—the cell growth retardation; III—the stationary phase.

The data of the accumulation of biomass and CA by *Y. lipolytica* NG40/UV5 grown on other substrates compared to rapeseed oil are shown in Table 1.

Table 1. Citric acid production by *Y. lipolytica* on various carbon sources.

Substrates	Biomass (g L^{-1})	CA (g L^{-1})	ICA (g L^{-1})	CA:ICA	Productivity (g $(L{\cdot}h)^{-1}$)	Y_{CA} (g g^{-1})
Rapeseed oil	17.0 ± 1.1	140.0 ± 5.0	5.3 ± 0.8	26.2:1	1.46	1.50
Glucose	18.7 ± 1.3	100.8 ± 9.2	4.9 ± 0.9	20.6:1	1.05	0.80
Glucose-containing aspen waste	5.6 ± 0.8	31.2 ± 2.1	7.84 ± 0.9	4:1	0.325	0.50
Glycerol	16.8 ± 1.1	87 ± 6.4	13 ± 1.1	6.7:1	0.906	0.64
Glycerol waste of biodiesel industry	20.0 ± 1.8	100 ± 3.4	15 ± 1.2	7.7:1	1.04	0.90
Ethanol	15.3 ± 1.4	106.7 ± 2.7	15 ± 1.4	7.1:1	1.32	0.87

As seen in Table 1, at the end of cultivation (144 h), *Y. lipolytica* NG40/UV5 produced 100.8 g L^{-1} CA and 4.9 g L^{-1} ICA with CA:ICA ratio of 20.6:1 in the medium containing glucose. The CA production yield Y_{CA} was 0.80 g g^{-1}; the fermenter productivity was calculated to be 1.05 g $(L{\cdot}h)^{-1}$.

As seen in Table 1, *Y. lipolytica* NG40/UV5 only produced 31.2 g L^{-1} CA and 7.84 g L^{-1} ICA with CA:ICA ratio of 4:1 in the medium containing glucose-containing aspen waste. The CA production yield Y_{CA} was 0.50 g g^{-1}; the fermenter productivity was calculated to be 0.325 g $(L{\cdot}h)^{-1}$.

As it can be seen from the data in Table 1, the mutant grows perfectly and synthesizes CA both in a medium with pure glycerol and in a medium with biodiesel-derived glycerol. *Y. lipolytica* NG40/UV5 produced 87 g L^{-1} CA with a ratio of CA to ICA of 6.7:1. The application of waste glycerol for *Y. lipolytica* NG40/UV5 cultivation increased CA production by 15% (up to 100 g L^{-1}) compared to that obtained from pure glycerol; the CA to ICA ratio was 7.7:1. The fermenter productivity was high and reached 0.906 and 1.04 (g $(L{\cdot}h)^{-1}$) in the media with pure- and biodiesel-derived glycerol, respectively. The mass yield (Y_{CA}) reached 0.64 and 0.9 g/g in the media with pure- and biodiesel-derived glycerol, respectively.

As seen in Table 1, *Y. lipolytica* NG40/UV5 produced 106.7 g L^{-1} CA and 15 g L^{-1} ICA with CA:ICA ratio of 7.1:1 in the medium containing ethanol. The CA production yield Y_{CA} was 0.87 g g^{-1}; the fermenter productivity was calculated to be 1.32 g/L·h.

Comparative data on the most efficient processes of CA production by yeasts *Y. lipolytica* from various carbohydrate-containing substrates are given in Table 2. As seen in this table, in the experiments with wild and mutant strains, *Y. lipolytica* produced CA in industrially sufficient amounts. For instance, in the mutant strain *Saccharomycopsis lipolytica* NTG9 grown on rapeseed oil, the CA concentration reached 152.3 g L^{-1} with the yield (Y_{CA}) of 1.5 g g^{-1} [25]. Aurich et al. [26] obtained a CA concentration of 198 g L^{-1} with the yield (Y_{CA}) of 1.16 g g^{-1}, which was achieved after a 300 h fed-batch cultivation of the wild strain *Y. lipolytica* H181. The wild strains *Y. lipolytica* H222 and *Y. lipolytica* W29, grown on glucose, produced 41 g L^{-1} [4] and 49 g L^{-1} of CA, respectively [27]. Recently, we found that the wild strain *Y. lipolytica* VKM Y 2373, cultivated in a medium with glucose under cell growth limitation using nitrogen, phosphorus and sulfur, produced CA at a level of 80–85 g L^{-1} with a yield of 0.70–0.75 g g^{-1} [7]. The overexpression of gene *PYR* encoding pyruvate carboxylase in *Y. lipolytica*, resulted in the production of CA at a level of 95–111.1 g L^{-1} with the yield (Y_{CA}) of 0.75–0.93 g g^{-1} [5,6]. Strain *Y. lipolytica* ACA-DS 50109 cultivated on glucose and olive mill wastewaters produced CA (28.9 g L^{-1}) with the product yield (Y_{CA}) of 0.53 g g^{-1} [28]. Later, the authors of the last article improved the process of CA production up to 52.0 g L^{-1} with the product yield (Y_{CA}) of 0.64 g g^{-1} using strain *Y. lipolytica* ACA-YC 5033, which was also able to remove harmful phenolic compounds from olive mill wastewaters [29]. The glycerol-grown yeast *Y. lipolytica* NRRL YB-423 produced 21.6 g L^{-1} of CA with mass yield of 0.54 g g^{-1} [12]; strain *Y. lipolytica* ACA-DC 50109 synthesized 62.5 g L^{-1} of CA with mass yield of 0.56 g g^{-1} from raw glycerol [20], while the recombinant strain *Y. lipolytica* NCYC3825 was able to produce 58 g L^{-1} of CA [30]. Earlier, we indicated that the other mutant *Y. lipolytica* NG40/UV7 synthesized, CA (122.2 g L^{-1}) with the yield of 0.95 g g^{-1}. The high CA production (up to 140 g L^{-1}) has also been reported for acetate-negative mutant *Y. lipolytica* Wratislavia AWG7, grown on crude glycerol [13] and recombinant strain *Y. lipolytica* H222-S4 (p67ICL1), harboring the invertase encoding *ScSUC2* gene of *Saccharomyces cerevisiae* under inducible XPR2 promoter control and multiple ICL1 copies, cultivated on sucrose [31].

Table 2. Comparative data of the processes of CA production from various substrates using *Y. lipolytica* strains.

Strain	Substrate	Characteristics of Strain	CA ($g\ L^{-1}$)	Y_{CA} ($g\ g^{-1}$)	References
S. lipolytica NTG9	canola oil	mutant/nitrosoguanidine	152.3	1.50	[25]
Y. lipolytica H181	sunflower oil	wild type	198.0	1.16	[26]
Y. lipolytica H222		wild type	41.0	0.55	[4]
Y. lipolytica W291		wild type	49.0	0.85	[27]
Y. lipolytica VKM Y 2373	glucose	wild type	80–85	0.70–0.75	[7]
Y. lipolytica PG86		*PYC* gene expression	95.0	0.75	[5]
Y. lipolytica PR32		*PYC* gene expression	111.1	0.93	[6]
Y. lipolytica ACA-DS 50109	glucose + olive mill	wild type	28.9	0.53	[28]
Y. lipolytica ACA-YC 5033	wastewaters	wild type	52.0	0.64	[29]
Y. lipolytica NRRL YB-423	glycerol	wild type	21.6	0.54	[12]
Y. lipolytica ACA-DC 50109	raw glycerol	wild type	62.5	0.56	[20]
Y. lipolytica NCYC 3825	raw glycerol	multigene expression	58.8	0.17	[30]
Y. lipolytica NG40/UV7	raw glycerol	mutant/nitrosoguanidine/UV	122.2	0.95	[17]
Y. lipolytica Wratislavia AWG7	raw glycerol	mutant/acetate^{-}	139.0	0.70	[13]
Y.lipolytica H222-S4 (p67ICL1)	sucrose	ScSUC2/ICL1	127–140	0.75–0.82	[31]
Y. lipolytica XYL+	xylose	*XYL* gene expression	80.0	0.53	[19]
Y. lipolytica Wratislavia K1	inulin	*INU1* gene expression	105.2	0.53	[21]
Y. lipolytica NG40/UV5	rapeseed oil	mutant/nitrosoguanidine/UV	140.0	1.5	Present study

It should be noted that plant raw materials, such as wood, straw, and agricultural products processing waste, are inexpensive, accessible, renewable, and environmentally friendly substrates for microbiological synthesis of practically valuable compounds. However, the effective conversion of these substrates into easily assimilable carbohydrates (glucose, xylose, and higher glucose-containing polymers) is a difficult task. The traditional technologies of hydrolysis of plant raw materials with the use of strong acids and alkalis are associated with the formation of by-products that inhibit the growth of microorganisms and biosynthesis of target substances. Therefore, recombinant producers, effectively assimilating plant raw materials, were developed. For instance, Ledesmo-Amaro et al. (2016) engineered *Y. lipolytica* able to metabolize xylose to produce CA and lipids. The overexpression of xylose reductase, xylitol dehydrogenase and xylulokinase resulted in a production of 80 g L^{-1} of CA from xylose by mutant strain *Y. lipolytica* [19].

Rakicka et al. (2016) reported the production from inulin by engineered strain *Y. lipolytica* Wratislavia K1. The overexpression of the *INU1* gene from *Kluyveromyces marxianus* coding inulinase resulted in the effective hydrolysis of inulin by mutant and the production of a high amount of CA (105.2 g L^{-1} from 200 g L^{-1} inulin) [21].

4. Conclusions

The results of the experiments indicated that the mutant *Y. lipolytica* NG40/UV5 was able to grow and synthesize CA on media containing all types of the investigated renewable raw materials. However, it should be noted that the accumulation of the by-product of fermentation—ICA—was even very high in four of the six substrates investigated (glucose-containing aspen waste, glycerol, glycerol waste of biodiesel industry and ethanol). The best results were obtained using rapeseed oil (140 g L^{-1} CA; CA:ICA ratio of 26.2:1; mass yield of CA production (Y_{CA}) of 1.5 g g^{-1}, and fermenter productivity of 1.46 g $(L{\cdot}h)^{-1}$). However, economic considerations may lead to the fact that using a cheaper and less pure substrate (glycerol waste of biodiesel industry; 100 g L^{-1} CA; CA:ICA ratio of 7.7:1; mass yield of CA production (Y_{CA}) of 0.9 g g^{-1}, and fermenter productivity of 1.04 g $(L{\cdot}h)^{-1}$) would be preferable on a production scale.

Author Contributions: I.G.M. conceived and designed the experiments, wrote the paper; S.V.K. and J.N.L. performed the experiments.

References

1. Cavallo, E.; Charreau, H.; Cerrutti, P.; Foresti, M.L. *Yarrowia lipolytica*: A model yeast for citric acid production. *FEMS Yeast Res.* **2017**, *17*. [CrossRef] [PubMed]
2. Anastassiadis, S.; Morgunov, I.G.; Kamzolova, S.V.; Finogenova, T.V. Citric acid production patent review. *Recent Pat. Biotechnol.* **2008**, *2*, 107–123. [CrossRef] [PubMed]
3. Finogenova, T.V.; Morgunov, I.G.; Kamzolova, S.V.; Chernyavskaya, O.G. Organic acid production by the yeast *Yarrowia lipolytica*: A review of prospects. *Appl. Biochem. Microbiol.* **2005**, *41*, 418–425. [CrossRef]
4. Moeller, L.; Strehlitz, B.; Aurich, A.; Zehnsdorf, A.; Bley, T. Optimization of citric acid production from glucose by *Yarrowia lipolytica*. *Eng. Life Sci.* **2007**, *7*, 504–511. [CrossRef]
5. Tan, M.J.; Chen, X.; Wang, Y.K.; Liu, G.L.; Chi, Z.M. Enhanced citric acid production by a yeast *Yarrowia lipolytica* over-expressing a pyruvate carboxylase gene. *Bioprocess Biosyst. Eng.* **2016**, *39*, 1289–1296. [CrossRef] [PubMed]
6. Fu, G.Y.; Lu, Y.; Chi, Z.; Liu, G.L.; Zhao, S.F.; Jiang, H.; Chi, Z.M. Cloning and characterization of a pyruvate carboxylase from *Penicillium rubens* and overexpression of the gene in the yeast *Yarrowia lipolytica* for enhanced citric acid production. *Mar. Biotechnol.* **2016**, *18*, 1–14. [CrossRef] [PubMed]
7. Kamzolova, S.V.; Morgunov, I.G. Metabolic peculiarities of the citric acid overproduction from glucose in yeasts *Yarrowia lipolytica*. *Bioresour. Technol.* **2017**, *243*, 433–440. [CrossRef] [PubMed]
8. Kamzolova, S.V.; Morgunov, I.G.; Aurich, A.; Perevoznikova, O.A.; Shishkanova, N.V.; Stottmeister, U.; Finogenova, T.V. Lipase secretion and citric acid production in *Yarrowia lipolytica* yeast grown on animal and vegetable fat. *Food Technol. Biotechnol.* **2005**, *43*, 113–122.
9. Kamzolova, S.V.; Lunina, J.N.; Morgunov, I.G. Biochemistry of citric acid production from rapeseed oil by *Yarrowia lipolytica* yeast. *J. Am. Oil Chem. Soc.* **2011**, *88*, 1965–1976. [CrossRef]
10. Aurich, A.; Specht, R.; Müller, R.A.; Stottmeister, U.; Yovkova, V.; Otto, C.; Holz, M.; Barth, G.; Heretsch, P.; Thomas, F.A.; et al. Microbiologically Produced Carboxylic Acids Used as Building Blocks in Organic Synthesis. In *Reprogramming Microbial Metabolic Pathways*; Wang, X., Chen, J., Quinn, P., Eds.; Springer: Dordrecht, The Netherlands, 2012; pp. 391–424. Available online: https://link.springer.com/chapter/10.1007%2F978-94-007-5055-5_19 (accessed on 28 September 2012).
11. Rymowicz, W.; Rywińska, A.; Żarowska, B.; Juszczyk, P. Citric acid production from raw glycerol by acetate mutants of *Yarrowia lipolytica*. *Chem. Pap.* **2006**, *60*, 391–395. [CrossRef]
12. Levinson, W.E.; Kurtzman, C.P.; Kuo, T.M. Characterization of *Yarrowia lipolytica* and related species for citric acid production from glycerol. *Enzym. Microb. Technol.* **2007**, *41*, 292–295. [CrossRef]
13. Rywińska, A.; Rymowicz, W.; Źarowska, B.; Wojtatowicz, M. Biosynthesis of citric acid from glycerol by acetate mutants of *Yarrowia lipolytica* in fed-batch fermentation. *Food Technol. Biotechnol.* **2009**, *47*, 1–6.
14. Makri, A.; Fakas, S.; Aggelis, G. Metabolic activities of biotechnological interest in *Yarrowia lipolytica* grown on glycerol in repeated batch cultures. *Bioresour. Technol.* **2010**, *101*, 2351–2358. [CrossRef] [PubMed]
15. Morgunov, I.G.; Kamzolova, S.V.; Lunina, J.N. The citric acid production from raw glycerol by *Yarrowia lipolytica* yeast and its regulation. *Appl. Microbiol. Biotechnol.* **2013**, *97*, 7387–7397. [CrossRef] [PubMed]
16. Rywinska, A.; Juszczyk, P.; Wojtatowicz, M.; Robak, M.; Lazar, Z.; Tomaszewska, L.; Rymowicz, W. Glycerol as a promising substrate for *Yarrowia lipolytica* biotechnological applications. *Biomass Bioenergy* **2013**, *48*, 148–166. [CrossRef]
17. Morgunov, I.G.; Kamzolova, S.V. Physiologo-biochemical characteristics of citrate-producing yeast *Yarrowia lipolytica* grown on glycerol-containing waste of biodiesel industry. *Appl. Microbiol. Biotechnol.* **2015**, *99*, 6443–6450. [CrossRef] [PubMed]
18. Wojtatowicz, M.; Rymowicz, W.; Kautola, H. Comparison of different strains of the yeast *Yarrowia lipolytica* for citric acid production from glucose hydrol. *Appl. Biochem. Biotechnol.* **1991**, *31*, 165–174. [CrossRef] [PubMed]
19. Ledesma-Amaro, R.; Lazar, Z.; Rakicka, M.; Guo, Z.; Fouchard, F.; Coq, A.C.; Nicaud, J.M. Metabolic engineering of *Yarrowia lipolytica* to produce chemicals and fuels from xylose. *Metab. Eng.* **2016**, *38*, 115–124. [CrossRef] [PubMed]

20. Papanikolaou, S.; Fakas, S.; Fick, M.; Chevalot, I.; Galiotou-Panayotou, M.; Komaitis, M.; Marc, I.; Aggelis, G. Biotechnological valorisation of raw glycerol discharged after biodiesel (fatty acid methyl-esters) manufacturing process: Production of 1,3-propanediol, citric acid and single oil. *Biomass Bioengergy* **2008**, *32*, 60–71. [CrossRef]
21. Rakicka, M.; Lazar, Z.; Rywinska, A.; Rymowicz, W. Efficient utilization of inulin and glycerol as fermentation substrates in erythritol and citric acid production using expressing inulinase. *Chem. Pap.* **2016**, *70*, 1452–1459. [CrossRef]
22. Stephanopoulos, G. Challenges in engineering microbes for biofuels production. *Science* **2007**, *315*, 801–804. [CrossRef] [PubMed]
23. Weusthuis, R.A.; Aarts, J.M.M.J.G.; Sanders, J.P.M. From biofuel to bioproduct: Is bioethanol a suitable fermentation feedstock for synthesis of bulk chemicals? *Biofuels Bioprod. Biorefin.* **2011**, *5*, 486–494. [CrossRef]
24. Finogenova, T.V.; Puntus, I.F.; Kamzolova, S.V.; Lunina, I.N.; Monastyrskaia, S.E.; Morgunov, I.G.; Boronin, A.M. Obtaining of the mutant *Yarrowia lipolytica* strains producing citric acid from glucose. *Prikl. Biokhimiia Mikrobiol.* **2008**, *44*, 219–224.
25. Good, D.W.; Droniuk, R.; Lawford, R.G.; Fein, J.E. Isolation and characterization of a *Saccharomycopsis lipolytica* mutant showing increased production of citric acid from canola oil. *Can. J. Microbiol.* **1985**, *31*, 436–440. [CrossRef]
26. Aurich, A.; Förster, A.; Mauesberger, S.; Barth, G.; Stottmeister, U. Citric acid production from renewable resources by *Yarrowia lipolytica*. *Biotechnol. Adv.* **2003**, *21*, 454–455.
27. Papanikolaou, S.; Chatzifragkou, A.; Fakas, S.; Galiotou-Panayotou, M.; Komaitis, M.; Nicaud, J.M.; Aggelis, G. Biosynthesis of lipids and organic acids by *Yarrowia lipolytica* strains cultivated on glucose. *Eur. J. Lipid Sci. Technol.* **2009**, *111*, 1221–1232. [CrossRef]
28. Papanikolaou, S.; Galiotou-Panayotou, M.; Fakas, S.; Komaitis, M.; Aggelis, G. Citric acid production by *Yarrowia lipolytica* cultivated on olive-mill wastewater-based media. *Bioresour. Technol.* **2008**, *99*, 2419–2428. [CrossRef] [PubMed]
29. Sarris, D.; Stoforos, N.G.; Mallouchos, A.; Kookos, I.K.; Koutinas, A.A.; Aggelis, G.; Papanikolaou, S. Production of added-value metabolites by *Yarrowia lipolytica* growing in olive mill wastewater-based media under aseptic and non-aseptic conditions. *Eng. Life Sci.* **2017**, *17*, 695–709. [CrossRef]
30. Celińska, E.; Grajek, W. A novel multigene expression construct for modification of glycerol metabolism in *Yarrowia lipolytica*. *Microb. Cell Fact.* **2013**, *12*, 102. [CrossRef] [PubMed]
31. Förster, A.; Aurich, A.; Mauersberger, S.; Barth, G. Citric acid production from sucrose using a recombinant strain of the yeast *Yarrowia lipolytica*. *Appl. Microbiol. Biotechnol.* **2007**, *75*, 1409–1417. [CrossRef] [PubMed]

9

Gravity-Driven Adaptive Evolution of an Industrial Brewer's Yeast Strain towards a Snowflake Phenotype in a 3D-Printed Mini Tower Fermentor

Andreas Conjaerts and Ronnie G. Willaert *

Alliance Research Group VUB-UGent NanoMicrobiology (NAMI), IJRG VUB-EPFL, BioNanotechnology & NanoMedicine (NANO), Structural Biology Brussels (SBB), Vrije Universiteit Brussel, Brussels 1050, Belgium; Andreas.Conjaerts@vub.ac.be

* Correspondence: Ronnie.Willaert@vub.ac.be

Academic Editors: Maurizio Ciani and Badal C. Saha

Abstract: We designed a mini tower fermentor that is suitable to perform adaptive laboratory evolution (ALE) with gravity imposed as selective pressure, and suitable to evolve a weak flocculating industrial brewers' strain towards a strain with a more extended aggregation phenotype. This phenotype is of particular interest in the brewing industry, since it simplifies yeast removal at the end of the fermentation, and many industrial strains are still not sufficiently flocculent. The flow of particles (yeast cells and flocs) was simulated, and the theoretical retainment advantage of aggregating cells over single cells in the tower fermentor was demonstrated. A desktop stereolithography (SLA) printer was used to construct the mini reactor from transparent methacrylic acid esters resin. The printed structures were biocompatible for yeast growth, and could be sterilised by autoclaving. The flexibility of 3D printing allowed the design to be optimized quickly. During the ALE experiment, yeast flocs were observed within two weeks after the start of the continuous cultivation. The flocs showed a "snowflake" morphology, and were not the result of flocculin interactions, but probably the result of (a) mutation(s) in gene(s) that are involved in the mother/daughter separation process.

Keywords: 3D printing; mini tower fermentor; *Saccharomyces cerevisiae*; industrial brewer's yeast; adaptive laboratory evolution (ALE); snowflake phenotype

---

1. Introduction

The use of evolutionary methods is a more "natural" approach to enhance the attributes of microorganisms, in contrast to genetic modification which has so far precluded its commercial use due to the low consumer tolerance for genetically modified organisms [1,2]. Evolutionary methods can be applied even before the genetic elements and their global interactions required for optimal performance by an organism are understood. Recently, artificial laboratory selection of microbial cells has been introduced to generate potentially robust and optimised microbial production systems [3,4]. Adaptive laboratory evolution (ALE) strategies allow for the metabolic engineering of microorganisms by combining genetic variation with the selection of beneficial mutations in an unbiased fashion [5]. A number of investigations have demonstrated the feasibility of directing evolution in natural *Saccharomyces pastorianus* hybrid stains in order to create variant strains with improved functional properties [6]. Such investigations have focused on adaptation to very high-gravity brewing conditions [7–9], associated stresses (such as osmotic stress and ethanol toxicity) [10,11], or the modification of the production of flavour compounds [12]. ALE has also been utilised to enhance the fermentation rate of *S. cerevisiae* with decreased formation of acetate and greater production of aroma compounds [13,14].

Continuous culture provides many benefits over classical batch-style cultivation to perform experimental evolution [15]. Steady-state cultures allow for precise control of growth rate and environment, and cultures can be propagated for weeks or months in these controlled environments, which is important for the study of experimental evolution. Continuous mini bioreactors have been successfully used as multiplexed chemostat arrays for adaptive evolution experiments with yeast cells [15,16]. The use of mini bioreactors has several advantages, such as reduced costs for media and labour, and the ability to perform a large number of fermentations in parallel [17–19].

Recently, three-dimensional (3D) printing has been used for medical applications [20], such as (1) medical models; (2) medical aids, orthoses, splints, and prostheses; (3) tools, instruments, and parts for medical devices; (4) inert implants; and (5) biomanufacturing [21–24]. Bioprocess applications include tissue engineering scaffolds and corresponding bioreactors [25]. 3D printing has recently been used to fabricate fullerene-type biocarriers for biofilm growth that can be used in bioreactors for wastewater treatment [26]. 3D printing technology shows great potential for the easy development of mini-scale bioreactors. Here, we report the use of 3D printing for the construction of a continuous mini tower fermentor.

The concept of 3D printing—also referred to as additive manufacturing, rapid prototyping, or solid-freeform fabrication (SFF)—was developed by Charles Hull in the early 1980's [27]. 3D printing is used for rapid prototyping of 3D models originally generated by a computer aided design (CAD) program. The 3D model is sliced into 2D horizontal cross sections, which are printed in consecutive layers. There are several well-established methods of 3D-printing, such as stereolithography (SLA), fused deposition modelling, selective laser sintering, multi jet fusion, and selective laser melting. SLA became the first commercialised 3D-printing technique (invented by Charles Hull in 1983), and remains one of the most powerful and versatile of all SFF techniques [28]. SLA works by exposing a layer of photosensitive liquid resin to a UV-laser beam so that the resin hardens and becomes solid. Once the laser has swept a layer of resin in the desired pattern and it begins to harden, the model-building platform in the liquid tank of the printer steps down the thickness of a single layer, and the laser begins to form the next layer. SLA is capable of printing at high resolutions (up to 25 μm) and relatively high production rates (1.5 cm/h).

In this contribution, we designed a continuous mini bioreactor that is suitable to perform ALE of a weak flocculating industrial brewers' strain towards a strain with a more extended aggregation (flocculation) phenotype. Yeast strains that aggregate or flocculate are of particular interest to the brewer, since it simplifies yeast removal at the end of the primary fermentation [29]. Since many industrial brewing strains still show no adequate flocculation, we applied ALE as a non-genetic engineering method. The design of the fermentor was based on a continuous mini tower fermentor; i.e., the continuous A.P.V. tower fermentor that was used in the 1960s to produce beer on an industrial scale in a British brewery [30,31]. In this type of fermentor, gravity is the selective pressure to enhance the aggregation phenotype during evolution. We demonstrate that SLA 3D printing can be used to construct the mini tower fermentor, and that the design is suitable to obtain a yeast cell aggregation ("snowflake") phenotype by performing ALE.

2. Materials and Methods

2.1. Yeast Strains and Media

An industrial *S. cerevisiae* brewer's strain was used; i.e., CMBSVM22 from the yeast collection of prof. em. Freddy Delvaux (Centre for Malting and Brewing Science (CMBS), Katholieke Universiteit Leuven, Leuven Belgium; Biercentrum Delvaux, Beekstraat 20, 3040 Neerijse, Belgium) [32]. The haploid *S. cerevisiae* BY4742 strain [33] was used for the biocompatibility assay, and the flocculent BY4742 [*FLO1*] strain [34] for the preliminary experiments during experimental design optimisation. These strains were precultured in YPD (Yeast extract-Peptone-Dextrose) medium (1% m/v yeast extract, 2% m/v meat peptone, 4% m/v D-glucose) overnight at 30 °C. Growth medium (100 g/L D-glucose,

4 g/L $(NH_4)SO_4$, 1.5 g/L KH_2PO_4, 1 g/L $MgSO_4.7H_2O$, and 5 g/L yeast extract) was used for the continuous adaptive evolution experiment in the mini tower bioreactor.

WL (Wallerstein Laboratory) nutrient agar was used to differentiate between colonies by morphology variations [35,36]. The medium contained 7.5 g WL nutrient agar per 100 mL deionized water. The cell suspensions were diluted to 10×10^6 cells/mL with sterile deionized water. Four droplets of 1 µL of the diluted suspension were spotted on each plate.

2.2. Determination of Cell Concentration and Growth Rate

The cell concentration was determined by cell dry mass (CDM) and optical density (OD) measurements. For CDM measurements, a sampled cell suspension of 3 mL was filtered with a 0.45-µm filter (Type HVLP, Millipore®, Darmstadt, Germany), and the filter was washed with deionised water. The filter was dried at 80 °C until the mass remained stable. The difference in mass between the filter before and after filtration gave the cell dry mass (CDM) concentration (g·CDM/L). For the OD measurements, absorption of the cell suspension was determined at a wavelength of 600 nm. Biomass measurements were performed in triplicate, and the standard deviation was calculated to represent the error bars. First-order growth kinetics was used to fit the exponential part of the growth curve and to determine the maximum specific growth rate.

2.3. Flocculation Assay

The assay was based on the method of D'Hautcourt and Smart [37], with slight modifications. Cells were grown for 24 h in 20 mL YPD. The cell suspension was centrifuged (4000 rpm, 3 min), and the medium was discarded. Subsequently, the cells were resuspended in an equal volume of EDTA-buffer to chelate the Ca^{2+}-ions and deflocculate the yeast flocs. The $OD_{600\ nm}$-value of the suspension was determined, and such a volume was transferred to an Eppendorf tube to ensure an $OD_{600\ nm}$-value of 10 in 1 mL. EDTA (ethylenediaminetetraacetic acid)-buffer (50 mM EDTA, pH 7) was added until a final volume of 1 mL was acquired. A sample of 50 µL was taken 0.5 mL below the meniscus, and the sample was diluted 20 times in a 1.5 mL cuvette with EDTA-buffer. The tubes were centrifuged (4000 rpm, 3 min), and the supernatant was discarded. The cells were resuspended in 1 mL flocculation buffer A (3 mM $CaSO_4$). The last step was repeated, but the cells were resuspended in flocculation buffer B (3 mM $CaSO_4$, 83 mM CH_3COONa, 4% *v*/*v* ethanol, pH 4.5). The tubes were shaken horizontally at 100 rpm for 10 min. Prior to taking 50 µL samples 0.5 mL below the meniscus, 3 min of sedimentation in a vertical position took place. The sample was diluted 20 times with EDTA-buffer in a 1.5 mL cuvette. The absorbance of both suspensions in the cuvettes was determined, and the related flocculation percentage was calculated:

$$\text{Flocculation percentage (\%)} = \frac{OD_{EDTA} - OD_{Flocculation\ buffer}}{OD_{EDTA}} \times 100$$

with OD_{EDTA} as the $OD_{600\ nm}$-value of cells in EDTA-buffer, and $OD_{Flocculation\ buffer}$ the $OD_{600\ nm}$-value of cells in flocculation buffer B.

2.4. Biocompatibility Assay

A commercial resin (Clear Resin GPCL02, FormLabs, Somerville, MA, USA) composed of a mixture of methacrylic acid esters and photoinitiators was used to print the mini reactor. To ensure that this resin had no negative effect on the growth of yeast cells, a biocompatibility assay was performed. Therefore, four sterile Erlenmeyer flasks were filled with 50 mL YPD. Prior to adding a piece of resin to two of the four flasks, the pieces were rinsed with soap water, ethanol, and sterile water. Afterwards, all flasks were inoculated with 4 mL of a preculture of the *S. cerevisiae* BY4742 strain in such a way that a final $OD_{600\ nm}$-value of 0.2 was reached. The flasks were incubated at 30 °C and shaken at 120 rpm. Every two hours, a 1 mL sample was taken from each flask, and the cell concentration was determined.

2.5. Mini Tower Fermentor Design and Construction

The mini tower reactor was designed in COMSOL Multiphysics 4.4 (Stockholm, Sweden). The bioreactor had a high-aspect-ratio (beer depth divided by the inner diameter) of around 8:1, which is based on the APV tower fermentor that was used to produce lager and ale beer in the 1960s [30,31,38,39]. The inlet tube had an inner radius of 1 mm and a length of 18 mm; the outlet tube had an inner radius of 2 mm and a length of 25 mm; both contained barbs to facilitate the attachment of the silicone tubing. The bottom of the vessel was conical with a bottom inner radius of 0.7 mm, a top radius of 5.5 mm, and a wall thickness of 1 mm (Figure 1F). The mid cylindrical part was 88 mm high and had an inner radius of 5.5 mm. Another cone connected the cylinder to the reactor head. This cone had a bottom inner radius of 5.5 mm, a top inner radius of 11 mm, and a height of 11 mm. A conical opening was incorporated in the reactor head with a bottom radius of 7.5 mm, a top inner radius of 9 mm, and a height of 20 mm. The dimensions of this opening fitted a silicone stopper that was pierced with a hypodermic needle (sharp 18G, 150 mm, Samco, Surrey, UK), allowing the reactor to be aerated at the bottom cone.

The benchtop SLA 3D printer Form 1+ (Formlabs, Somerville, MA, USA) was used to construct the bioreactor. The COMSOL Multiphysics design was exported as an "stl" file and imported in the printing software PreForm 2 (Formlabs, Somerville, MA, USA). A layer thickness of 0.05 mm was selected for the printing job. Supports settings were: a density of 1, point size of 0.6 mm, and selection of no internal supports. The reactor was oriented with the top of the reactor towards the base, under an angle of 20° around the *z*-axis. The printed structure was shaken gently in two consecutive isopropanol baths, and was sterilised by autoclaving (121 °C, 20 min).

In preliminary experiments, the design was optimised by performing short-term cultivations with a strong flocculating yeast strain (BY4742 [*FLO1*]). The design adaptations are illustrated in Figure 1A–F). The medium was fed through the bottom inlet, and was disposed through the middle outlet. To prevent flocs from sedimenting through the inlet tube, the inner radius of the inlet was decreased to obtain a higher medium inlet velocity (Figure 1B–F versus A). In the initial designs (Figure 1A–D), a gas outlet was present at the top of the reactor. To enhance the medium flow through the reactor, this outlet was removed in the final design, and gases were removed via the liquid outlet. The inlet and outlet were modified by printing of barbs at the outsides, which allowed the silicone tubing to be attached strongly using plastic straps (Figure 1G). The bottom of the vessel was conical, which facilitated sedimentation and fluidisation of the yeast flocs. The body of the reactor was a long cylinder. The wider cylindrical reactor head was connected to the body by another cone to avoid floc retention within the reactor head. Therefore, the length and slope of this cone was increased from the second design on. A conical opening was incorporated in the reactor head and fitted with a silicone stopper. This allowed the reactor to be gently aerated and fluidised from the bottom by piercing the silicone stopper with a long needle.

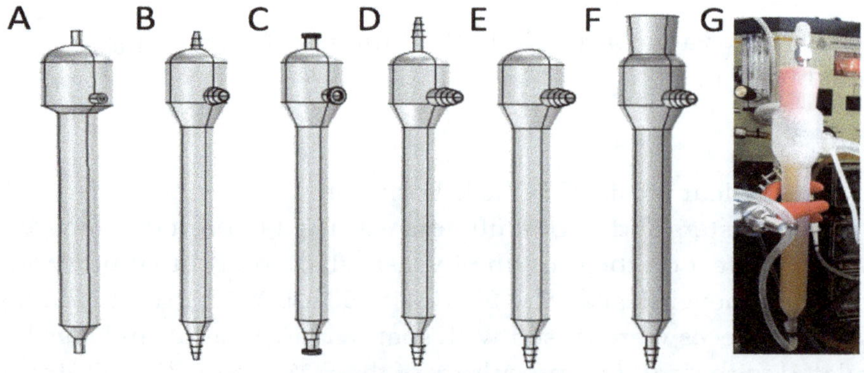

Figure 1. The chronological stages in the development of the 3D-printed mini tower fermentor (**A–F**); (**G**) the printed reactor **F** setup during the evolution experiment.

2.6. *Mini Tower Fermentor Mathematical Modelling*

To evaluate the effect of gravity as selective pressure within the mini tower reactor, fluid flow and particle tracing simulations were performed in COMSOL Multiphysics 4.4 by selecting the Fluid Flow (Single-Phase Flow, Laminar Flow) and the Particle Tracing for Fluid Flow physics. Water at 20 °C was selected as the working liquid (density 999.62 kg/m^3, dynamic viscosity 0.001 Pa·s). Laminar fluid flow through the reactor was simulated with laminar inflow as boundary condition at the inlet. An inlet flow rate of 1.47 mL/h (with an entrance length of 5 cm) was employed, which was in agreement with the initial flow rate of the actual experiment (see Section 2.7). Pressure (atmospheric pressure) was selected as the boundary condition at the outlet, and backflow was suppressed. Physics-controlled mesh was selected as mesh settings with normal element size. Particle Tracing for Fluid flow was used to calculate the motion of the particles in a background laminar fluid flow. Three thousand particles were released at the inlet, with a random diameter between 3 µm and 70 µm. This range includes the size of planktonic cells as well as small yeast flocs. The density of the particles was set at 1117 kg/m^3, which is the average of values found in the literature [40–43]. The particles were released at the inlet (bottom) of the reactor. The forces acting on the particles are the drag force of the fluid and the gravity force.

2.7. *Experimental Set-Up*

A fermentation vessel (5 L) (BioFlo III, New Brunswick Scientific, Edison, NJ, USA), which was aerated, agitated at 137 rpm, and heated to 32 °C, was used as medium reservoir, and connected to the mini tower fermentor inlet with silicone tubing (Figure 1G). The peristaltic pump of the BioFlo III fermentor was used as feeding pump. This pump was initially set at a flow rate of 1.47 mL/h, which corresponds to a dilution rate of 0.11 h^{-1}. A silicone stopper was inserted in the mini reactor head and was pierced by a needle to reach the bottom of the reactor. The needle was connected to an aquarium pump to aerate the reactor. The outlet of the reactor was connected to a four-port manifold. Each outlet of the manifold was connected to a needle head. This needle was pierced through a silicone stopper, which was put in a 15- or 50-mL Falcon tube. The manifold enabled fast and simple exchanging of the outlet reservoir. The four Falcon tubes were permanently kept on ice. Inoculation was performed by piercing a syringe through the silicone stopper and ejecting the inoculum directly into the mini reactor.

3. Results

3.1. *Simulation of Single Cell and Cell Aggregate Behaviour in the Mini Tower Fermentor*

A tower fermentor is suitable to exploit gravity as selective pressure in evolution experiments, since aggregates will be retained preferentially in the reactor compared to single cells. The selective pressure within the tower fermentor was evaluated by particle tracing simulations in COMSOL Multiphysics. Only particles with a diameter smaller than approximately 11 µm were able to reach the top of the reactor (Figure 2). This corresponds to single cells or possibly a complex of two or three cells. According to the calculations, it is very unlikely for larger particles (such as flocs) to reach the outlet of the reactor. Over a longer period of time, this would result in an amplification of flocculating cells within the fermentor. Cells are also retained at the locations of the conical structures. These simulations do not take into account the upward air flow or CO_2 release by the cells during cultivation, which will increase the flow through the reactor. As experimentally demonstrated (see further), these latter conditions also result in a higher probability of cell aggregate retention in the reactor.

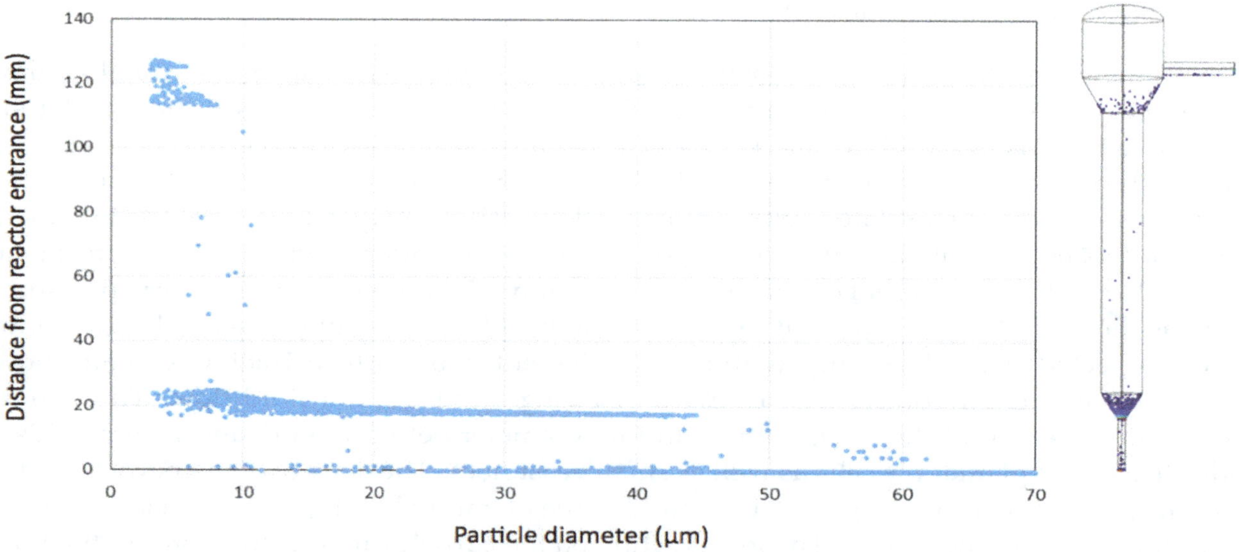

Figure 2. Particle tracing in a 3D-printed bioreactor using COMSOL Multiphysics. The distance travelled from the reactor entrance is plotted against the cell particle diameter.

3.2. Biocompatibility Assay

To ensure that the used resin for the construction of the reactor did not have any negative effects on the growth of the obtained yeast strains, a biocompatibility assay was performed. The *S. cerevisiae* BY4742 strain was grown in the presence and absence of a printed structure. The printed resin material had no effect on the growth of the yeast cells (see Appendix A: Figure A1). The maximum specific growth rate in the presence and absence of the resin material were comparable (0.21 h^{-1} and 0.22 h^{-1}, respectively), indicating the yeast growth biocompatibility of the polymeric material.

3.3. Adaptive Evolution during Continuous Operation

The mini tower fermentor was used for an evolution experiment with the intention of obtaining a more flocculent phenotype of an industrial brewer's strain. The strain was characterised by a very low flocculation percentage (4% ± 6%). The conditions inside the tower fermentor were monitored by sampling at the exit, and by measuring the pH and cell density (Figure 3). The cultivation was initiated with a dilution rate of approximately 0.1 h^{-1}. The cell concentration visibly increased inside the reactor over time. The dilution rate was increased in steps to avoid wash-out and selectively retain larger cell aggregates. The dilution rate was doubled after seven days, which resulted in an increased biomass concentration. The dilution rate was further increased up to 0.4 h^{-1}. At this rate, a dense yeast slurry containing cell aggregates was present in the reactor. The experiment was ended after 14 days of continuous culture when a dense yeast slurry could be visually observed in the reactor.

The yeast population was subsequently examined by optical microscopy (Figure 4). The presence of multicellular yeast flocs was observed. The morphology of the flocs can be described as "snowflake" aggregates. Yeast cell clusters with a comparable morphology have been recently described [44–46]. The evolved cells were plated out on WL nutrient agar, which can be used to detect morphological differences between strains [36]. The population at the start, and after seven and fifteen days were compared (Figure 5). The CMBSVM22 strain possesses the *FLO11* gene required for invasive growth [32,47–49], which could clearly be seen on the bottom of the plates. The morphology of the colonies differed after 14 days, the same time when the snowflake flocs were observed. Most likely, a mixture of multicellular clusters and single cells was spotted, causing the presence of radial differences in colony colour. A flocculation assay was attempted, but the aggregates could not be deflocculated by adding deflocculation buffer. Deflocculation was also not successful after sonication and heating up to 90 °C during 15 min in the presence of 8 M urea.

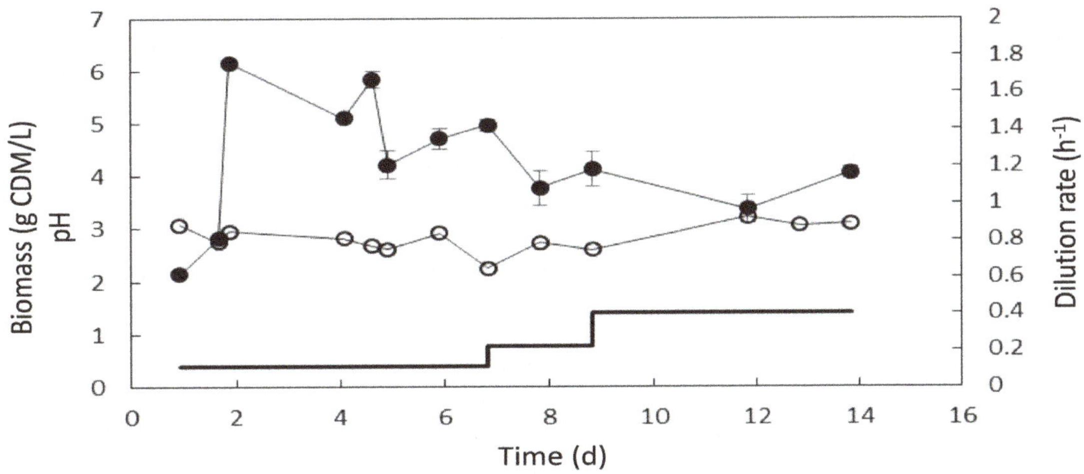

Figure 3. Cultivation of CMBSVM22 in the 3D-printed continuous mini bioreactor. The evolution of the biomass concentration (g·cell dry mass (CDM)/L) (●) of the effluent stream, the pH (○), and the dilution rate (—) during the evolution experiment.

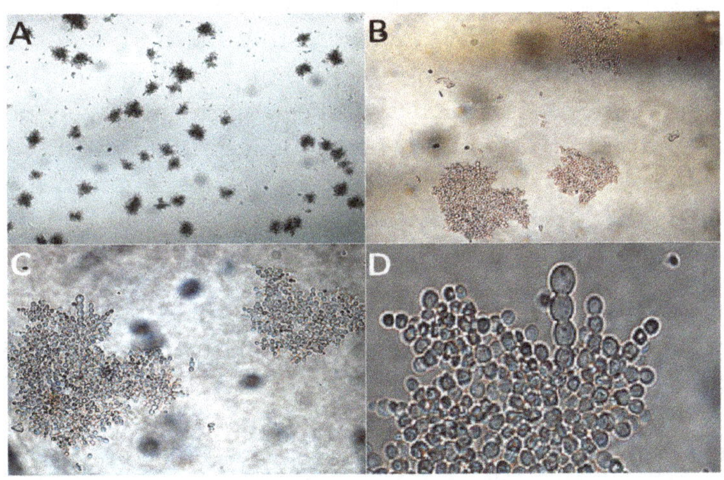

Figure 4. Microscopic observations of snowflake flocs at the end of the fermentation. Objectives: (**A**) 10×; (**B**) 20×; (**C**) zoom-in of 40×; (**D**) 10×.

Figure 5. Colonies of CMBSVM22 grown on WL nutrient agar plates. Top (upper panels) and bottom (lower panels) view of two colonies at (**A**) the start of the cultivation; (**B**) after 7 days of cultivation; and (**C**) after 14 days of cultivation.

4. Discussion

A 3D-printed reactor was created, based on an industrial tower fermentor. The reactor is characterized by a high aspect-ratio, which creates a gravity-based selective pressure: single cells have a higher probability to be washed out than cell aggregates, as the denser aggregates tend to sediment faster. Consequently, this reactor type ensures excellent conditions for strain evolution towards a more aggregative phenotype.

We used a commercial physics modelling software package that includes 3D modelling tools. The created 3D model can be exported as an "stl" file that was imported in the print program. The flow of particles such as yeast cells and flocs in the 3D model of the tower fermentor was calculated with the modelling software. In silico simulations confirmed the gravity-based selective pressure in the reactor, and the theoretical retainment advantage of aggregating cells in the reactor was demonstrated.

The employment of 3D-printing for the development of a bioprocess has several advantages. As demonstrated in this contribution, the flexibility of adapting the design allows the design to be optimized quickly. Depending on the selected print resolution, the mini reactor can be printed in a few hours (4–7 h). The used commercial resin is biocompatible, and the printed structure can be autoclaved. We selected transparent resin, which allowed visual observation of the content.

A long-term continuous cultivation experiment was performed to improve the flocculation characteristics of a brewer's ale strain. Cell flocculation is of particular interest to the brewing industry, since it simplifies yeast removal at the end of the fermentation [29]. Still, many industrial brewing strains show no adequate flocculation. ALE is an interesting method to adapt the strain, since genetic manipulation—which is not allowed in the manufacture of products for human consumption—is not involved. At the end of the evolution experiment, yeast cell aggregates were observed with "snowflake" morphology. These multicellular structures did not disaggregate in EDTA buffer nor urea buffer, which indicates that the Flo proteins are not involved in the aggregation mechanism [50,51]. A high temperature of 90 °C could also not destroy the aggregates, which confirms that the aggregation mechanism is not flocculin-based, since these flocs can be disaggregated above 60 °C [50].

A similar morphology of yeast cell clusters was observed when mother/daughter separation was impaired [46]. This phenotype was recently discovered by performing a gravity-based repeated batch evolution experiment with a diploid *S. cerevisiae* Y55 strain where sedimented cells were collected and cultivated further in a next batch culture [44]. It was demonstrated that these multicellular clusters are uniclonal and appeared as "snowflakes". This phenotype was also obtained in an ALE experiment with a haploid *S. cerevisiae* CEN.PK113-7D strain using long-term cultivation in sequential batch reactors [45]. It was shown that a frameshift mutation in *ACE2*—which encodes a transcriptional regulator involved in cell cycle control and mother–daughter cell separation (MDS)—caused the snowflake phenotype. The disruption of the transcription factor Ace2p prevented the splitting of mother and daughter cell [52], and affected the expression of its targets. One of the targets of Ace2p is Cts1p, which is a chitinase that is required for the degradation of the cell septum. Gene expression of *CTS1* was approximately 90% reduced in the evolved strain [45]. Six other genes (*DSE4, DSE2, SUN4, DSE1, SCW11*, and *AMN1*) that are regulated by *ACE2* were significantly downregulated in the snowflake yeast [46]. These seven downregulated genes are involved in daughter cell separation, many acting directly to degrade the bud neck septum as MDS enzymes [52–55]. It was shown that *ACE2* knockouts formed cellular clusters [45,53,56,57]. The knockout of *CBK1* (a serine/threonine protein kinase of the RAM signalling network that regulates localisation and activity of Ace2p and the Ssd1p translation repressor) also results in cell clusters [56,58,59]. These results indicate that a lack of the degradation of the septum between the mother and daughter cells could explain the observed multicellular clusters. Whole genome sequencing of the evolved strain that was obtained in the current experiment is necessary to find out if the observed aggregates are the result of (a) mutation(s) in genes involved in MDS, or if mutations in other genes can also result in the snowflake phenotype.

5. Conclusions

A continuous mini tower fermentor was designed and constructed using a 3D printer. The printed fermentor was biocompatible for yeast cultivation and could be sterilized by autoclaving. The tower fermentor was suitable to perform experimental evolution experiments where gravity acts as the selective pressure, and allowed yeast cells to evolve from planktonic single cells towards cell aggregates. After 14 days of continuous cultivation of a non-flocculating industrial brewer's yeast strain in the fermentor, yeast flocs with a "snowflake" morphology were observed. Yeast floc characterisation showed that the cell–cell interactions in the cell aggregates were not the result of flocculin interactions, but probably the result of (a) mutation(s) in gene(s) that are involved in the mother/daughter separation process.

Acknowledgments: The Belgian Federal Science Policy Office (Belspo) and the European Space Agency (ESA) PRODEX program supported this work. The Research Council of the Vrije Universiteit Brussel (Belgium) and the University of Ghent (Belgium) are acknowledged to support the Alliance Research Group VUB-UGent NanoMicrobiology (NAMI), and the International Joint Research Group (IJRG) VUB-EPFL BioNanotechnology & NanoMedicine (NANO).

Author Contributions: Ronnie G. Willaert and Andreas Conjaerts conceived and designed the experiments; Andreas Conjaerts performed the experiments; Andreas Conjaerts and Ronnie G. Willaert analysed the data; Ronnie G. Willaert contributed reagents/materials/analysis tools; Ronnie G. Willaert and Andreas Conjaerts wrote the paper.

Appendix A

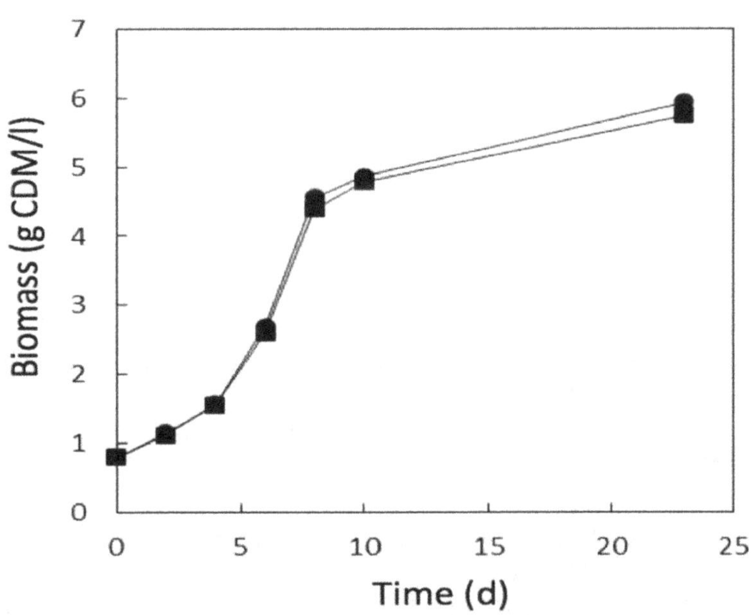

Figure A1. Biocompatibility assay of *S. cerevisiae* BY4742. Evolution of the biomass concentration in the presence (●) and absence (■) of the methacrylic acid ester resin printed material during 1 day of growth.

References

1. Dequin, S. The potential of genetic engineering for improving brewing, wine-making and baking yeasts. *Appl. Microbiol. Biotechnol.* **2001**, *56*, 577–588. [CrossRef] [PubMed]
2. Saerens, S.M.; Duong, C.T.; Nevoigt, E. Genetic improvement of brewer's yeast: Current state, perspectives and limits. *Appl. Microbiol. Biotechnol.* **2010**, *86*, 1195–1212. [CrossRef] [PubMed]

3. Conrad, T.M.; Lewis, N.E.; Palsson, B.Ø. Microbial laboratory evolution in the era of genome-scale science. *Mol. Syst. Biol.* **2011**, *7*, 509. [CrossRef] [PubMed]
4. Portnoy, V.A.; Bezdan, D.; Zengler, K. Adaptive laboratory evolution—Harnessing the power of biology for metabolic engineering. *Curr. Opin. Biotechnol.* **2011**, *22*, 590–594. [CrossRef] [PubMed]
5. Dragosits, M.; Mattanovich, D. Adaptive laboratory evolution—Principles and applications for biotechnology. *Microb. Cell Fact.* **2013**, *12*, 64. [CrossRef] [PubMed]
6. Gibson, B.; Liti, G. *Saccharomyces pastorianus*: Genomic insights inspiring innovation for industry. *Yeast* **2015**, *32*, 17–27. [CrossRef] [PubMed]
7. Blieck, L.; Toye, G.; Dumortier, F.; Verstrepen, K.J.; Delvaux, F.R.; Thevelein, J.M.; van Dijck, P. Isolation and characterization of brewer's yeast variants with improved fermentation performance under high-gravity conditions. *Appl. Environ. Microbiol.* **2007**, *73*, 815–824. [CrossRef] [PubMed]
8. Huuskonen, A.; Markkula, T.; Vidgren, V.; Lima, L.; Mulder, L.; Geurts, W.; Walsh, M.; Londesborough, J. Selection from industrial lager yeast strains of variants with improved fermentation performance in very-high-gravity worts. *Appl. Environ. Microbiol.* **2010**, *76*, 1563–1573. [CrossRef] [PubMed]
9. Yu, Z.; Zhao, H.; Li, H.; Zhang, Q.; Lei, H.; Zhao, M. Selection of *Saccharomyces pastorianus* variants with improved fermentation performance under very high gravity wort conditions. *Biotechnol. Lett.* **2012**, *34*, 365–370. [CrossRef] [PubMed]
10. James, T.C.; Usher, J.; Campbell, S.; Bond, U. Lager yeasts possess dynamic genomes that undergo rearrangements and gene amplification in response to stress. *Curr. Genet.* **2008**, *53*, 139–152. [CrossRef] [PubMed]
11. Ekberg, J.; Rautio, J.; Mattinen, L.; Vidgren, V.; Londesborough, J.; Gibson, B.R. Adaptive evolution of the lager brewing yeast *Saccharomyces pastorianus* for improved growth under hyperosmotic conditions and its influence on fermentation performance. *FEMS Yeast Res.* **2013**, *13*, 335–349. [CrossRef] [PubMed]
12. Strejc, J.; Siříštova, L.; Karabín, M.; Almeida e Silva, J.B.; Brányik, T. Production of alcohol-free beer with elevated amounts of flavouring compounds using lager yeast mutants. *J. Inst. Brew.* **2013**, *119*, 149–155. [CrossRef]
13. Cadière, A.; Ortiz-Julien, A.; Camarasa, C.; Dequin, S. Evolutionary engineered *Saccharomyces cerevisiae* wine yeast strains with increased in vivo flux through the pentose phosphate pathway. *Metab. Eng.* **2011**, *13*, 263–271. [CrossRef] [PubMed]
14. Cadière, A.; Aguera, E.; Caillé, S.; Ortiz-Julien, A.; Dequin, S. Pilot-scale evaluation the enological traits of a novel, aromatic wine yeast strain obtained by adaptive evolution. *Food Microbiol.* **2012**, *32*, 332–337. [CrossRef] [PubMed]
15. Dunham, M.J.; Kerr, E.O.; Miller, A.W.; Payen, C. Chemostat culture for yeast physiology and experimental evolution. *Cold Spring Harb. Protoc.* **2016**. [CrossRef]
16. Miller, A.W.; Befort, C.; Kerr, E.O.; Dunham, M.J. Design and use of multiplexed chemostat arrays. *J. Vis. Exp.* **2013**, *72*, e50262. [CrossRef] [PubMed]
17. Kumar, S.; Wittmann, C.; Heinzle, E. Minibioreactors. *Biotechnol. Lett.* **2004**, *26*, 1–10. [CrossRef] [PubMed]
18. Betts, J.I.; Baganz, F. Miniature bioreactors: Current practices and future opportunities. *Microb. Cell Fact.* **2006**, *5*, 21. [CrossRef] [PubMed]
19. Hortsch, R.; Weuster-Botz, D. Milliliter-scale stirred tank reactors for the cultivation of microorganisms. *Adv. Appl. Microbiol.* **2010**, *73*, 61–82. [PubMed]
20. Tuomi, J.; Paloheimo, K.S.; Vehviläinen, J.; Björkstrand, R.; Salmi, M.; Huotilainen, E.; Kontio, R.; Rouse, S.; Gibson, I.; Mäkitie, A.A. A novel classification and online platform for planning and documentation of medical applications of additive manufacturing. *Surg. Innov.* **2014**, *21*, 553–559. [CrossRef] [PubMed]
21. Mironov, V.; Boland, T.; Trusk, T.; Forgacs, G.; Markwald, R.R. Organ printing: Computer-aided jet-based 3D tissue engineering. *Trends Biotechnol.* **2003**, *21*, 157–161. Erratum in: *Trends Biotechnol.* **2004**, *22*, 265. [CrossRef]
22. Mironov, V.; Kasyanov, V.; Markwald, R.R. Organ printing: From bioprinter to organ biofabrication line. *Curr. Opin. Biotechnol.* **2011**, *22*, 667–673. [CrossRef] [PubMed]
23. Leukers, B.; Gülkan, H.; Irsen, S.H.; Milz, S.; Tille, C.; Schieker, M.; Seitz, H. Hydroxyapatite scaffolds for bone tissue engineering made by 3D printing. *J. Mater. Sci. Mater. Med.* **2005**, *16*, 1121–1124. [CrossRef] [PubMed]

24. Salmi, M. Possibilities of preoperative medical models made by 3D printing or additive manufacturing. *J. Med. Eng.* **2016**, *2016*, 6191526. [CrossRef] [PubMed]
25. Sears, N.A.; Seshadri, D.R.; Dhavalikar, P.S.; Cosgriff-Hernandez, E. A review of three-dimensional printing in tissue engineering. *Tissue Eng. Part B Rev.* **2016**, *22*, 298–310. [CrossRef] [PubMed]
26. Dong, Y.; Fan, S.Q.; Shen, Y.; Yang, J.X.; Yan, P.; Chen, Y.P.; Li, J.; Guo, J.S.; Duan, X.M.; Fang, F.; et al. A novel bio-carrier fabricated using 3D printing technique for wastewater treatment. *Sci. Rep.* **2015**, *5*, 12400. [CrossRef] [PubMed]
27. Gross, B.C.; Erkal, J.L.; Lockwood, S.Y.; Chen, C.; Spence, D.M. Evaluation of 3D printing and its potential impact on biotechnology and the chemical sciences. *Anal. Chem.* **2014**, *86*, 3240–3253. [CrossRef] [PubMed]
28. Melchels, F.P.; Feijen, J.; Grijpma, D.W. A review on stereolithography and its applications in biomedical engineering. *Biomaterials* **2010**, *31*, 6121–6130. [CrossRef] [PubMed]
29. Boulton, C.; Quain, D. Fermentation systems. In *Brewing Yeast and Fermentation*; Blackwell Science Ltd.: Oxford, UK, 2001; pp. 260–372.
30. Klopper, W.J.; Roberts, R.H.; Royston, M.G.; Ault, R.G. Continuous fermentation in a tower fermenter. *Proc. Eur. Brew. Conv.* **1965**, 238–259.
31. Ault, R.G.; Hampton, A.N.; Newton, R.; Roberts, R.H. Biological and biochemical aspects of tower fermentation. *J. Inst. Brew.* **1969**, *75*, 260–277. [CrossRef]
32. Van Mulders, S.E.; Ghequire, M.; Daenen, L.; Verbelen, P.J.; Verstrepen, K.J.; Delvaux, F.R. Flocculation gene variability in industrial brewer's yeast strains. *Appl. Microbiol. Biotechnol.* **2010**, *88*, 1321–1331. [CrossRef] [PubMed]
33. Brachmann, C.B.; Davies, A. Designer deletion strains derived from *Saccharomyces cerevisiae* S288C: A useful set of strains and plasmids for PCR-mediated gene disruption and other applications. *Yeast* **1998**, *14*, 115–132. [CrossRef]
34. Van Mulders, S.E.; Christianen, E.; Saerens, S.M.G.; Daenen, L.; Verbelen, P.J.; Willaert, R.; Verstrepen, K.J.; Delvaux, F.R. Phenotypic diversity of Flo protein family-mediated adhesion in *Saccharomyces cerevisiae*. *FEMS Yeast Res.* **2009**, *9*, 178–190. [CrossRef] [PubMed]
35. Richards, M. The use of giant-colony morphology for the differentiation of brewing yeasts. *J. Inst. Brew.* **1967**, *73*, 162–166. [CrossRef]
36. Pallmann, C.L.; Brown, J.A.; Olineka, T.L.; Cocolin, L.; Mills, D.A.; Bisson, L.F. Use of WL medium to profile native flora fermentations. *Am. J. Enol. Vitic.* **2001**, *52*, 198–203.
37. D'Hautcourt, O.; Smart, K.A. Measurement of brewing yeast flocculation. *J. Am. Soc. Brew. Chem.* **1999**, *57*, 123–128.
38. Den Blanken, J.G. The tower fermenter for lager beer production. *Brew. Gardian* **1974**, 35–39.
39. Seddon, A.W. Continuous tower fermentation—Experiences in establishing large scale commercial production. *Brewer* **1976**, 72–78.
40. Haddad, S.A.; Lindegren, C.C. A method for determining the weight of an individual yeast cell. *Appl. Microbiol.* **1953**, *1*, 153–156. [PubMed]
41. Davis, R.H.; Hunt, T.P. Modeling and measurement of yeast flocculation. *Biotechnol. Prog.* **1986**, *2*, 91–97. [CrossRef] [PubMed]
42. Fontana, A.; Bore, C.; Ghommidh, C.; Guiraud, J.P. Structure and sucrose hydrolysis activity of *Saccharomyces cerevisiae* aggregates. *Biotechnol. Bioeng.* **1992**, *40*, 475–482. [CrossRef] [PubMed]
43. Van Hamersveld, E.H.; Van Der Lans, R.G.; Luyben, K.C. Quantification of brewers' yeast flocculation in a stirred tank: Effect of physical parameters on flocculation. *Biotechnol. Bioeng.* **1997**, *56*, 190–200. [CrossRef]
44. Ratcliff, W.C.; Denison, R.F.; Borrello, M.; Travisano, M. Experimental evolution of multicellularity. *Proc. Natl. Acad. Sci. USA* **2012**, *109*, 1595–1600. [CrossRef] [PubMed]
45. Oud, B.; Guadalupe-Medina, V.; Nijkamp, J.F.; de Ridder, D.; Pronk, J.T.; van Maris, A.J.; Daran, J.M. Genome duplication and mutations in *ACE2* cause multicellular, fast-sedimenting phenotypes in evolved *Saccharomyces cerevisiae*. *Proc. Natl. Acad. Sci USA* **2013**, *110*, E4223–E4231. [CrossRef] [PubMed]
46. Ratcliff, W.C.; Fankhauser, J.D.; Rogers, D.W.; Greig, D.; Travisano, M. Origins of multicellular evolvability in snowflake yeast. *Nat. Commun.* **2015**, *6*, 6102. [CrossRef] [PubMed]
47. Guo, B.; Styles, C.A.; Feng, Q.; Fink, G.R. A *Saccharomyces* gene family involved in invasive growth, cell-cell adhesion, and mating. *Proc. Natl. Acad. Sci. USA* **2000**, *97*, 12158–12163. [CrossRef] [PubMed]

48. Goossens, K.V.; Willaert, R.G. The *N*-terminal domain of the Flo11 protein from Saccharomyces cerevisiae is an adhesin without mannose-binding activity. *FEMS Yeast Res.* **2012**, *12*, 78–87. [CrossRef] [PubMed]
49. Kraushaar, T.; Brückner, S.; Veelders, M.; Rhinow, D.; Schreiner, F.; Birke, R.; Pagenstecher, A.; Mösch, H.U.; Essen, L.O. Interactions by the fungal Flo11 adhesin depend on a fibronectin Type III-like adhesin domain girdled by aromatic bands. *Structure* **2015**, *23*, 1005–1017. [CrossRef] [PubMed]
50. Mill, P.J. The nature of the interactions between flocculent cells in the flocculation of *Saccharomyces cerevisiae*. *J. Gen. Microbiol.* **1964**, *35*, 61–68. [CrossRef] [PubMed]
51. Stratford, M.; Keenan, M.H. Yeast flocculation: Quantification. *Yeast* **1988**, *4*, 107–115. [CrossRef] [PubMed]
52. Weiss, E.L. Mitotic exit and separation of mother and daughter cells. *Genetics* **2012**, *192*, 1165–1202. [CrossRef] [PubMed]
53. King, L.; Butler, G. Ace2p, a regulator of *CTS1* (chitinase) expression, affects pseudohyphal production in *Saccharomyces cerevisiae*. *Curr. Genet.* **1998**, *34*, 183–191. [CrossRef] [PubMed]
54. O'Conallain, C.; Doolin, M.T.; Taggart, C.; Thornton, F.; Butler, G. Regulated nuclear localisation of the yeast transcription factor Ace2p controls expression of chitinase (*CTS1*) in *Saccharomyces cerevisiae*. *Mol. Gen. Genet.* **1999**, *262*, 275–282. [CrossRef] [PubMed]
55. McBride, H.J.; Yu, Y.; Stillman, D.J. Distinct regions of the Swi5 and Ace2 transcription factors are required for specific gene activation. *J. Biol. Chem.* **1999**, *274*, 21029–21036. [CrossRef] [PubMed]
56. Bidlingmaier, S.; Weiss, E.L.; Seidel, C.; Drubin, D.G.; Snyder, M. The Cbk1p pathway is important for polarized cell growth and cell separation in *Saccharomyces cerevisiae*. *Mol. Cell. Biol.* **2001**, *21*, 2449–2462. [CrossRef] [PubMed]
57. Voth, W.P.; Olsen, A.E.; Sbia, M.; Freedman, K.H.; Stillman, D.J. *ACE2*, *CBK1*, and *BUD4* in budding and cell separation. *Eukaryot Cell* **2005**, *4*, 1018–1028. [CrossRef] [PubMed]
58. Versele, M.; Thevelein, J.M. Lre1 affects chitinase expression, trehalose accumulation and heat resistance through inhibition of the Cbk1 protein kinase in *Saccharomyces cerevisiae*. *Mol. Microbiol.* **2001**, *41*, 1311–1326. [CrossRef] [PubMed]
59. Weiss, E.L.; Kurischko, C.; Zhang, C.; Shokat, K.; Drubin, D.G.; Luca, F.C. The *Saccharomyces cerevisiae* Mob2p-Cbk1p kinase complex promotes polarized growth and acts with the mitotic exit network to facilitate daughter cell-specific localization of Ace2p transcription factor. *J. Cell Biol.* **2002**, *158*, 885–900. [CrossRef] [PubMed]

10

The Xylose Metabolizing Yeast *Spathaspora passalidarum* is a Promising Genetic Treasure for Improving Bioethanol Production

Khaled A. Selim [1,*], Saadia M. Easa [2] and Ahmed I. El-Diwany [1]

1 Pharmaceutical and Drug Industries Research Division, National Research Centre, 33-El-Bohouth St. (former El Tahrir St.), Dokki, P.O. Box, Giza 12622, Egypt; ahmed_eldiwany@yahoo.com

2 Microbiology Department, Faculty of Science, Ain Shams University, Cairo 11566, Egypt; elkas_1111@yahoo.com

* Correspondence: Khaled.a.selim@gmail.com

Abstract: Currently, the fermentation technology for recycling agriculture waste for generation of alternative renewable biofuels is getting more and more attention because of the environmental merits of biofuels for decreasing the rapid rise of greenhouse gas effects compared to petrochemical, keeping in mind the increase of petrol cost and the exhaustion of limited petroleum resources. One of widely used biofuels is bioethanol, and the use of yeasts for commercial fermentation of cellulosic and hemicellulosic agricultural biomasses is one of the growing biotechnological trends for bioethanol production. Effective fermentation and assimilation of xylose, the major pentose sugar element of plant cell walls and the second most abundant carbohydrate, is a bottleneck step towards a robust biofuel production from agricultural waste materials. Hence, several attempts were implemented to engineer the conventional *Saccharomyces cerevisiae* yeast to transport and ferment xylose because naturally it does not use xylose, using genetic materials of *Pichia stipitis*, the pioneer native xylose fermenting yeast. Recently, the nonconventional yeast *Spathaspora passalidarum* appeared as a founder member of a new small group of yeasts that, like *Pichia stipitis*, can utilize and ferment xylose. Therefore, the understanding of the molecular mechanisms regulating the xylose assimilation in such pentose fermenting yeasts will enable us to eliminate the obstacles in the biofuels pipeline, and to develop industrial strains by means of genetic engineering to increase the availability of renewable biofuel products from agricultural biomass. In this review, we will highlight the recent advances in the field of native xylose metabolizing yeasts, with special emphasis on *S. passalidarum* for improving bioethanol production.

Keywords: fermentation; xylose metabolism; genetic engineering; biofuel; *Spathaspora passalidarum*; *Pichia stipitis*

---

1. Fermentation Technology and Challenges

The modern biotechnological applications for generation of alternative and renewable sources of biofuels are receiving more attention due to global worries over the climate change, rapid global warming, and the rising of fossil fuel costs. One of such growing biotechnological trends is the fermentation technology to convert the sugar-rich agriculture waste into bioethanol by conventional or non-conventional yeasts [1–4]. In general, yeasts have advantages over bacteria for commercial fermentation due to the thickness of their cell walls, less stringent nutritional requirements, large sizes, utmost resistance to contamination, and better growth at acidic pH of bioreactor fermenters.

In nature, the second most abundant hemicellulosic sugar in fast-growing hardwoods and agricultural biomass is xylose. Xylose sugar forms up to 15–25% of all angiosperm biomass, and it could supply an alternative fuel source for its ability to be commercially fermented into ethanol. Several approaches have been employed to engineer xylose assimilation metabolism into conventional fermenting yeasts, such as *Saccharomyces cerevisiae* [4–6]. Therefore, efficient hemicellulosic sugar fermentation is crucial for the economic conversion of lignocellulose biomass to renewable biofuels [4,6–8]. The discovery of xylose-fermenting yeasts in new niches and genetic engineering of yeasts to be capable of rapid fermentation of xylose and other sugars to recoverable concentrations of bioethanol could provide alternative biofuel sources for the future (Figure 1) [4,9].

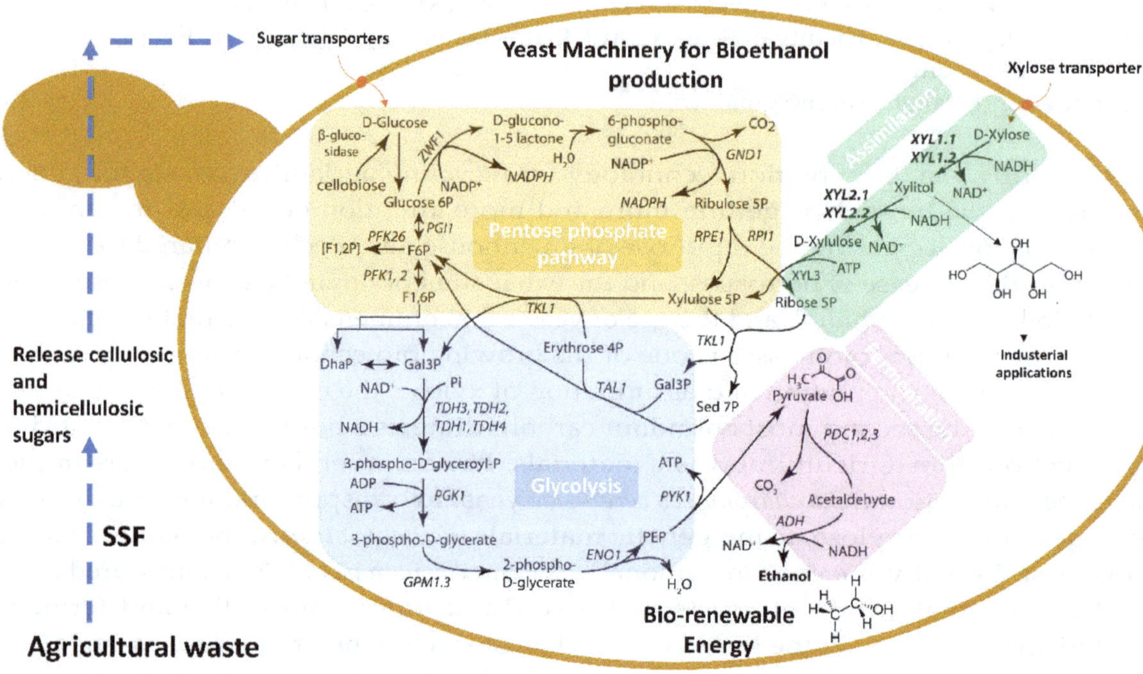

Figure 1. Model of yeast fermentation machinery for bioethanol production using the agriculture waste to feed yeasts [4]. The metabolic pathways for xylose and glucose assimilation and fermentation are indicated including the pentose phosphate pathway and glycolysis. The agriculture waste is treated through the enzymatic and chemical simultaneous saccharification and fermentation (SSF) processes to release the cellulosic and hemicellulosic sugars. The hexose and pentose sugars are transported by specific hexose and pentose sugar transporters into yeast for further metabolizing processes.

As a rule of thumb for metabolizing the xylose in most of xylose-fermenting yeasts [4], firstly the xylose is reduced by xylose reductase (XR) to xylitol. In the second step, the xylitol is oxidized by xylitol dehydrogenase (XDH) to xylulose. Afterward, the xylulose passes into the pentose phosphate pathway being metabolized into glyceraldehyde-3-P which is further reduced to pyruvate. Finally, the pyruvate is decarboxylated to acetaldehyde which is further reduced to ethanol by alcohol dehydrogenase (Figure 1). Notably, most xylose reductase enzymes have dual cofactor specificity, using both NADH and NADPH, but typically favor NADPH. However, xylitol dehydrogenase enzymes use NAD^+ specifically as a cofactor, which could cause imbalance between the cofactor's source for the XR-XDH pathway and xylitol accumulation under uncontrolled oxygen conditions (Figure 2) [10].

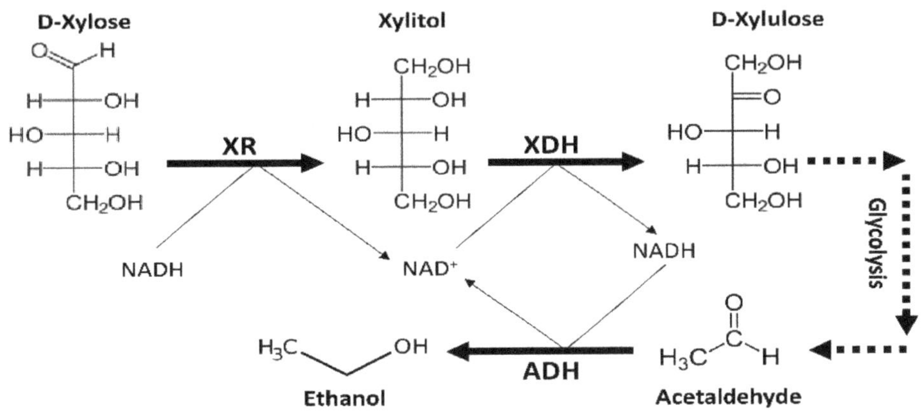

Figure 2. Schematic model of the central metabolism for bioethanol production from xylose indicating the rate limiting steps (XR: xylose reductase; XDH: xylitol dehydrogenase, and ADH: alcohol dehydrogenase) and cofactors demand/balance in most of native xylose metabolizing yeasts. Xylose assimilation reactions starts with XR to produce xylitol. The xylitol is further metabolized by XDH to produce xylulose, which further metabolized to xylulose-phosphate to enter the glycolysis (indicate by black dotted arrow and summarized in Figure 1) to produce acetaldehyde. The acetaldehyde finally converted to ethanol by ADH.

One of pioneer xylose fermenting yeasts is *Pichia stipites*. *P. stipitis* is heterothallic ascomycetous yeast, predominantly haploid and related to pentose fermenting yeasts, such as *Candida shehatae* [1,4,11–15]. *P. stipitis* was recently renamed to be *Scheffersomyces stipitis* [15] and it is natively one of the highest xylose-utilizing and fermenting yeasts. In type culture collections, the *P. stipitis* strains are among the best xylose-metabolizing microbes [16]. Under controlled low O_2 conditions, *P. stipitis* is able to consume xylose and produce up to 57 g/L of bioethanol at 30°C [4,13,14,17]. *Pichia* uses an alternative nuclear genetic code (ANGC) in which CUG encodes for Ser rather than Leu [17], which makes the genetic manipulation of *Pichia* with the commercial drug resistance markers unusually problematic because essentially all of these markers are derived from bacteria that use the universal codon system. Moreover, one of classical challenges in fermentation technology is that some of key enzymes of bioethanol production pathway are expressed relatively in low levels [1,4,13,14]. Therefore, the metabolic engineering of the bioethanol pathway in yeasts, which can ferment the sugars of the agriculture biomass with considerable and recoverable bioethanol concentrations, could enhance the productivity and sustainability of renewable biofuel sources [1,13,14]. To improve bioethanol production and xylose metabolism, a stable and manipulatable genetic system that enables overexpression or deletion of one or more of key enzymes and sugar transporters in xylose-fermenting yeast *P. stipitis*, was developed [1,3,18]. This approach comprises modelling, metabolic and flux analysis, quantitative metabolomics and transcriptomics followed by the targeted overexpression or deletion of genes of the rate-limiting steps [1,3,4,13,14]. Since there is reasonable information about the metabolic capacity of *P. stipitis* to ferment xylose on various omics levels, this makes it an attractive model system for metabolic engineering.

Recently, a new xylose-fermenting yeast *Spathaspora passalidarum* was discovered, which naturally co-ferments xylose, glucose, and cellobiose and demonstrates potentials in the effective conversion of mixed sugars from hemicellulosic hydrolysates into ethanol [9]. *S. passalidarum* was initially isolated from extremely O_2-limited and hemicellulosic sugar rich environments from the gut of a wood-boring beetle (as will be discussed below). Although the anaerobic fermentation of glucose is broadly known, the xylose fermentation typically needs a controlled oxygen condition. Uncontrolled oxygen conditions in another xylose-fermenting yeast, such as *P. stipitis*, leads to accumulation of xylitol due to insufficient amounts of NAD^+, and as consequence the xylose metabolism will be blocked (Figure 2) [9,13,14]. To solve this problem, precise controlled O_2 (very low O_2 concentrations) during the xylose fermentation is required in *P. stipitis* to generate NAD^+ from NADH.

The bioethanol production from the bioconversion of lignocellulose biomass must be achieved at high rates and yields for economically recoverable concentrations. The achieving of such targets for efficient bioethanol production are more difficult with cellulose and hemicellulose. The major barrier for cellulose utilization is enzymatic saccharification, while for hemicellulose it is the utilization of mixed sugars (hexose sugars: glucose, galactose, mannose, and rhamnose; and pentose sugars: xylose and arabinose) in the presence of ferulic and acetic acids along with other byproducts of the thermochemical pretreatment of the hydrolysates [13,14]. However, *S. passalidarum* and *P. stipitis* yeasts possess a set of unique physiological merits that make them very useful biodegradable organisms for bioconversion of lignocellulosic biomass [4,9,13,14]. *Pichia* can utilize and ferment effectively cellobiose, glucose, galactose, and mannose along with xylan high oligomeric sugars xylan and mannan, in addition to its extensively studied ability to metabolize and ferment the xylose [4,13,14]. The primary sugar released in enzymatic hydrolysis is cellobiose and, remarkably, *P. stipitis* and *S. passalidarum* have the capability to utilize the cellobiose, which make such yeasts potent organisms for simultaneous saccharification and fermentation (SSF) or hydrolysate, because most commercially available cellulase products are often deficient in β-glucosidase enzyme so the accumulation of cellobiose inhibits cellulose activities. Since *P. stipitis* and *S. passalidarum* can directly metabolize the cellobiose, they have the potential to improve SSF processes [4,9,10,13,14,19,20].

Collectively, the native ability of *P. stipitis* and *S. passalidarum* to metabolize the oligomeric sugars is of high importance as the mild acidic pretreatments of agriculture waste biomass can prevent the formation of the sugar degradation byproducts, which could inhibit significantly the fermentation process, but can release about 15–55% of soluble oligomeric sugars. Therefore, with low cost and high yield, the hemicellulosic sugars can be more readily recovered and underutilized from cellulose biomass than glucose. Although such easy recoverable sugars can be utilized for formation of a number of useful products such as xylitol, butanol, lactic acid, and other chemicals, bioethanol is still the major product with the largest potential market. Hence, bioethanol production from the lignocellulosic biomasses is receiving a lot of attention as a consequence of agriculture policies and energy demands to improve the production of alternative renewable biofuels and to reduce CO_2 emissions [2,4,13,14].

2. *Spathaspora passalidarum* a Promising Genetic Source

The Spathaspora clade contains many bioethanol producer yeasts, including *Spathaspora arborariae, Spathaspora brasiliensis, Spathaspora gorwiae, Spathaspora hagerdaliae, Spathaspora passalidarum, Spathaspora roraimanensis, Spathaspora suhii, Spathaspora xylofermentans*. They are usually endosymbioticly associated with wood-boring-beetles that occupy rotting wood. *Spathaspora passalidarum* (Figure 3), the first identified species of genus Spathaspora, was isolated from the gut of passalid beetle *Odontotaenius disjunctus* [9,21–24]. Notably, *S. arborariae, S. gorwiae, S. hagerdaliae,* and *S. passalidarum* ferment xylose to produce bioethanol, whereas the rest within the Spathaspora clade are thought to be xylitol producers [9].

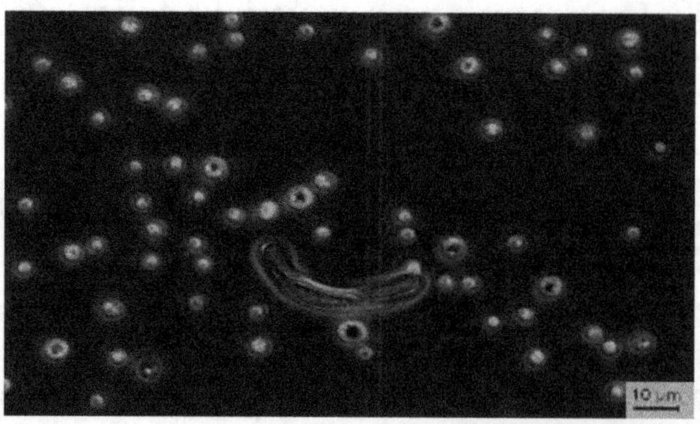

Figure 3. *Spathaspora passalidarum* budding cells with characteristic curved and elongated ascospore.

S. passalidarum was firstly described in 2006 by Nguyen et al. [21]. The authors speculated that *Spathaspora* mainly exists in the beetle's biosphere rather than the beetle's gut microbiota, and it may be only by coincidence that *O. disjunctus* beetles ingested decaying wood contaminated by yeasts. Later in 2012 and 2017, another 12 strains were described in two independent studies from wood-boring beetles and wood samples of Amazonian forest in Brazil [23,25]. Among these strains, only one isolate was obtained from the gut of *Popilus marginatus* beetle, while the rest of the strains were obtained from the woody samples inhabited by the beetles [23,25]. In 2014, two more strains were isolated from rotted wood in China [26]. Additionally, Rodrussamee and colleges in 2018 reported a new thermotolerant strain, named *S. passalidarum* CMUWF1–2, which was isolated from Thailand soil [27]. The frequency of finding *S. passalidarum* mainly among the woody samples supports the notion that those yeasts are probably associated with decaying wood niches rather than with the gut microbiota of wood-boring beetles. However, the fact of the low frequency of finding *S. passalidarum* among other yeast species keeps an open possibility that they inhabit mainly the wood-related beetles [9].

3. Fermentation Capability of *Spathaspora passalidarum*

It is believed that the beetle's gut is truly anaerobic or microaerobic, therefore it was speculated that *S. passalidarum* possess a unique adaptation capability to survive under oxygen-depleted conditions on mixtures of hemicellulosic sugars in the midgut of wood-boring beetles [10,19]. Currently, *S. passalidarum* is among the best xylose-utilizing and fermenting yeasts. Under anaerobic or microaerobic conditions, *S. passalidarum* possess rapid utilization and consumption rates for xylose and produces up to 0.48 g/g bioethanol (near to the maximum theoretical bioethanol production of 0.51 g/g), in contrast to *P. stipitis* which can hardly metabolize xylose anaerobically, accumulating xylitol and a very low yield of bioethanol [10,19,20,28,29]. Under anaerobic conditions, Hou in 2012 showed that *S. passalidarum* has a high growth rate with rapid consumption rate of sugars and can ferment xylose into a high yield of bioethanol with higher production efficiency than *P. stipitis* [10]. Similarly, Veras and colleges in 2017 showed that under anaerobic conditions, *S. passalidarum* accumulates 1.5 times more bioethanol than *S. stipitis*, while both stains accumulate around 0.44 g/g under O_2 limiting conditions [30]. The previous work by Hou (2012) defined strictly that *S. passalidarum* can metabolize and ferment xylose in tightly capped flasks [10]. In contrast to the previous report by Hou (2012) [10], under stringent O_2 limiting conditions, the *S. passalidarum* was not able virtually to utilize the sugars, indicating that native wild-type *S. passalidarum* does not ferment sugars under truly anaerobic conditions [19,20]. Therefore, it is still under debate whether *S. passalidarum* can ferment xylose truly anaerobically or whether it requires a controlled microoxygenic condition similar to *P. stipitis*.

One of the major challenges in fermentation technology is the inability of the majority of known microbes to co-ferment xylose and glucose, since glucose usually inhibits the metabolization of the other sugars in lignocellulose hydrolysate, as in the case of *P. stipitis* [13,14]. Astonishingly, in a recent study to address the metabolic profiling and fermentation capacity of *S. passalidarum*, *S. passalidarum* was found to co-ferment xylose, cellobiose, and glucose simultaneously with high bioethanol yields ranging from 0.31 to 0.42 g/g [19,20]. Moreover, an adapted *S. passalidarum* strain was found to accumulate up to 39 g/L bioethanol with a 0.37 g/g yield from a lignocellulosic hydrolysate. The specific production rate of bioethanol on xylose as a carbon source was superior with three times more than the corresponding rate on glucose, where the flux of glycolytic intermediates was meaningfully lower on glucose than on xylose and its xylose reductase enzyme had a higher affinity for NADH than NADPH [19,20]. Thus, the allosteric activation of glycolytic routes associated with the xylose utilization and the NADH-dependent xylose reductase are most likely the causes for such unique ability of *S. passalidarum* to co-ferment mixed sugars [19,23]. Later, such results were confirmed in a metabolic flux study, where *S. passalidarum* showed about 1.5–2 times high flux rate in the NADH-dependent xylose reductase reaction [31], which caused continuous recycling and reduction of xylitol levels. Such directed high flux rates to glycolytic routes and pentose phosphate pathway was the cause for high levels of bioethanol production in *S. passalidarum* [31]. In large scale fed-batch fermentation study, *S.*

passalidarum was able metabolize around 90% of xylose sugar and all of glucose of sugarcane bagasse hydrolysate, even so glucose had approximately three-fold higher xylose content; and produced a high ethanol yield of 0.46 g/g with volumetric productivity of 0.81 g/L/h in contrast to *P. stipitis* which produced 0.32 g/g ethanol with productivity of 0.36 g/L/h [32]. In follow up study, *S. passalidarum* UFMG-CM-Y473 strain was able to simultaneously utilize and co-ferment about 78% of the released sugars (xylose, glucose, and cellobiose) of pretreated sugarcane bagasse hydrolysate (delignified and enzymatically hydrolyzed) to yield up to 0.32 g/g bioethanol with productivity of 0.34 g/L/h without any nutritional supplementation [33]. Moreover, the new thermotolerant strain, *S. passalidarum* CMUWF1–2, was able to co-ferment various sugars (mannose, galactose, xylose, and arabinose) of lignocellulosic biomass, even in presence of glucose, to accumulate considerable amounts of bioethanol and low amounts of xylitol at higher temperatures. For example, it was able to accumulate 0.43, 0.40, and 0.20 g/g ethanol per xylose at 30, 37, and 40 °C, respectively [27]. Constant with absence of the glucose repression effect on the utilization of other sugars, *S. passalidarum* CMUWF1–2 exhibited a resistance to 2-deoxy glucose, the nonmetabolizable glucose analog, and tolerance to elevated levels of glucose (35.0% of *w*/*v*) and ethanol (8.0% of *v*/*v*) [27]. In contrast, the first discovered *S. passalidarum* NRRL Y-27907 strain was sensitive to 2-deoxy glucose, as 2-deoxy glucose suppressed the xylose consumption under anaerobic conditions. While under aerobic conditions, the 2-deoxy glucose inhibited, only partially, *S. passalidarum* NRRL Y-27907 [10]. Therefore, the author speculated that xylose uptake in *S. passalidarum* NRRL Y-27907 may take place by different xylose transport systems under aerobic and anaerobic conditions. Under aerobic conditions, xylose is taken up by means of ATP-dependent/high affinity xylose-proton symporter and low affinity transporter via facilitated diffusion driven only by the sugar gradient. While, under anaerobic condition, the yeasts are most likely to use only the low affinity xylose pump, as the active transport via xylose-proton symporter will deplete the ATP levels. The inhibitory effect of 2-deoxy glucose on *S. passalidarum* NRRL Y-27907 can be therefore explained by (1) blocking of the low-affinity-facilitated diffusion transporters which are occupied with transporting 2-deoxy glucose, and (2) the inhibition xylose active transport due to the depletion of the intracellular ATP levels to actively phosphorate the 2-deoxy glucose into the non-metabolizable phospho-2-deoxy glucose [10].

4. Genetic and Physiological Features of *Spathaspora passalidarum* Emphasis Special Roles for Xylose Reductase and Xylitol Dehydrogenase

These unusual unique traits of *S. passalidarum* are very attractive for studying on a molecular level. The complete genome sequence of xylose-fermenting yeast *S. passalidarum* was therefore necessary and it was accomplished and published for first time in 2011 [34]. The comparative genomic, transcriptomic, and metabolomic analysis between two of the native xylose-fermenting yeasts, the relatively newly discovered *S. passalidarum*, and the deeply studied *P. stipitis*, allowed a better understanding of the regulatory mechanisms of lignocellulose utilization, and identified the target key genes involved in xylose metabolism [9,10,19,31,34]. The comparative genomic and phylogenetic analysis clearly revealed that *S. passalidarum* is one of the CUG yeast clades, similar to *P. stipites* [34]. In addition, the transcriptome analysis indicated upregulation of the genes implicated in transporting carbohydrate and xylose- and carbohydrate-metabolisms under xylose growth. Several of genes, which are involved in regulation of redox balance and recycling of $NAD(P)H/^{+}$, were upregulated to probably keep the redox balance during xylose utilization. Additionally, the genes encoding for cellulases and β-glucosidases were also upregulated, which suggests a positive feedback of xylose on the upstream genes to activate its own liberation from the higher oligomeric sugars of hemicelluloses by means of the catalytic activities of cellulases and β-glucosidases [34].

The previously mentioned capabilities of *S. passalidarum* to co-ferment different sugars and accumulate high levels of bioethanol with very low concentrations of xylitol, can be explained by the presence of a set of physiological characters encoded by unique set of genes [10,19,34]. Thus, the high capacity of xylose fermentation and low levels of xylitol accumulation by *S. passalidarum* was speculated to be due to the cofactor's equilibrium between the intracellular demand and supply of the cofactors via NADH-favored xylose reductase enzyme and NAD^+-specific xylitol dehydrogenase enzyme, the key enzymes of the xylose utilization pathway [10,28]. Normally, the xylose metabolization occurs through the reduction of xylose to xylitol with xylose reductase, which requires NADPH or NADH as a cofactor with preference for NADPH. Only few NADH-favored xylose reductase enzymes have been described so far [10,35–37]. Then the xylitol is metabolized further by xylitol dehydrogenase which is strictly NAD^+ dependent (Figure 2). The unbalance between NAD^+ supplement and requirement can block the xylose metabolization and leads to accumulation of xylitol. Later, *S. passalidarum* was found to harbor two genes encoding for xylose reductase (*SpXYL1.1* and *SpXYL1.2*) [28]. The *SpXYL1.1* gene product is more equivalent to *XYL1* found in other yeasts. The expression levels of *SpXYL1.2* were found to be higher than *SpXYL1.1* and bioethanol production in *S. passalidarum* was attributed to higher xylose reductase activity with NADH than with NADPH [28]. The *SpXYL1.2* was found to use both NADH and NADPH with preference for NADH, while *SpXYL1.1* was stringently NADPH-dependent. Furthermore, the transformation of *S. cerevisiae* with *SpXYL1.2* of *S. passalidarum* enabled the overexpressing *S. cerevisiae*::*Sp*XYL1.2 strain to grow anaerobically on xylose and to ferment it to higher ethanol yield than the isogenic *S. cerevisiae* TMB 3422 strain, which overexpresses *P. stipitis XYL1*. While, the *S. cerevisiae*::*SpXYL1.1* overexpressing strain was not able to grow on xylose [28]. Similarly, in the yeast-like fungus *Aureobasidium pullulans*, the overexpression of *SpXYL1.2* xylose reductase along with *S. passalidarum* xylitol dehydrogenase encoded by *SpXYL2.2* enhanced the xylose metabolization by 17.76% and improved the fermentation capability and the pullulan production by 97.72% of the overexpressing mutants compared with the parental strain [38].

Finally, a metabolic analysis of *S. passalidarum* speculated that NADH-preferred xylose reductase and NAD^+-dependent xylitol dehydrogenase would tend to drive both of xylose assimilation via the oxidoreductase pathway and the acetaldehyde reduction to ethanol by the alcohol dehydrogenase enzyme [19]. Recently, a metabolic flux analysis of different xylose-fermenting yeasts confirmed a better cofactors balance within *S. passalidarum* cells during xylose catabolism to bioethanol production than within *P. stipitis* cells [31], which further supports the growth characteristics of *S. passalidarum*.

Collectively, those unique and unusual traits of *S. passalidarum* encourage using it as a source for genes to improve xylose utilization and bioethanol production from lignocellulosic biomass in the current xylose fermenting yeasts, such as *P. stipites*, or to introduce xylose metabolism genes to develop industrial strains of *Saccharomyces cerevisiae* capable of co-fermentation of pentose and hexose sugars. Or alternatively, to domesticate it given the excellent results already accomplished by wild-type representatives of that species for co-fermentation of mixed sugars. To facilitate that purpose, Li et al. (2017) developed a stable genetic expression system compatible with the CUG yeasts clade for genomic integration of Gene Of Interest (GOI) into several yeasts [39]. The developed multi-host integrative system was functional in several of the xylose-fermenting yeasts including *S. passalidarum*, *P. stipitis*, and *Candida jeffriesii* and *Candida amazonensis*, as well as in a hexose metabolizing yeast *Saccharomyces cerevisiae*, for heterologous expression of green fluorescent protein (GFP) or lactate dehydrogenase. For lactate dehydrogenase overexpressing strains, all the engineered yeast strains were able to metabolize either glucose (in case of *S. cerevisiae*) or xylose (in case of xylose-fermenting yeasts) to produce lactate [39].

5. New Adaptive Strains of *Spathaspora passalidarum* for Potential Industrial Applications

One of the unique features of those xylose metabolizing yeasts, is the ability to use not only the monomeric hexose and pentose sugars but also the high oligomeric disaccharide sugar in mixed co-fermentation [9,19], which can be an advantage for large scale industrial applications. The mild acid pretreatment of agriculture waste biomasses is relatively cheap and prevents the accumulation of harmful compounds, which inhibits the fermentation processes, but releases the sugars in higher oligomeric stats. Therefore, the ability of such native xylose fermenting strains, *P.*

stipitis, and *S. passalidarum*, to use the high oligomeric sugar can be a great advantage for various biotechnological applications.

One of the major problems that hinders the use of *S. passalidarum* for industrial bioethanol production, even with its remarkable ability for bioethanol production, is the high sensitivity of *S. passalidarum* to the chemical inhibitors, such as ferulic and acetic acids, which are released in preparation of hemicellulosic hydrolysates [9]. Several elaborative studies have focused mainly on improving the tolerability of *S. passalidarum* to the hydrolysates inhibitors with keeping in mind the bioethanol productivity of the strains [19,40–43]. Hou and Yao in (2012) reported a strong strain [40], which is able to grow on furfurals and many other inhibitors of wheat straw hydrolysate (75%) and able to accumulate up to 0.40 g/g ethanol. Such strain was generated through hybridization of a *S. cerevisiae* and a UV-mutagenized *S. passalidarum* [40]. In 2012 also, another resistant strain was developed under O_2 limiting conditions through several passage of the wild-type *S. passalidarum* NRRL Y-27907 on wood hydrolysate, followed by adaptive growth of the strain on corn stover AFEX (ammonia fiber expansion) hydrolysate [19]. Even with such efforts, the strain was not able to accumulate significant amounts of ethanol during the fermentation of the AFEX hydrolysate, despite its ability to grow in AFEX hydrolysate media. When the acetic acid was depleted from AFEX hydrolysate media, ethanol production was surprisingly observed with a yield of 0.45 g/g and most of the xylose content was consumed [19]. Later in 2017, Morales and colleagues developed an evolutionary adapted strain [41] with high tolerance toward the classical inhibitor of the fermentation processes, acetic acid, and that produces ethanol with a yield of 0.48 g/g. In a non-detoxified hydrolysate of *Eucalyptus globulus*, the authors reported also the ability of this strain to co-utilize mixed sugars of xylose, glucose, and cellobiose under microaerobic conditions [41]. This strain was generated by UV irradiation followed by successive growing of the strain under elevated acetic acid concentrations [41]. In similar way, another group also obtained a mutated *S. passalidarum* strain but via plasma mutagenesis and continuous cultivation in alkaline liquor pretreated corncob [42]. Under a simultaneous saccharification and co-fermentation, the obtained strain produced bioethanol with efficiency of 75% [42]. Finally, Su et al. in 2018 developed an adaptive *S. passalidarum* strain (named YK208-E11) [43], which is designated for resistance to AFEX hydrolysate inhibitors, from the wild-type NRRL Y-27907 through high-throughput screen via combining several approaches of batch adaptation, cell recycling, and cell mating [43]. The *S. passalidarum* YK208-E11 strain produced less biomass (about 40% compared to the wild-type), co-metabolized mixed sugars of xylose, glucose, and cellobiose, and exhibited a three-fold improvement in the ethanol production rate with a yield of 0.45 g/g. The whole genome sequence of *S. passalidarum* YK208-E11 strain revealed a deletion of about 11 kb in this strain. The ORF, which was deleted in *S. passalidarum* YK208-E11, is encoding for proteins predicted to be involved in cell division and respiration. Therefore, the authors speculated that this deletion may account for those unique adaptive/physiological features of this AFEX-acclimatized *S. passalidarum* YK208-E11 strain [43].

6. Future Perspective for Engineering New Strains for Better Bioethanol Production

The metabolic engineering approaches involve targeted overexpression and/or deletion of fermentative key genes that facilities quick and efficient conversion of sugars into bioethanol with high recoverable yields [1,3]. As we discussed above, *S. passalidarum* xylose reductase and xylitol dehydrogenase are among the promising candidates for targeted overexpression. The cumulative knowledge of the transcriptomics, metabolomics, and comparative genomics studies for *P. stipitis* and *S. passalidarum*, identified other key enzymes controlling the xylose assimilation, rather than XDH and XR (Figure 2). One of such promising key genes is *adh* that encodes for fermentative isozyme alcohol dehydrogenase (ADH), which is vital for production and/or assimilation of ethanol (Figure 2). Generally, ADH catalyzes the final (rate limiting) step in the yeast glycolytic pathway, the reduction of acetaldehyde to ethanol and NAD^+, and therefore it accepts NADH as a co-factor [44,45]. However, ADH enzymes are also able to perform the reverse reaction from ethanol to acetaldehyde, enabling the yeasts to oxidize and grow on ethanol as a carbon source. In *P. stipitis*, the ADH fermentative activities is crucial not only for ethanol production and/or consumption but also for maintenance redox balance within the yeast cell, so it is considered to be a part of the cofactor balance system in *P. stipitis* [44].

The sequencing projects of *S. passalidarum* NRRL Y-27907 and *P. stipitis* CBS6054 (JGI-MycoCosm) revealed the presence of several/different alcohol dehydrogenase (ADH) encoding genes. For example, in *P. stipitis*, sevem genes were predicted to encode for alcohol dehydrogenase (*Ps*ADH1 to *Ps*ADH7) enzymes [13,17]. Among them *Ps*ADH1 and *Ps*ADH2 were found to be essential for xylose assimilation and ethanol production [44,46]. Each of the ADH proteins in *S. passalidarum* and *P. stipitis* are supposed to have different kinetic properties. Some of the enzymes could be mainly responsible for producing ethanol while others might be responsible for oxidizing it. In *S. passalidarum*, the gene encoding for *Sp*ADH1 was found to be expressed at a very high level during xylose metabolization [34]. In addition, metabolic analysis and metabolic flux analysis revealed that alcohol dehydrogenase is one of key enzymes driving ethanol production in *S. passalidarum* [19,31]. Notably, owing to relative *Sp*ADH1 abundance, the *Sp*ADH1 promotor was used to develop a multi-host integrative system for xylose-fermenting yeast [39].

While in *P. stipitis*, the function of some ADH enzymes are better understood, in particular *Ps*ADH1 and *Ps*ADH2 [13,17,44,46–49]. Transcriptomic studies of the *P. stipites adh* system indicated that the *Ps*ADH activities are correlated with and induced under O_2 limited/microaerobic conditions [46,48]. Under xylose fermentation, the *Ps*ADH1 was found to be the primary key enzyme among the *Ps*ADH system. The deletion of *Ps*ADH1 caused a reduction in *P. stipites* growth rate and a notable increase in xylitol accumulation accompanied with a dramatic decrease in ethanol production, due to intracellular cofactors imbalance [44]. The *Ps*ADH2 is not expressed under microaerobic or aerobic conditions unless *Ps*ADH1 is deleted [44,46], which further confirms that the significant role of *PsADH1* is in sugar assimilation and ethanol production. The levels of *Ps*ADH1 and *Ps*ADH2 transcripts were observed, however, to be low through xylose metabolism relative to the transcript levels of other fermentative and glycolytic enzymes [13,17]. In addition, *Ps*ADH1 and *Ps*ADH2 were able to complement the growth of the *S. cerevisiae* Δ*adh* mutant on ethanol as a sole carbon source [47]. Moreover, *Ps*ADH3 to *Ps*ADH7 were speculated to keep the balance between the cofactors NADPH and NADH [17]. However, the expression patterns of the other *Ps*ADHs on xylose and glucose under microaerobic conditions, in particular, for *Ps*ADH7 and *Ps*ADH4 are not fully understood [13]. *Ps*ADH5 was found in proximity to NADPH dehydrogenase, implying a function in maintenance the intracellular cofactors balance, however, it is not proven yet. Notably, *Ps*ADH7 was found to be upregulated under aerobic growth on xylose [50]. *Ps*ADH7 was described as a strictly NADP(H) dependant enzyme with broad spectrum for substrates-specificity, including variety of aromatic and linear aldehydes (e.g., acetaldehyde, butanal, propanal, and furfural) and alcohols (e.g., ethanol, butanol, pentanol, hexanol, and octanol) for forward and reverse reactions, respectively [50]. Surprisingly, *Ps*ADH7 was able to utilize xylitol as a substrate too with moderate activity. In the same context, the overexpression of *Ps*ADH7 into a *P. stipites* xylitol dehydrogenase mutant (Δ*Ps*XDH) [18], which cannot metabolize xylitol and therefore cannot grow on xylose as a sole carbon source, was able exclusively to complement the growth of Δ*Ps*XDH on xylose, in contrast to *Ps*ADH1, 2, 4, and 5 [50]. Hence, there is a need to understand the kinetic characteristics of each of *Ps*ADH and *Sp*ADH enzymes in order to target the correct genes for overexpression and/or deletion. Finally, we would like to state that genes encoding for *adh* isozymes are worth studying, especially of *S. passalidarum*, owning to their significant functions in bioethanol production/consumption and/or intracellular cofactor balance.

7. Conclusions

Taken together, the advances in fermentation performance by *S. passalidarum* pave the way for engineering the conventional and the nonconventional fermenting yeasts, such as *S. cerevisiae* and *P. stipites*, for economical fermentation of hexose and pentose sugars in hemicellulosic hydrolysates on industrial scales. Keeping in mind that the efficient metabolization and fermentation of xylose is essential for the bioconversion of lignocellulosic biomasses into biofuels and chemicals, but the conventional wildtype strains like *S. cerevisiae* cannot use the xylose. Therefore, researchers keep trying to engineer the xylose utilization pathway into the conventional yeast. The genomes of the natural xylose-fermenting yeasts, in particular of *P. stipitis* and *S. passalidarum,* are of huge importance, as their genomics features and regulatory patterns can serve as guides and genomic resources for further genetic engineering development in those native xylose-metabolizing yeasts or to engineer non-xylose fermenting yeasts. Therefore, *S. passalidarum* and *P. stipitis* can be considered as genomic treasure sources for various genes to engineer the xylose metabolism and to improve the bioethanol production [1,24,34].

Author Contributions: K.A.S. undertook extensive literature review and wrote the manuscript. S.M.E. and A.I.E.-D. supervised the research. All authors have read and agreed to the published version of the manuscript.

Acknowledgments: The authors would like to thank Thomas Jeffries (University of Wisconsin Madison), Laura Willis (Forest Products Laboratory, USDA), Amal rabeá, Ali Selim, and Katerina Peros for the invaluable and unlimited support to complete this project. K.A.S. received the Parwon scholarship at Institute for Microbial and Biochemical Technology, FPL, USDA.

References

1. Jeffries, T.W. Engineering yeasts for xylose metabolism. *Curr. Opin. Biotechnol.* **2006**, *17*, 320–326. [CrossRef] [PubMed]
2. Balat, M.; Balat, H. Recent trends in global production and utilization of bio-ethanol fuel. *Appl. Energy* **2009**, *86*, 2273–2282. [CrossRef]
3. Jarboe, L.R.; Zhang, X.; Wang, X.; Moore, J.C.; Shanmugam, K.T.; Ingram, L.O. Metabolic Engineering for Production of Biorenewable Fuels and Chemicals: Contributions of Synthetic Biology. *J. Biomed. Biotechnol.* **2010**, *2010*, 761042. [CrossRef] [PubMed]
4. Selim, K.A.; El-Ghwas, D.E.; Easa, S.M.; Hassan, M.I.A. Bioethanol a Microbial Biofuel Metabolite; New Insights of Yeasts Metabolic Engineering. *Fermentation* **2018**, *4*, 16. [CrossRef]
5. Jeffries, T.W.; Shi, N.Q. Genetic engineering for improved xylose fermentation by yeasts. *Adv. Biochem. Eng. Biotechnol.* **1999**, *65*, 117–161.
6. Ruchala, J.; Kurylenko, O.O.; Dmytruk, K.V.; Sibirny, A.A. Construction of advanced producers of first- and second-generation ethanol in *Saccharomyces cerevisiae* and selected species of non-conventional yeasts (*Scheffersomyces stipitis*, *Ogataea polymorpha*). *J. Ind. Microbiol. Biotechnol.* **2020**, *47*, 109–132. [CrossRef]
7. Hinmann, N.D.; Wright, J.D.; Hoagland, W.; Wyman, C.E. Xylose fermentation—An economic analysis. *Appl. Biochem. Biotechnol.* **1989**, *20*, 391–401. [CrossRef]
8. Saha, B.C.; Dien, B.S.; Bothast, R.J. Fuel ethanol production from corn fiber—Current status and technical prospects. *Appl. Biochem. Biotechnol.* **1998**, *70*, 115–125. [CrossRef]
9. Cadete, R.M.; Rosa, C.A. The yeasts of the genus *Spathaspora*: Potential candidates for second-generation biofuel production. *Yeast* **2018**, *35*, 191–199. [CrossRef]
10. Hou, X. Anaerobic xylose fermentation by *Spathaspora passalidarum*. *Appl. Microbiol. Biotechnol.* **2012**, *94*, 205–214. [CrossRef]
11. Kurtzman, C.P. *Candida shehatae*—Genetic diversity and phylogenetic relationships with other xylose-fermenting yeasts. *Antonie Leeuwenhoek* **1990**, *57*, 215–222. [CrossRef] [PubMed]
12. Melake, T.; Passoth, V.V.; Klinner, D. Characterization of the genetic system of the xylose-fermenting yeast *Pichia stipitis*. *Curr. Microbial.* **1996**, *33*, 237–242. [CrossRef] [PubMed]
13. Jeffries, T.W.; Van Vleet, J.R.H. *Pichia stipitis* genomics, transcriptomics, and gene clusters. *FEMS Yeast Res.* **2009**, *9*, 793–807. [CrossRef] [PubMed]
14. Van Vleet, J.H.; Jeffries, T.W. Yeast metabolic engineering for hemicellulosic ethanol production. *Curr. Opin. Biotechnol.* **2009**, *20*, 300–306. [CrossRef] [PubMed]
15. Kurtzman, C.P.; Suzuki, M. Phylogenetic analysis of the ascomycete yeasts that form coezyme Q-9 and the proposal of the new genera *Babjeviella*, *Meyerozyma*, *Millerozyma*, *Priceomyces* and *Scheffersomyces*. *Mycoscience* **2010**, *51*, 2–14. [CrossRef]

16. Van Dijken, J.P.; van den Bosch, E.; Hermans, J.J.; de Miranda, L.R.; Scheffers, W.A. Alcoholic fermentation by 'nonfermentative' yeasts. *Yeast* **1986**, *2*, 123–127. [CrossRef]
17. Jeffries, T.W.; Grigoriev, I.V.; Grimwood, J.; Laplaza, J.M.; Aerts, A.; Salamov, A.; Schmutz, J.; Lindquist, E.; Dehal, P.; Shapiro, H.; et al. Genome sequence of the lignocellulose-bioconverting and xylose fermenting yeast *Pichia stipitis*. *Nat. Biotechnol.* **2007**, *25*, 319–326. [CrossRef]
18. Laplaza, J.M.; Torres, B.R.; Jin, Y.S.; Jeffries, T.W. *Sh ble* and Cre adapted for functional genomics and metabolic engineering of *Pichia stipitis*. *Enzym. Microb. Technol.* **2006**, *38*, 741–747. [CrossRef]
19. Long, T.M.; Su, Y.K.; Headman, J.; Higbee, A.; Willis, L.B.; Jeffries, T.W. Cofermentation of glucose, xylose, and cellobiose by the beetle-associated yeast, *Spathaspora passalidarum*. *Appl. Environ. Microbiol.* **2012**, *78*, 5492–5500. [CrossRef]
20. Su, Y.K.; Willis, L.B.; Jeffries, T.W. Effects of aeration on growth, ethanol and polyol accumulation by *Spathaspora passalidarum* NRRL Y-27907 and *Scheffersomyces stipitis* NRRL Y-7124. *Biotechnol. Bioeng.* **2015**, *112*, 457–469. [CrossRef]
21. Nguyen, N.H.; Suh, S.O.; Marshall, C.J.; Blackwell, M. Morphological and ecological similarities: Wood-boring beetles associated with novel xylose-fermenting yeasts, *Spathaspora passalidarum* gen. sp. nov. and *Candida jeffriesii* sp. nov. *Mycol. Res.* **2006**, *110*, 1232–1241. [CrossRef] [PubMed]
22. Cadete, R.M.; Santos, R.O.; Melo, M.A.; Mouro, A.; Gonc¸alves, D.L.; Stambuk, B.U.; Gomes, F.C.O.; Lachance, M.A.; Rosa, C.A. *Spathaspora arborariae* sp. nov., a D-xylose-fermenting yeast species isolated from rotting wood in Brazil. *FEMS Yeast Res.* **2009**, *9*, 1338–1342. [PubMed]
23. Cadete, R.M.; Melo, M.A.; Dussán, K.J.; Rodrigues, R.C.; Silva, S.S.; Zilli, J.E.; Vital, M.J.; Gomes, F.C.; Lachance, M.A.; Rosa, C.A. Diversity and physiological characterization of D-xylose-fermenting yeasts isolated from the Brazilian Amazonian Forest. *PLoS ONE* **2012**, *7*, e43135. [CrossRef] [PubMed]
24. Cadete, R.M.; Melo, M.A.; Zilli, J.E.; Vital, M.J.; Mouro, A.; Prompt, A.H.; Gomes, F.C.; Stambuk, B.U.; Lachance, M.A.; Rosa, C.A. *Spathaspora brasiliensis* sp. nov., *Spathaspora suhii* sp. nov., *Spathaspora roraimanensis* sp. nov. and *Spathaspora xylofermentans* sp. nov., four novel (D)-xylose-fermenting yeast species from Brazilian Amazonian forest. *Antonie Leeuwenhoek* **2013**, *103*, 421–431. [CrossRef] [PubMed]
25. Souza, G.F.L.; Valentim, L.T.C.N.; Nogueira, S.R.P.; Abegg, M.A. Efficient production of second generation ethanol and xylitol by yeasts from Amazonian beetles (Coleoptera) and their galleries. *Afr. J. Microbiol. Res.* **2017**, *11*, 814–824.
26. Ren, Y.; Chen, L.; Niu, Q.; Hui, F. Description of *Scheffersomyces henanensis* sp. nov., a new D-xylose-fermenting yeast species isolated from rotten wood. *PLoS ONE* **2014**, *9*, e92315. [CrossRef]
27. Rodrussamee, N.; Sattayawat, P.; Yamada, M. Highly efficient conversion of xylose to ethanol without glucose repression by newly isolated thermotolerant *Spathaspora passalidarum* CMUWF1-2. *BMC Microbiol.* **2018**, *18*, 73. [CrossRef]
28. Cadete, R.M.; de las Heras, A.M.; Sandström, A.G.; Ferreira, C.; Gírio, F.; Gorwa-Grauslund, M.-F.; Rosa, C.A.; Fonseca, C. Exploring xylose metabolism in *Spathaspora* species: *XYL1.2 from Spathaspora passalidarum as the key for efficient anaerobic xylose fermentation in metabolic engineered Saccharomyces cerevisiae. Biotechnol. Biofuels* **2016**, *9*, 167.
29. Kwak, S.; Jin, Y.S. Production of fuels and chemicals from xylose by engineered *Saccharomyces cerevisiae*: A review and perspective. *Microb. Cell Fact.* **2017**, *16*, 82. [CrossRef]
30. Veras, H.C.T.; Parachin, N.S.; Almeida, J.R.M. Comparative assessment of fermentative capacity of different xylose-consuming yeasts. *Microb. Cell Fact.* **2017**, *16*, 153. [CrossRef]
31. Veras, H.C.T.; Campos, C.G.; Nascimento, I.F.; Abdelnur, P.V.; Almeida, J.R.M.; Parachin, N.S. Metabolic flux analysis for metabolome data validation of naturally xylose-fermenting yeasts. *BMC Biotechnol.* **2019**, *19*, 58. [CrossRef] [PubMed]
32. Nakanishi, S.C.; Soares, L.B.; Biazi, L.E.; Nascimento, V.M.; Costa, A.C.; Rocha, G.J.M.; Ienczak, J.L. Fermentation strategy for second generation ethanol production from sugarcane bagasse hydrolyzate by *Spathaspora passalidarum* and *Scheffersomyces stipitis*. *Biotechnol. Bioeng.* **2017**, *114*, 2211–2221. [CrossRef] [PubMed]
33. De Souza, R.D.F.R.; Dutra, E.D.; Leite, F.C.B.; Cadete, R.M.; Rosa, C.A.; Stambuk, B.U.; Stamford, T.L.M.; de Morais, M.A., Jr. Production of ethanol fuel from enzyme-treated sugarcane bagasse hydrolysate using d-xylose-fermenting wild yeast isolated from Brazilian biomes. *3 Biotech.* **2018**, *8*, 312. [CrossRef] [PubMed]

34. Wohlbach, D.J.; Kuo, A.; Sato, T.K.; Potts, K.M.; Salamov, A.A.; Labutti, K.M.; Sun, H.; Clum, A.; Pangilinan, J.L.; Lindquist, E.A.; et al. Comparative genomics of xylose-fermenting fungi for enhanced biofuel production. *Proc. Natl. Acad. Sci. USA* **2011**, *108*, 13212–13217. [CrossRef] [PubMed]
35. Jeppsson, M.; Bengtsson, O.; Franke, K.; Lee, H.; Hahn-Hagerdal, B.; Gorwa-Grauslund, M.F. The expression of a *Pichia stipitis* xylose reductase mutant with higher K(m) for NADPH increases ethanol production from xylose in recombinant *Saccharomyces cerevisiae*. *Biotechnol. Bioeng.* **2006**, *93*, 665–673. [CrossRef]
36. Petschacher, B.; Nidetzky, B. Altering the coenzyme preference of xylose reductase to favor utilization of NADH enhances ethanol yield from xylose in a metabolically engineered strain of Saccharomyces cerevisiae. *Microb. Cell* **2008**, *7*, 9. [CrossRef]
37. Bengtsson, O.; Hahn-Hagerdal, B.; Gorwa-Grauslund, M.F. Xylose reductase from *Pichia stipitis* with altered coenzyme preference improves ethanolic xylose fermentation by recombinant Saccharomyces cerevisiae. *Biotechnol. Biofuels* **2009**, *2*, 9. [CrossRef]
38. Guo, J.; Huang, S.; Chen, Y.; Guo, X.; Xiao, D. Heterologous expression of *Spathaspora passalidarum* xylose reductase and xylitol dehydrogenase genes improved xylose fermentation ability of *Aureobasidium pullulans*. *Microb. Cell Fact.* **2018**, *17*, 64. [CrossRef]
39. Li, H.; Fan, H.; Li, Y.; Shi, G.Y.; Ding, Z.Y.; Gu, Z.H.; Zhang, L. Construction and application of multi-host integrative vector system for xylose-fermenting yeast. *FEMS Yeast Res.* **2017**, *17*, 6. [CrossRef]
40. Hou, X.; Yao, S. Improved inhibitor tolerance in xylosefermenting yeast *Spathaspora passalidarum* by mutagenesis and protoplast fusion. *Appl. Microbiol. Biotechnol.* **2012**, *93*, 2591–2601. [CrossRef]
41. Morales, P.; Gentina, J.C.; Aroca, G.; Mussatto, S.I. Development of an acetic acid tolerant *Spathaspora passalidarum* strain through evolutionary engineering with resistance to inhibitors compounds of autohydrolysate of Eucalyptus globulus. *Ind. Crop Prod.* **2017**, *106*, 5–11. [CrossRef]
42. Yu, H.; Guo, J.; Chen, Y.; Fu, G.; Li, B.; Guo, X.; Xiao, D. Efficient utilization of hemicellulose and cellulose in alkali liquor-pretreated corncob for bioethanol production at high solid loading by *Spathaspora passalidarum* U1–58. *Bioresour. Technol.* **2017**, *232*, 168–175. [CrossRef] [PubMed]
43. Su, Y.K.; Willis, L.B.; Rehmann, L.; Smith, D.R.; Jeffries, T.W. *Spathaspora passalidarum* selected for resistance to AFEX hydrolysate shows decreased cell yield. *FEMS Yeast Res.* **2018**, *18*. [CrossRef] [PubMed]
44. Cho, J.Y.; Jeffries, T.W. *Pichia stipitis* genes for alcohol dehydrogenase with fermentative and respiratory functions. *Appl. Environ. Microbiol.* **1998**, *64*, 1350–1358. [CrossRef]
45. Lin, Y.; He, P.; Wang, Q.; Lu, D.; Li, Z.; Wu, C.; Jiang, N. The alcohol dehydrogenase system in the xylose-fermenting yeast *Candida maltosa*. *PLoS ONE* **2010**, *5*, e11752. [CrossRef]
46. Cho, J.Y.; Jeffries, T.W. Transcriptional control of ADH genes in the xylose fermenting yeast *Pichia stipitis*. *Appl. Environ. Microbiol.* **1999**, *65*, 2363–2368. [CrossRef]
47. Passoth, V.; Schafer, B.; Liebel, B.; Weierstall, T.; Klinner, U. Molecular cloning of alcohol dehydrogenase genes of the yeast *Pichia stipitis* and identification of the fermentative ADH. *Yeast* **1998**, *14*, 1311–1325. [CrossRef]
48. Passoth, V.; Cohn, M.; Schafer, B.; Hahn-Hagerdal, B.; Klinner, U. Molecular analysis of the hypoxia induced ADH2-promoter in the respiratory yeast *Pichia stipitis*. *Yeast* **2003**, *20*, S199. [CrossRef]
49. Passoth, V.; Cohn, M.; Schafer, B.; Hahn-Hagerdal, B.; Klinner, U. Analysis of the hypoxia-induced ADH2 promoter of the respiratory yeast *Pichia stipitis* reveals a new mechanism for sensing of oxygen limitation in yeast. *Yeast* **2003**, *20*, 39–51. [CrossRef]
50. Lu, C.; Chapter, V. In Elucidating Physiological Roles of *Pichia stipitis* Alcohol Dehydrogenases in Xylose Fermentation and Shuffling Promoters for Multiple Genes in *Saccharomyces cerevisiae* to Improve Xylose Fermentation. Ph.D. Thesis, University of Wisconsin, Madison, WI, USA, 2007.

11

Impact of Glucose Concentration and NaCl Osmotic Stress on Yeast Cell Wall β-D-Glucan Formation during Anaerobic Fermentation Process

Vassileios Varelas [1,*], Evangelia Sotiropoulou [1], Xara Karambini [1], Maria Liouni [1] and Elias T. Nerantzis [2]

1 Laboratory of Industrial Chemistry, Department of Chemistry, School of Science, National and Kapodistrian University of Athens, PO Box 157 71, Zografou, Athens 157 72, Greece; evasot210@hotmail.com (E.S.); xarakarambini@gmail.com (X.K.); mLiouni@chem.uoa.gr (M.L.)
2 Laboratory of Biotechnology & Industrial Fermentations, Department of Oenology and Beverage Technology, School of Food Technology and Nutrition, Technological Educational Institute (TEI) of Athens, Egaleo 122 43, Greece; enerat@teiath.gr
* Correspondence: vavarelas@chem.uoa.gr

Abstract: Yeast β-glucan polysaccharide is a proven immunostimulant molecule for human and animal health. In recent years, interest in β-glucan industrial production has been increasing. The yeast cell wall is modified during the fermentation process for biomass production. The impact of environmental conditions on cell wall remodelling has not been extensively investigated. The aim of this research work was to study the impact of glucose and NaCl stress on β-glucan formation in the yeast cell wall during alcoholic fermentation and the assessment of the optimum fermentation phase at which the highest β-glucan yield is obtained. VIN 13 *Saccharomyces cerevisiae* (*S. cerevisiae*) strain was pre-cultured for 24 h with 0% and 6% NaCl and inoculated in a medium consisting of 200, 300, or 400 g/L glucose. During fermentation, 50 mL of fermented medium were taken periodically for the determination of Optical Density (OD), cell count, cell viability, cell dry weight, β-glucan concentration and β-glucan yield. Next, dry yeast cell biomass was treated with lytic enzyme and sonication. At the early stationary phase, the highest β-glucan concentration and yield was observed for non-NaCl pre-cultured cells grown in a medium containing 200 g/L glucose; these cells, when treated with enzyme and sonication, appeared to be the most resistant. Stationary is the optimum phase for cell harvesting for β-glucan isolation. NaCl and glucose stress impact negatively on β-glucan formation during alcoholic fermentation. The results of this work could comprise a model study for yeast β-glucan production on an industrial scale and offer new perspectives on yeast physiology for the development of antifungal drugs.

Keywords: yeast fermentation; glucose stress; NaCl stress; β-glucan production; cell wall remodelling

---

1. Introduction

Saccharomyces cerevisiae is the model microorganism most commonly used in the study of yeast and higher eukaryotic cell physiology [1]. A major substrate for its growth is sugars, which are degraded through an aerobic or anaerobic fermentation process with the production of CO_2 or alcohol and CO_2, respectively [2]. Nowadays, the biotechnological research and industrial applications of *S. cerevisiae* have extended far beyond food and beverage production to medicine development, molecular biology and genetics, and environmental protection technologies [3].

In yeast, the plasma membrane is surrounded by the cell wall, a structure of 100–200 nm width that consists of consecutive layers. The main structural elements of the yeast cell wall are constructed of three groups of polysaccharides: (a) mannoproteins (or mannans): α-mannose polymers, which account for around 40% of the cell wall dry weight [4]; (b) α- and β-glucans: D-glucose homopolymers with α- or β-glycosidic bonds between C-1 and C-2, C-3, C-4, and C-6 of glucopyranose rings; they account for approximately 30–60% of the cell wall dry weight (β-D-glucans are the main cell wall glucans in which two different types are found: a major β-1,3-D-glucan (85%), which is the main β-glucan, representing more than 50–55% of cell wall, and a minor β-1,6-D-glucan (15%), representing 5–10% of cell wall dry weight [5–8]); and (c) chitin: β-1,4-D-*N*-acetyl-glucosamine homopolymers, equates to around 2% of the yeast cell wall dry weight [8,9]. The plasma membrane is an extension of the cell wall to the interior of the yeast cell. Its width is 7.5 nm and it consists of lipids and proteins that are found in much lower proportions compared to the cell wall components [10,11]. Yeast β-glucans (β-1,3-D-glucan and β-1,6-D-glucan) belong to biological response modifiers (BRMs) due to their ability to enhance and stimulate the human immune system presenting antitumor, anti-inflammatory, antimicrobial, wound healing, coronary heart disease prevention, hepatoprotective, weight loss, and antidiabetic properties [12–14].

Depending on the yeast growth and cultivation conditions, the cell wall dry weight may vary between 10% and 25% of the total cell dry weight [15]. The cell wall has four main functions: to (1) protect the cell from osmotic stress by maintaining its turgidity [10]; (2) protect the cell from mechanical stress; (3) maintain the cell's shape; and (4) act as a hold for the cell's surface proteins [15,16].

During yeast cell batch fermentation, the cell wall is modified. These changes are affected by the yeast strain and cell growth conditions [17–20]. The conditions of cell growth, such as pH, temperature, aeration, carbon sources, limitation of nitrogen, and the mode of cultivation, affect the cell wall polysaccharide composition, structure and morphology [21]. Also, the addition of supplements (SDS, ethylenediaminetetraacetic acid (EDTA), NaCl) in the yeast culture medium and the conditions during alcoholic fermentation for beer production impact variously on the cell wall β-glucan concentration at the end of the fermentation process [22] as it also modifies the polysaccharide structure of produced spent yeast biomass [23].

The remodelling of the yeast cell wall, as a response to various environmental stresses, is controlled by: (a) the cell wall integrity (CWI) and (b) high osmotic glycerol (HOG) through mitogen-activated protein kinase (MAPK) signalling pathways that modulate the cell wall gene expression and lead to the construction of a newly modified cell wall [24,25]. In addition to the HOG pathway, the cell growth, carbon storage and stress response are regulated by the RAS-cAMP PKA pathway [26], while during rapid hyperosmotic shocks the yeast cell sensitivity is regulated by the cell wall elasticity increase through the inactivation of the Crh family of cell wall cross-linking enzymes [25].

The impact of glucose concentration and NaCl osmotic stress on yeast performance during anaerobic fermentation has been well studied [27–35], but their impact on cell wall β-glucan production has not been studied until now.

The aim of this work was the investigation of yeast cell wall β-glucan production during the anaerobic fermentation process under the influence of hyperosmotic stress of fermented media glucose concentration and NaCl pre-cultured cells used for defined media inoculation. The motivations for this research study were: (a) the fact that there are few reports concerning yeast growth conditions and cell wall polysaccharide composition; (b) the lack of studies of the impact of different environmental stresses on yeast cell wall structure during the fermentation process and the yeast cell osmoadaptation responses against an osmotic shock; (c) the lack of knowledge of the entire physiological role of β-glucan in the yeast cell; (d) the fact that the monitoring of the yeast cell wall is a dynamic and unexplored research field that offers potential biotechnological applications and purposes; (e) the increasing interest in the industrial production of β-glucan from various yeast sources (medium cultures, brewery and winery waste yeast biomass) due to its proven immunostimulant properties and its incorporation in medicines, cosmetics and functional foods; and (f) the potential benefits from the study of the yeast cell wall in the creation of new antifungal drugs. Furthermore, the study of the yeast cell wall under various conditions can offer new perspectives on the biochemical pathways involved in the alcoholic fermentation process and the biology of the stressed yeast cell.

2. Materials and Methods

2.1. Yeast Cells and Culture Growth Conditions

Yeast cells *S. cerevisiae* VIN 13 strain (commercial dry yeast) used for laboratory experiments were provided by Anchor Yeast (Cape Town, South Africa). Yeast cells were grown in a medium containing (per litre of deionised water): 100 g D-glucose, 1 g K_2HPO_4, 1 g KH_2PO_4, 0.2 g $ZnSO_4$, 0.2 g $MgSO_4$, 2 g yeast extract and 2 g NH_4SO_4. All the media components were purchased from the Sigma Chemical Company (St. Louis, MO, USA).

2.2. Yeast Cell Preconditioning and Inoculum Preparation

One gram dry weight of yeast was resuspended in 100 mL of deionised water in an Erlenmeyer flask of 250 mL volume, at 30–35 °C, for 30 min with NaCl 6% *w/v*. The inoculum for experimental fermentations was prepared as follows: after 24 h of pre-culturing, 10 mL was collected and centrifuged at 5000 rpm for 15 min. Cells were resuspended in deionised water and re-centrifuged. This was repeated twice prior to the determination of total cell number and cell viability in the final washed inoculum; 2×10^5 of living cells were used to inoculate 1000 mL of medium substrate [27].

2.3. Preparation of Yeast Fermentation Medium

The medium for laboratory batch fermentations consisted of the following: 200, 300, 400 g/L glucose, 1 g/L K_2HPO_4, 1 g/L KH_2PO_4, 0.2 g/L $ZnSO_4$, 0.2 g/L $MgSO_4$, 2 g/L yeast extract and 2 g/L NH_4SO_4 (Sigma-Aldrich Co., St. Louis, MO, USA). Mineral components and glucose were sterilised separately at 120 °C, and 2 atm pressure for 20 min. The pH was set to 4.0 with the use of 1 N HCl. Batch fermentations were carried out in 2000 mL volume flasks containing 1000 mL of growth medium with NaCl-preconditioned cells without shaking at 25 °C [36].

2.4. Cell Wall Treatment with Lytic Enzyme and Sonication

Yeast cells were treated with lytic enzyme Glucanex®200G (β-1,3-glucanase with some β-1,6-glucanase) (Novozymes, Kobenhavn, Denmark) for the cell wall lysis (pH 4.64/50 °C/5 h/10 times higher than the normal dose of 0.015 g/L, optical density of the resuspended cells with 10 mM citrate buffer adjusted to approximately 1.0 OD at 600 nm, 15% *w/v* cell concentration) [18,37,38] and sonication for the cell disruption at 35 kHz/70 W/6 min (u = 230 V/AC, I = 1.6 A, f = 50/60 Hz) in an ice bath with a Transsonic 570/H sonicator (Elma, Singen, Germany) [39].

2.5. Determination of Cell Concentration

Yeast cell number was determined by counting using a hemocytometer (Neubauer type) [36].

2.6. Cell Viability Determination

Cell viability was determined using a haemocytometer (Neubauer type) and according to the methylene blue method: 1 mL of sample medium was taken and diluted in 9 mL of deionised water. 1 mL of this solution was dissolved with 1 mL of 10% *v/v* methylene blue solution and left for 10 min. Aliquots of 1 µL were placed on the haemocytometer using a Pasteur pipette. The haemocytometer was then microscopically observed with an optical microscope (Olympus model CHK2-F-GS microscope). Yeast cell viability was calculated and expressed as follows:

$$\text{Viability (\%)} = a/n \times 100,$$

where

a: number of metabolically active cells
n: total cell number [27].

2.7. Monitoring of Fermentation Kinetics

OD at 600 nm was determined spectrometrically (phasmatophotometer mini 1240, SHIMADZU, Beijing, China) and was used for the monitoring of fermentation kinetics and the discrimination of exponential, lag and death phase of cells growth during the fermentation process [18].

2.8. Determination of Cell Dry Weight

The sediment of yeast cells was frozen (−80 °C/24 h), lyophilised (−50 °C /vacuum/24 h) using a Thermo Fischer (Waltham, MA, USA) drying digital unit. The weight of each yeast cells freeze dried sample was measured gravimetrically with an analytical balance (Kern, Kern and Sohn GmbH, Balingen, Germany) with an accuracy of four decimal places.

2.9. Determination of β-Glucan Concentration (%)

The determinations of total yeast β-glucan concentration in lyophilised yeast biomass were performed with the use of Enzymatic K-EBHLG Yeast β-Glucan Assay Kit (Megazyme, Bray, Ireland).

2.10. Determination of β-Glucan Yield (%)

The yield of β-glucan was calculated as the product of yeast cells' dry weight with β-glucan purity [40].

2.11. Statistical Analysis

Each experiment was carried out in triplicate and the reproducibility was within the range of ±5%. Results are displayed as means of three determinations in all methods with standard deviation. Experiments were set up in a completely randomised design while for the determination of significant differences between the different treatments and the tested parameters, all results were analysed using two-way analysis of variance (ANOVA) run on XLSTAT software (Addinsoft Co., New York, NY, USA), with significant differences indicated at $p \leq 0.05$.

3. Results

3.1. Experimental Design

Yeast cells *S. cerevisiae* VIN 13 strain were inoculated in a medium containing three different glucose concentrations: 200, 300, or 400 g/L. The inoculation of each sample of the fermentation batch was performed according to the instructions of the yeast company (Anchor Yeast, Cape Town, South Africa) for wine production with 2×10^5 living cells of 0.1 g dry yeast preconditioned in 0% and 6% NaCl for 24 h before the inoculation according to the procedure described by Logothetis et al. (2013) [41]. The volume of the inoculated substrate was 1000 mL for each sample. During alcoholic fermentation, for the monitoring of fermentation kinetics, samples of 50 mL from each fermented medium were collected every 8 h after homogenisation of the fermented medium for the direct determination of cell viability and optical absorption (OD). In a next step, these 50 mL samples were centrifuged (5000 rpm/10 min) and the sediment (yeast biomass) was freeze dried for the determination of cell dry weight and the yeast cell β-glucan concentration. Subsequently, the freeze-dried yeast cells were treated with lytic enzyme and sonication for the estimation of cell wall resistance and its correlation with cell wall β-glucan amount. After the centrifugation of the samples and before their freeze drying, they were washed three times with distilled water for the removal of the possible glucose that may have remained in the sample. This was applied in order to avoid glucose weight being taken into account in the determination of the dry weight of the sample, which might cause subsequent data errors to arise. After 192 h for the completion of the alcoholic fermentation, the suspension was separated from the yeast biomass sediment (centrifugal/5000 rpm/10 min) for the determination of various oenological analytical parameters, while the sediment was freeze-dried for β-glucan determination in yeast cell biomass.

3.2. Cell Viability

For all the samples, cell viability rose during the early exponential phase of fermentation and reached a first peak after 16 h. At this point, cells of samples (200 g/L glucose, 0% NaCl), (200 g/L glucose, 6% NaCl) and (300 g/L, 6% NaCl) showed the highest viability (%) (97.78 ± 2.04%, 97.36 ± 1.59% and 97.28 ± 1.02% respectively) while NaCl-stressed cells of samples (300 g/L glucose, 6% NaCl) and (400 g/L glucose, 6% NaCl) had higher viability portions (97.28 ± 1.02% and 93.43 ± 1.44% respectively) compared to the corresponding (300 g/L, 0% NaCl) and (400 g/L glucose, 0% NaCl) ones (93.94 ± 1.42% and 89.89 ± 1.13% respectively) (Figure 1). The cell viability reached a second, lower peak during the end of the exponential and the beginning of the stationary phase for all samples. The more viable cells were counted for samples (200 g/L glucose, 0% NaCl) and (200 g/L glucose, 6% NaCl) (97.84 ± 2.03% and 93.54 ± 1.88% respectively) (Figure 1). At the death phase (48–120 h), the cell viability decreased for all the samples, with a higher decrease for samples (300 g/L glucose, 6% NaCl) and (400 g/L glucose, 6% NaCl) (Figure 1). From 120 to 192 h, an increase in cell viability occurred for all samples, but this was more abrupt for NaCl-stressed cells of samples (300 g/L glucose, 6% NaCl) and (400 g/L glucose, 6% NaCl). Despite this increase, for all samples, after 192 h of fermentation, cell viability was lower when compared to values of the second peak. Finally, the viability of NaCl preconditioned and non-preconditioned yeast cells growing in a medium containing 200 g/L glucose was similar at the end of the fermentation (85.34 ± 1.3% and 83.44 ± 0.8% respectively) (Figure 1).

3.3. Optical Density (OD) and Cell Count

The results of OD (600 nm) measurements showed that the lag phase was 0–8 h for all the fermented samples; the exponential phase was between 0–40 h for the samples (200 g/L glucose, 0% NaCl) and (200 g/L glucose, 6% NaCl) and 0–48 h for the others. The stationary phase for the samples (200 g/L glucose, 0% NaCl) and (200 g/L glucose, 6% NaCl) was short (40–48 h), while for the other samples it was longer (48–72 h). The death phase was determined to be between 48 and 192 h for the samples (200 g/L glucose, 0% NaCl) and (200 g/L glucose, 6% NaCl)) and 72–192 h for the samples (300 g/L glucose, 0% NaCl), (300 g/L glucose, 6% NaCl), (400 g/L glucose, 0% NaCl), and (400 g/L glucose, 6% NaCl) (Figure 2).

The cell count was used in order to determine the total yeast cell number during the fermentation process. Our results revealed that during the death phase, the OD (600 nm) and the cell number decreased for all the samples but, surprisingly, between 120 and 192 h the number of cells grown in 400 g/L glucose (0% and 6% NaCl) increased considerably (198.75×10^6 and 186.44×10^6 cells/mL respectively) compared with the other media (200 g/L glucose, 0% NaCl), (200 g/L glucose, 6% NaCl), (300 g/L glucose, 0% NaCl) and (200 g/L, 6% NaCl) (24.15×10^6, 30.12×10^6, 52.13×10^6 and 45.54×10^6 cells/mL respectively) (Figure 3). The samples with 200 g/L glucose showed two peaks, at 24 and 48 h, with the first peak a little higher than the second. The other samples (200 g/L glucose, 0% NaCl), (200 g/L glucose, 6% NaCl), (300 g/L glucose, 0% NaCl), (200 g/L glucose, 6% NaCl) showed one peak at 24 h, which was significantly higher compared to that of samples (200 g/L glucose, 0% NaCl) and (200 g/L glucose, 6% NaCl), thus indicating a higher cell number (Figure 3).

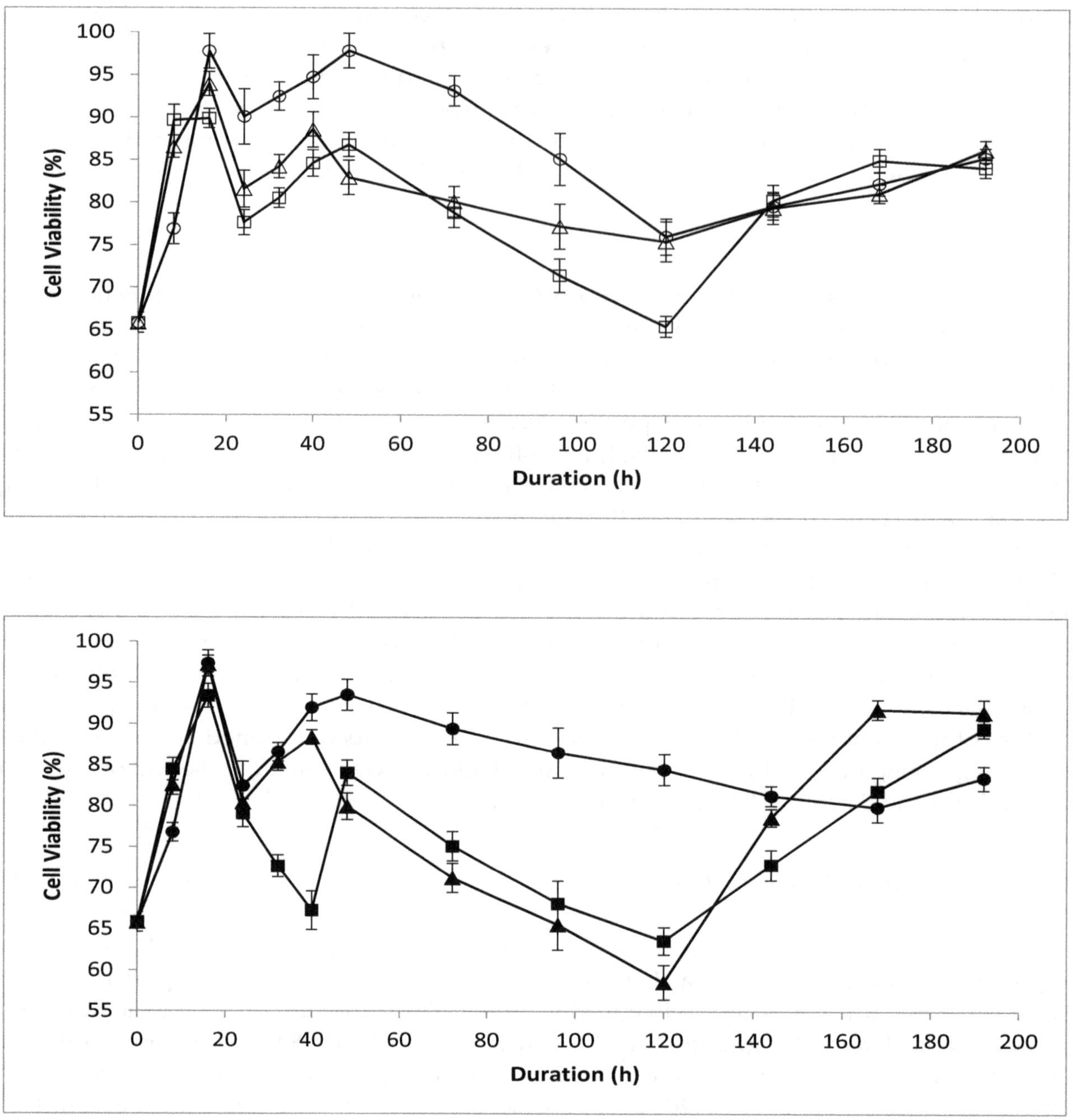

Figure 1. (**A**) Yeast cell viability (%) of non-NaCl-stressed cells during various fermentation times (h) with three different media containing 200, 300, or 400 g/L glucose ($n = 3$ replications for each time). (**B**) Yeast cell viability (%) of NaCl-stressed cells during the fermentation process (h) with three different media containing 200, 300, or 400 g/L glucose ($n = 3$ replications for each time) (○): (200 g/L glucose, 0% NaCl), (Δ): (300 g/L glucose, 0% NaCl), (□): (400 g/L glucose, 0% NaCl), (●): (200 g/L glucose, 6% NaCl), (▲): (300 g/L glucose, 6% NaCl), (■): (400 g/L glucose, 6% NaCl).

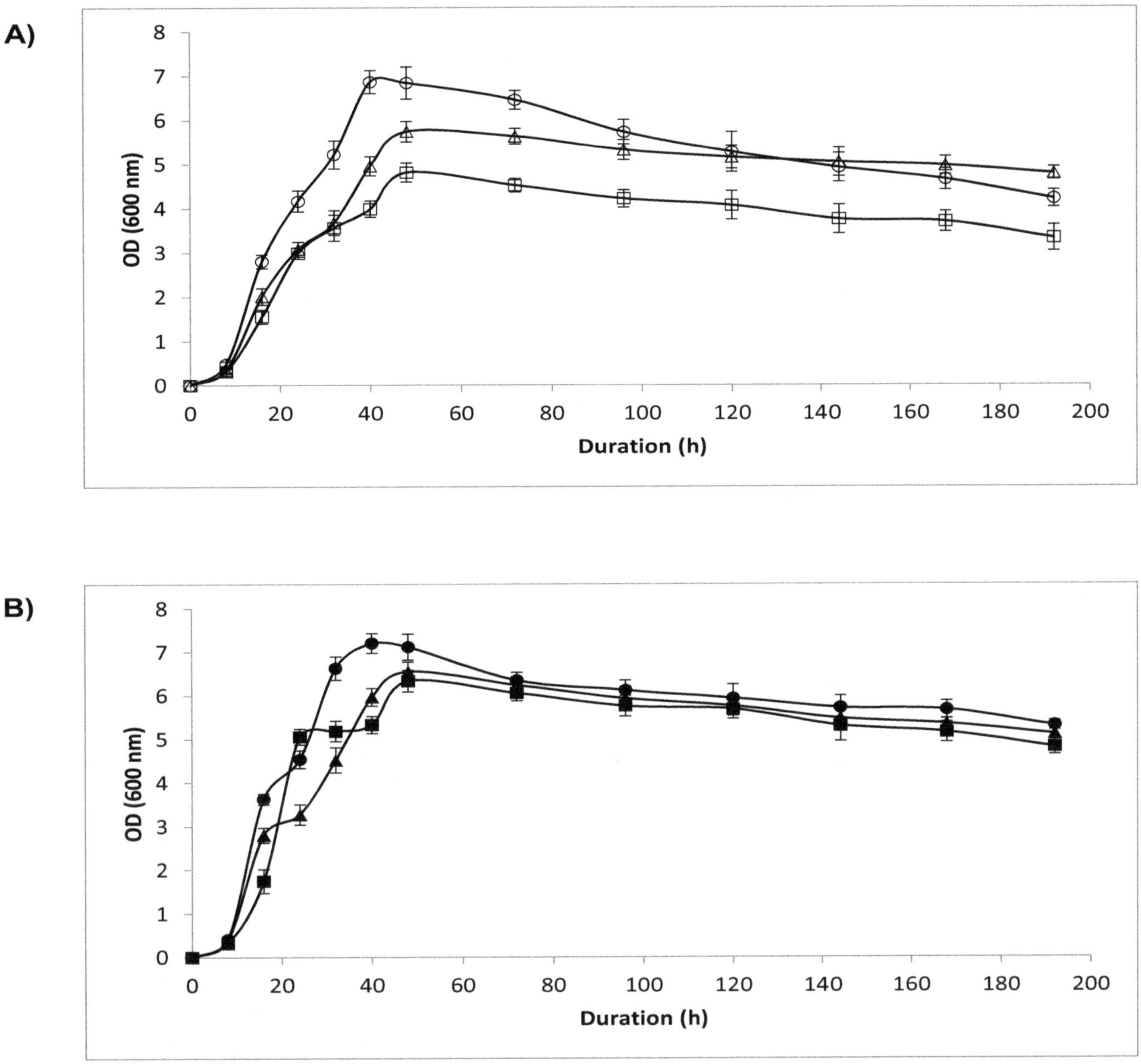

Figure 2. (**A**) OD (600 nm) of non-NaCl-stressed cells during various fermentation times (h) with three different media containing 200, 300, or 400 g/L glucose ($n = 3$ replications for each time); (**B**) OD (600 nm) of NaCl-stressed cells during the fermentation process (h) with three different media containing 200, 300, or 400 g/L glucose ($n = 3$ replications for each time) (○): (200 g/L glucose, 0% NaCl), (Δ): (300 g/L glucose, 0% NaCl), (□): (400 g/L glucose, 0% NaCl), (●): (200 g/L glucose, 6% NaCl), (▲): (300 g/L glucose, 6% NaCl), (■): (400 g/L glucose, 6% NaCl).

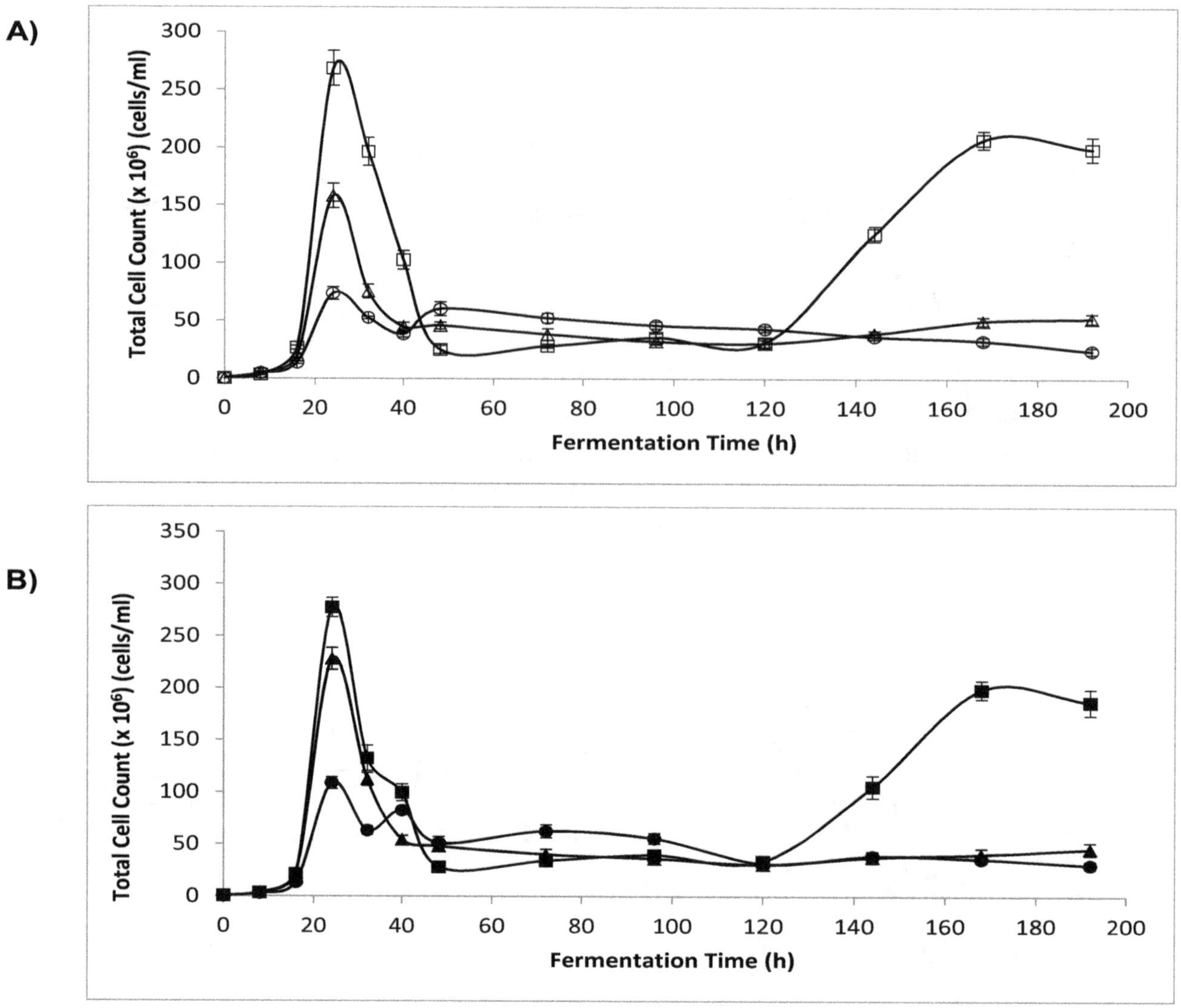

Figure 3. (**A**) Total cell count (cells/mL) of non-NaCl-stressed cells during various fermentation times (h) with three different media containing 200, 300, or 400 g/L glucose after cell treatment with sonication ($n = 3$ replications for each time and error bars represent standard deviation of the average value of all replications with each range of fermentation time); (**B**) Total cell count (cells/mL) of NaCl-stressed cells during various fermentation times (h) with three different media containing 200, 300, or 400 g/L glucose after cell treatment with sonication ($n = 3$ replications for each time and error bars represent standard deviation of the average value of all replications with each range of fermentation time) (○): (200 g/L glucose, 0% NaCl), (Δ): (300 g/L glucose, 0% NaCl), (□): (400 g/L glucose, 0% NaCl), (●): (200 g/L glucose, 6% NaCl), (▲): (300 g/L glucose, 6% NaCl), (■): (400 g/L glucose, 6% NaCl).

3.4. Dry Weight and Sample β-Glucan Yield

The freeze-dried yeast cell biomass obtained from a 50-mL sample of fermented medium at various times was determined gravimetrically. For each of the fermented media, the dry weight of yeast biomass rose constantly until the early stationary phase and then decreased constantly during the death phase only for media (200 g/L glucose, 0% NaCl), (200 g/L glucose, 6% NaCl) and (300 g/L glucose, 0% NaCl). The fermented media (300 g/L glucose, 6% NaCl), (400 g/L glucose, 6% NaCl) and (400 g/L glucose, 6% NaCl) showed a smaller decline, with a slight increase in yeast biomass for (300 g/L glucose, 6% NaCl) and (400 g/L glucose, 0% NaCl) and a more significant one for (400 g/L glucose, 6% NaCl) at the end of the fermentation (144–192 h) (Figure 4). The highest value of yeast dry weight (0.1900 ± 0.0066 g) was obtained with 200 g/L glucose, NaCl-stressed cells, with 40 h fermentation time (Figure 4).

A)

B)

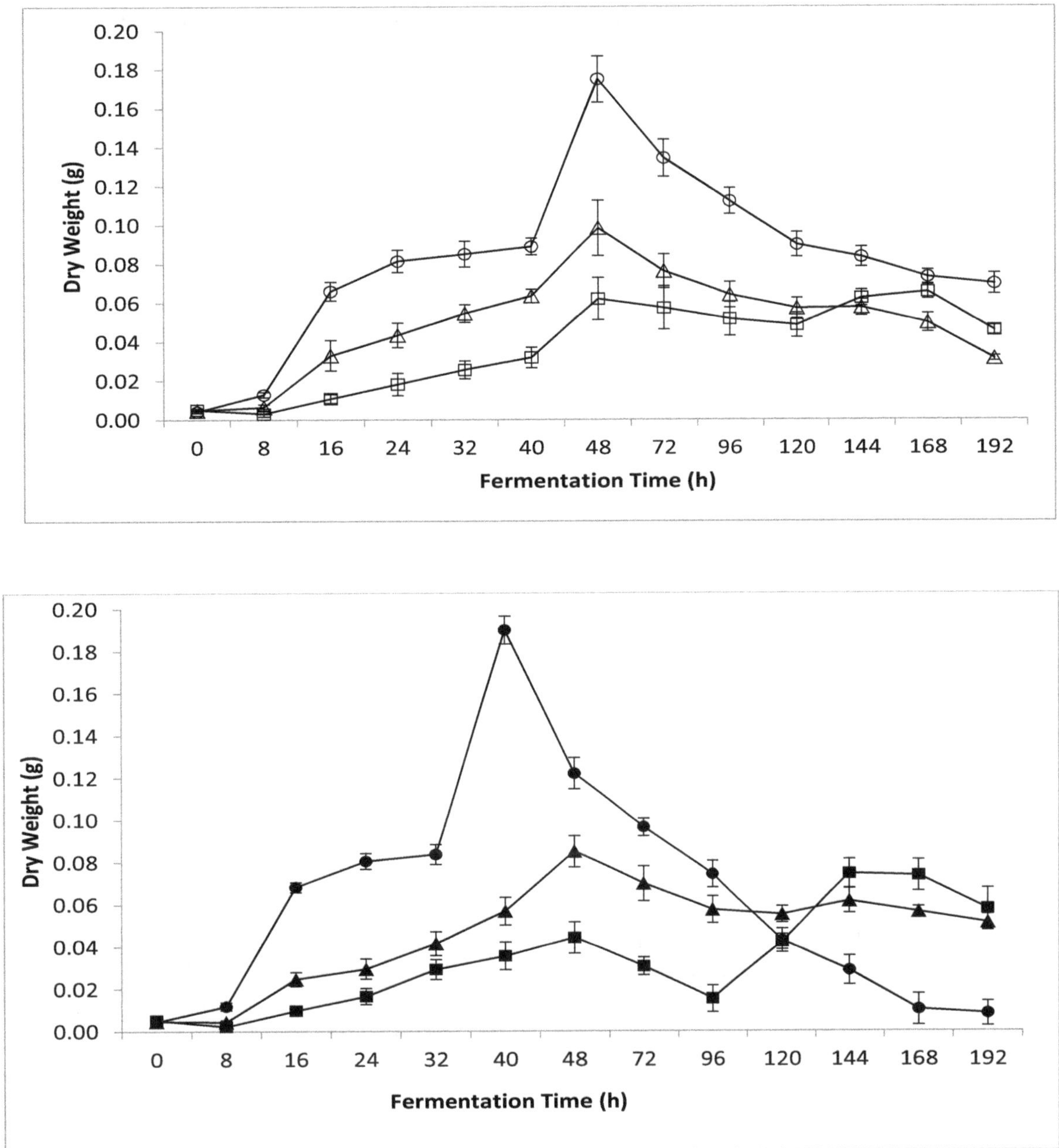

Figure 4. (**A**) Dry weight (g) of a 50-mL sample of non-NaCl-stressed cells during various fermentation times (h) with three different media containing 200, 300, or 400 g/L glucose ($n = 3$ replications for each time and error bars represent standard deviation of the average value of all replications with each range of fermentation time); (**B**) Dry weight (g) of a 50-mL sample of non-NaCl stressed cells during various fermentation times (h) with three different media containing 200, 300, or 400 g/L glucose ($n = 3$ replications for each time and error bars represent standard deviation of the average value of all replications with each range of fermentation time) (○): (200 g/L glucose, 0% NaCl), (Δ): (300 g/L glucose, 0% NaCl), (□): (400 g/L glucose, 0% NaCl), (●): (200 g/L glucose, 6% NaCl), (▲): (300 g/L glucose, 6% NaCl), (■): (400 g/L glucose, 6% NaCl).

The β-Glucan yield (g) was optimum for media (200 g/L glucose, 0% NaCl), (200 g/L glucose, 6% NaCl), (300 g/L glucose, 0% NaCl) and (300 g/L glucose, 6% NaCl) at the stationary phase

($p \leq 0.05$), while for medium (400 g/L glucose, 0% NaCl) it reached a peak at the stationary phase and remained constant until the end of fermentation and for medium (400 g/L glucose, 6% NaCl) it was optimum at the end of fermentation (Figure 5). The highest β-Glucan yield (0.1038 ± 0.0073 g) was obtained with fermented medium (200 g/L glucose/0% NaCl (non-stressed cells)/48 h fermentation time) at the stationary phase (Figure 5).

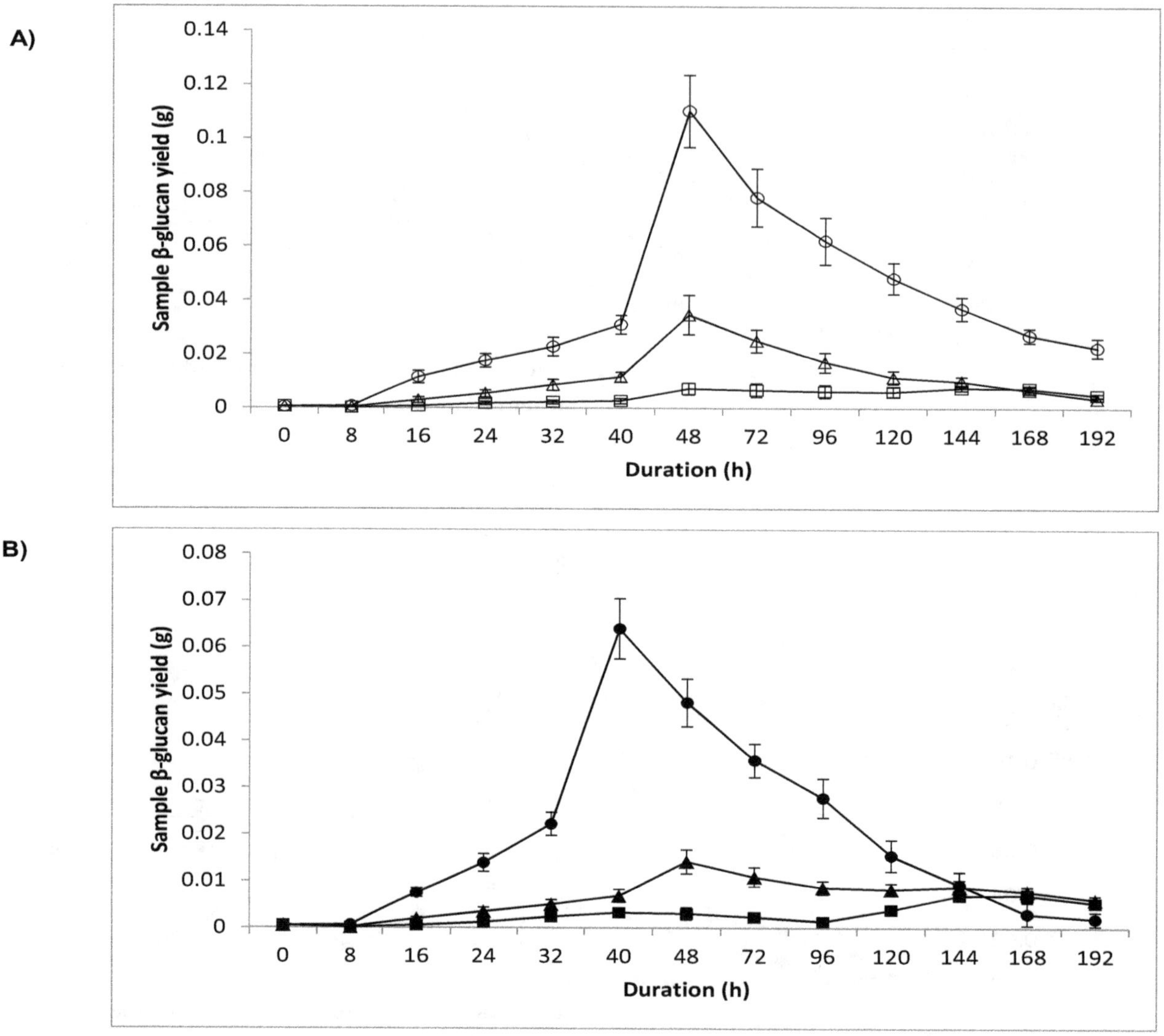

Figure 5. (**A**) β-Glucan yield (%) of a 50-mL sample of non-NaCl-stressed cells during various fermentation times (h) with three different media containing 200, 300, or 400 g/L glucose (n = 3 replications for each time and error bars represent standard deviation of the average value of all replications with each range of fermentation time). Mean values between 0–144 h were significantly different at $p \leq 0.05$ level. (**B**) β-Glucan yield (%) of a 50-mL sample of non-NaCl-stressed cells during various fermentation times (h) with three different media containing 200, 300, or 400 g/L glucose (n = 3 replications for each time and error bars represent standard deviation of the average value of all replications with each range of fermentation time). Mean values between 0–144 h were significantly different at $p \leq 0.05$ level. (○): (200 g/L glucose, 0% NaCl), (Δ): (300 g/L glucose, 0% NaCl), (□): (400 g/L glucose, 0% NaCl), (●): (200 g/L glucose, 6% NaCl), (▲): (300 g/L glucose, 6% NaCl), (■): (400 g/L glucose, 6% NaCl).

3.5. β-Glucan Concentration

For the fermented media (200 g/L glucose, 0% NaCl), (200 g/L glucose, 6% NaCl), (300 g/L glucose, 0% NaCl) and (300 g/L glucose, 6% NaCl), the highest yeast cell wall β-glucan concentration was observed at the stationary phase ($p \leq 0.05$) (Figure 6). For media (400 g/L glucose, 0% NaCl) and (400 g/L glucose, 6% NaCl), the β-glucan concentration rose until the cell culture entered the stationary phase and remained stable during the death phase (Figure 6). The highest β-glucan yeast cell concentration was observed for non-NaCl-stressed and NaCl-stressed cells grown in a medium containing 200 g/L glucose (62.95 ± 3.39 % and 39.49 ± 1.78 % respectively). Also, cells grown in these media showed a β-glucan concentration higher at the end of the alcoholic fermentation compared to the cells grown with the other media (Figure 6).

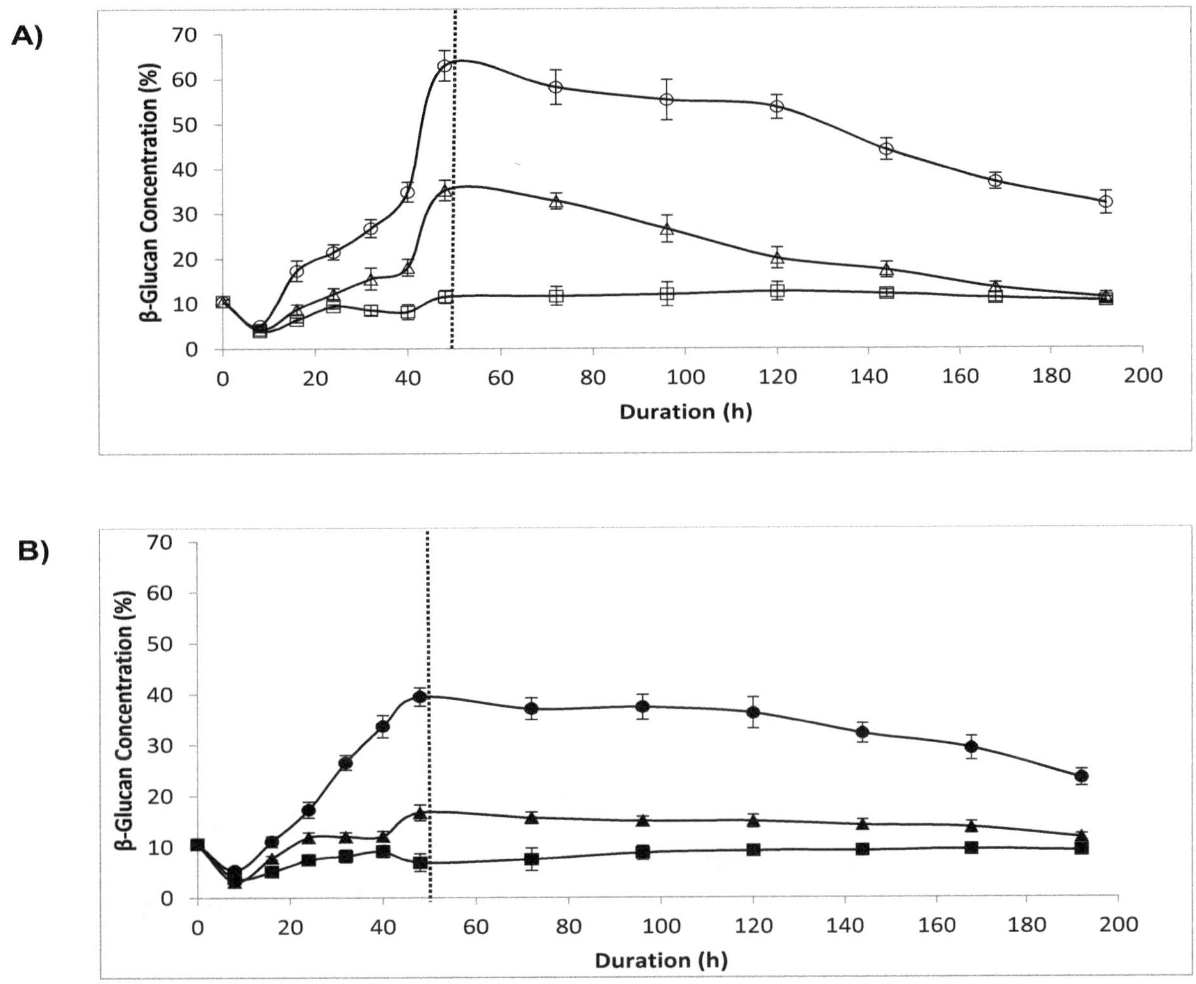

Figure 6. (**A**) Yeast cell wall β-glucan (%) of non-NaCl-stressed cells during various fermentation times (h) with three different media containing 200, 300, or 400 g/L glucose ($n = 3$ replications for each time and error bars represent standard deviation of the average value of all replications with each range of fermentation time). Mean values between 0–144 h were significantly different at $p \leq 0.05$ level. (**B**) Yeast cell wall β-glucan (%) of NaCl-stressed cells during various fermentation times (h) with three different media containing 200, 300, or 400 g/L glucose ($n = 3$ replications for each time and error bars represent standard deviation of the average value of all replications with each range of fermentation time). Mean values between 0–144 h were significantly different at $p \leq 0.05$ level. The stationary phase is indicated with the vertical line. (○): (200 g/L glucose, 0% NaCl), (Δ): (300 g/L glucose, 0% NaCl), (□): (400 g/L glucose, 0% NaCl), (●): (200 g/L glucose, 6% NaCl), (▲): (300 g/L glucose, 6% NaCl), (■): (400 g/L glucose, 6% NaCl).

3.6. Cell Wall Treatment with Lytic Enzyme and Sonication

The highest cell wall resistance to the end of exponential phase either to lytic enzyme or to sonication was observed for sample (200 g/L glucose, 0% NaCl), which had the highest β-glucan concentration. A decline in these results was observed for the fermented media (400 g/L glucose, 0% NaCl) and (400 g/L glucose, 6% NaCl) during the death phase (120–192 h of fermentation); despite their lowest β-glucan cell wall content, they appeared to be more resistant to the action of lytic enzyme but not to ultrasounds when compared to cells of samples (300 g/L glucose, 0% NaCl) and (300 g/L glucose, 6% NaCl) (Figures 7 and 8).

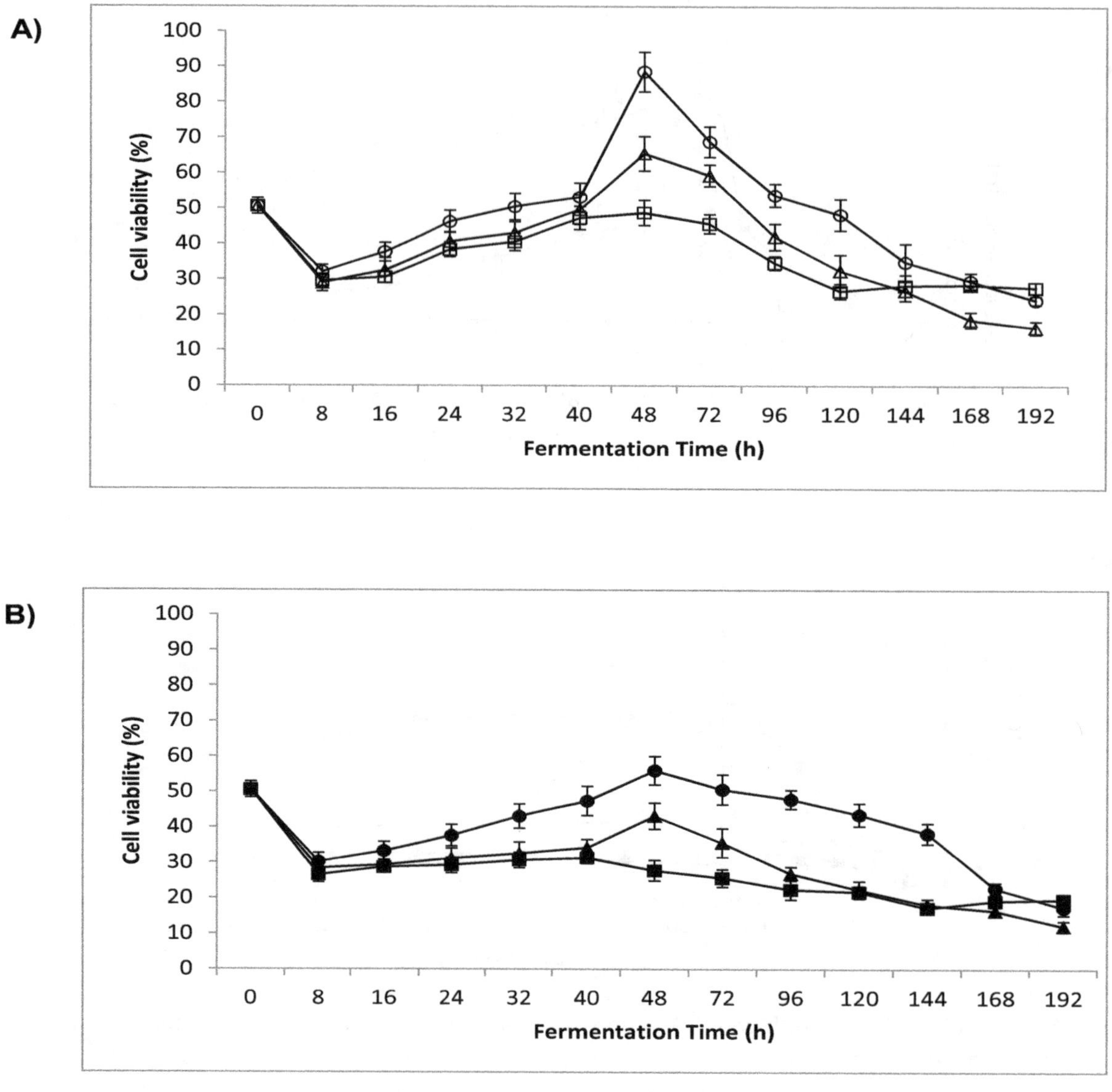

Figure 7. (**A**) Yeast cell viability (%) of non-NaCl-stressed cells during various fermentation times (h) with three different media containing 200, 300, or 400 g/L glucose after cell treatment with Glucanex 200G lytic enzyme (n = 3 replications for each time and error bars represent standard deviation of the average value of all replications with each range of fermentation time); (**B**) Yeast cell viability (%) of NaCl-stressed cells during various fermentation times (h) with three different media containing 200, 300, or 400 g/L glucose after cell treatment with Glucanex 200G lytic enzyme (n = 3 replications for each time and error bars represent standard deviation of the average value of all replications with each range of fermentation time) (○): (200 g/L glucose, 0% NaCl), (Δ): (300 g/L glucose, 0% NaCl), (□): (400 g/L glucose, 0% NaCl), (●): (200 g/L glucose, 6% NaCl), (▲): (300 g/L glucose, 6% NaCl), (■): (400 g/L glucose, 6% NaCl).

A)

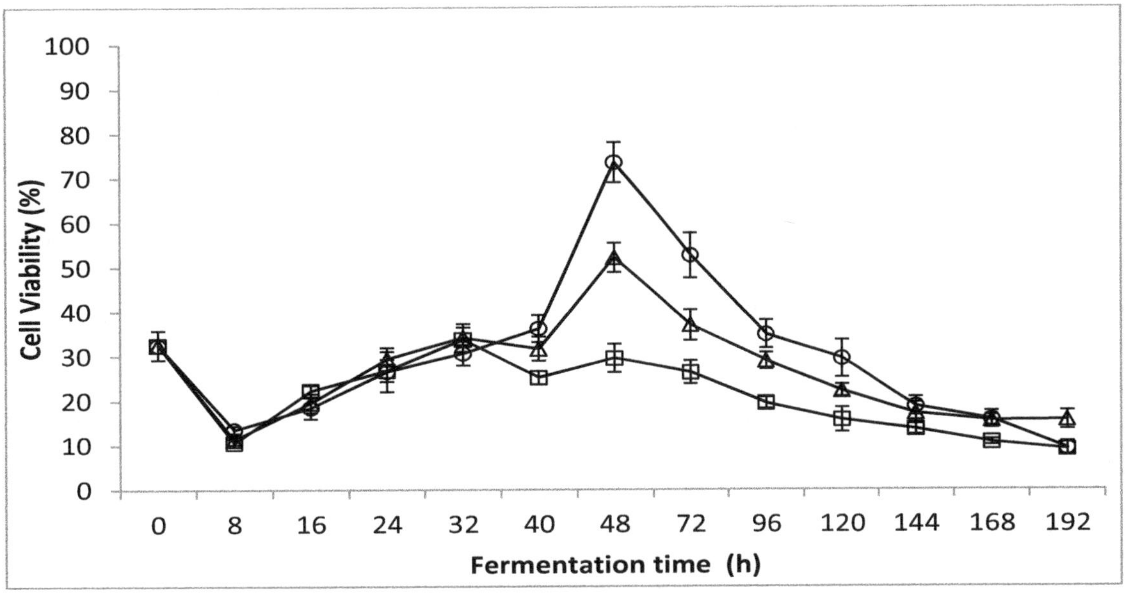

B)

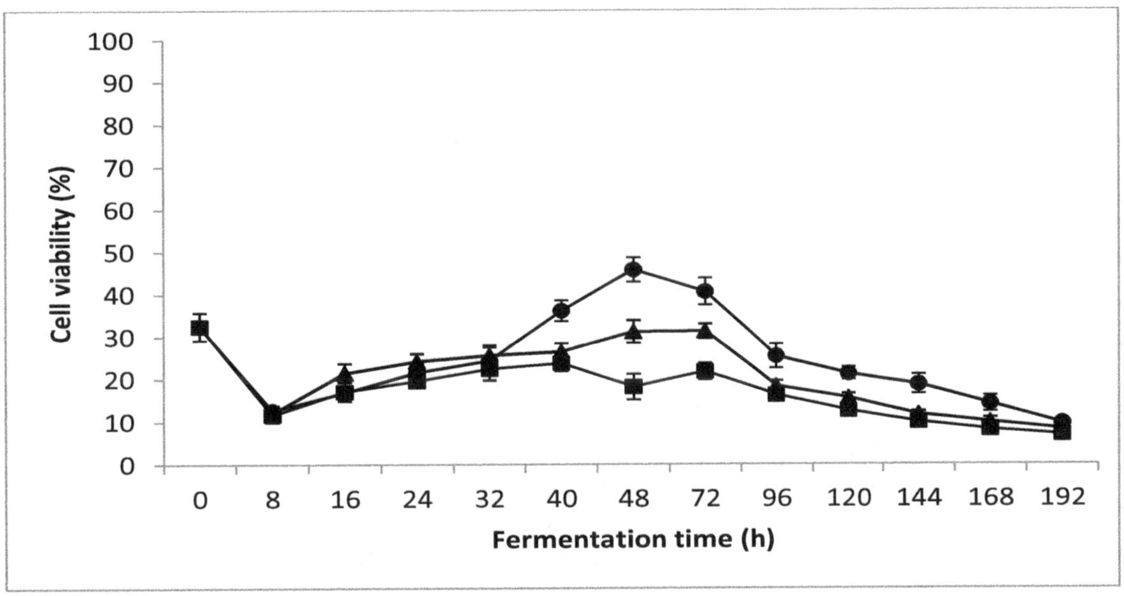

Figure 8. (**A**) Yeast cell viability (%) of non-NaCl-stressed cells during various fermentation times (h) with three different media containing 200, 300, or 400 g/L glucose after cell treatment with sonication ($n = 3$ replications for each time and error bars represent standard deviation of the average value of all replications with each range of fermentation time); (**B**) Yeast cell viability (%) of NaCl-stressed cells during various fermentation times (h) with three different media containing 200, 300, or 400 g/L glucose after cell treatment with sonication ($n = 3$ replications for each time and error bars represent standard deviation of the average value of all replications with each range of fermentation time) (○): (200 g/L glucose, 0% NaCl), (Δ): (300 g/L glucose, 0% NaCl), (□): (400 g/L glucose, 0% NaCl), (●): (200 g/L glucose, 6% NaCl), (Δ): (300 g/L glucose, 6% NaCl), (■): (400 g/L glucose, 6% NaCl).

4. Discussion

The impact of various environmental stresses on the yeast fermentation performance has been well studied and reviewed before [27–32,42,43], but the study of the yeast cell wall composition during the fermentation process is insufficient, with only a few published works referring to the impact of fermentation growth conditions and mode of cultivation on the yeast cell wall structure [20,21,44], the mode of fermentation on yeast cell wall β-glucan content [18], and the role of additives in the culture medium in β-glucan production [22,45]. The impact of glucose and NaCl stress on yeast cell wall remodelling via the study of the fluctuation of β-glucan concentration during the fermentation process has not been studied until now.

4.1. Cell Viability

Cell viability measurement with the methylene blue method is an established technique for yeast cell staining [46], used for the estimation of clear viable yeast cells [36,38]. Cell viability measurements were conducted: (a) for the estimation of the impact of stress conditions on cell viability during the fermentation process, and (b) for the correlation of yeast β-glucan concentration with cell biomass dry weight and total cell number, as non-viable and lysed cells were expected to have a higher β-glucan concentration (expressed as % of the yeast dry weight). In our experiments, cell viability reached two peaks and then decreased for all the fermented samples (Figure 1). At this point, the glucose concentration stress seems to be critical as cell viability decreases as glucose increases but the corresponding values for NaCl-stressed and non-stressed cells were similar for the same glucose concentration for all fermented samples (Figure 1). At the death phase (48–120 h), the synergistic action of glucose and NaCl stress seems to impact negatively on the cell viability, with a larger decrease in viable cell count for fermented samples with NaCl-stressed cells grown in media containing 300 and 400 g/L glucose (Figure 1).

Our results differed from the observations of other researchers under the same glucose (200 g/L) and NaCl (6%) stress conditions with the use of the *S. cerevisiae* VIN 13 strain, which observed a constant decrease of cell viability during the fermentation process [27]. These differences might be due to the different cell NaCl pre-culturing time (24 h in our experiments, 48 h in their work), the different volume of fermented medium (1000 mL in our experiments, 250 mL in their work) and the different package containing dry yeast used for carrying out the fermentations.

Also, at the end of death phase, the NaCl-preconditioned cells grown in the denser glucose medium had the highest viable cell values. These results cannot be compared with other researchers' results that used the same yeast strain with salt pre-culturing (NaCl 6% *w*/*v*) but in an even denser glucose medium (550 g/L), as cell viability values are not given [41]. Our results were in accordance with the observations of other researchers who found that NaCl pre-cultured yeast cells showed an increased cell viability at the end of the fermentation process [36]. Despite this increase, for all samples, after 192 h of fermentation, cell viability was lower compared to the values of the second peak. Lower cell viability was possibly caused by cell autolysis during the death phase of the alcoholic fermentation and consequently to a rise in the cell's dry weight due to cytoplasm leaking out from the cell to the supernatant [43]. Concerning the impact of NaCl 6% (*w*/*v*) concentration on yeast cells, our results with NaCl pre-cultured and non-pre-cultured yeast cells growing in a medium containing 200 g/L glucose differed from results of other researchers that indicated that higher osmotic shock conditions (>5% *w*/*v* NaCl) resulted in a higher cell viability at the end of the fermentation compared to the non-NaCl-stressed cells, while the viability value (83.44 ± 0.8%) for NaCl-stressed cells was in accordance with the corresponding one that previous researchers reported [27,47] (Figure 1). Concerning all the fermented media with the same glucose concentration, they appeared to have almost the same cell number at the end of fermentation (Figure 1), thus indicating that NaCl stress for the same glucose concentration had almost no impact on the cell viability at the end of the fermentation (Figure 1). As a general rule, the viability results were in accordance with the results reported by Logothetis et al. (2014), which suggested that salt pre-culturing of yeast *S. cerevisiae* impacts positively on cell viability [36].

4.2. Optical Density and Cell Count

Optical density (OD) is used for the estimation of yeast growth and the determination of exponential, stationary and lag phase during batch fermentations [18,21]. For media (300 g/L glucose,

0% NaCl), (300 g/L glucose, 6%), (400 g/L glucose, 0% NaCl) and (400 g/L glucose, 6% NaCl) the three fermentation phases coincided while for media (200 g/L glucose, 0% NaCl) and (200 g/L glucose, 6% NaCl) the fermentation phases also coincided but the exponential and stationary phases were shorter compared to the other media, thus indicating that NaCl stress had no impact on fermentation kinetics, while glucose concentration had a slight impact (Figure 2). Our OD results cannot be compared with those of other researchers who used the same yeast strain and stress conditions, as OD measurements are not given [27,47]. Also, our OD results differed from the report of Kim et al. (2006), but these researchers used a different *S. cerevisiae* strain with no glucose and NaCl-stressed yeast cells, and also used different media and fermentation times [18].

The cell count was used in order to determine the total yeast cell number during the fermentation process and correlate it with cell viability [47], but also with dry yeast mass weight (g) and β-glucan yield (%). During the death phase, the cell number decreased, with a simultaneous decrease in OD (600 nm) for all the fermented media; surprisingly, though, between 120 and 192 h, the number of cells grown in 400 g/L glucose increased spectacularly compared to the other samples. This cell number increase of glucose- and NaCl-stressed cells was not accompanied by a cell dry mass increase. This probably indicates a smaller yeast cell but further observations with electron microscopy must be performed in order to validate such a hypothesis (Figure 3). These results differed from other researchers' studies that used the same yeast strain, glucose and NaCl stress conditions, which suggests that NaCl-induced osmotic stress caused growth arrest in yeast cells, while an increase in osmotic stress with elevated NaCl concentration caused a decrease in yeast growth and total cell number over time [27]. During the first 40 h of fermentation, the cell count of media (200 g/L glucose, 0% NaCl) and (200 g/L glucose, 6% NaCl) was lower than that of the other media, while their dry weights appeared to be higher and their cell walls contained much more β-glucan (Figures 3 and 6), thus indicating fewer cells but larger ones, those with wider cell walls or both.

4.3. *Dry Weight and Sample β-Glucan Yield*

The determination of yeast dry weight at the various fermentation phases was done in order to calculate the β-glucan yield (g) at the different fermentation phases and optimise the phase at which the cells must be harvested. The highest value of dry weight (g) and β-glucan yield (g) was obtained with 200 g/L glucose/NaCl-non-stressed and 200 g/L glucose/stressed cells at 48 and 40 h, respectively, thus indicating that the stationary phase is the most appropriate time for cell harvesting, a result that is in accordance with the observations of Kim et al. (2006) [18]. From the obtained results, it is shown that glucose and NaCl stress impacts negatively on yeast dry weight and β-glucan yield (Figures 4 and 5). Also, for media (400 g/L glucose, 0% NaCl) and (400 g/L glucose, 6% NaCl), the highest value for yeast dry weight and β-glucan yield appeared at the end of the death phase and, despite the fact that β-glucan concentration was lowered, the total cell number increased. It seems that, in this way, the yeast cell maintains homeostasis in order to cope with the conditions of glucose and NaCl stress during the fermentation process, but this needs further investigation and is proposed as future research.

4.4. *β-Glucan Concentration, Lytic Enzyme and Sonication*

The determination of yeast β-glucan concentration was done for the estimation of polysaccharide and its accumulation in the yeast cell, while furthering the understanding of yeast cell wall β-glucan modification during the fermentation process under stress conditions, the estimation of cell wall lysis and disruption resistance against lytic enzymes and sonication was tested with cell viability measurements.

Generally, for all the fermented media, the β-glucan yeast cell wall concentration rose constantly during the exponential phase and rose to a peak value at the stationary phase and then decreased during the death phase (Figure 6). Also, the cell treatment with lytic enzyme and sonication revealed that the cell's resistance against lysis and disruption increased as the cell entered the stationary phase.

Next, as the cell entered the death phase, its resistance against lytic enzymes and sonication lowered, thus supporting the observations of a constant accumulation of β-glucan in the cell wall during the exponential phase, with its maximum quantity at the stationary phase and consequently a decrease during the death phase (Figures 7 and 8). During yeast alcoholic fermentation, the accumulation of ethanol causes an increase in yeast cell membrane permeability [42]. During the death phase, all the samples appeared to be more susceptible to lytic enzyme and sonication treatment (Figures 7 and 8) but for NaCl-stressed cells grown in a medium containing 300 g/L glucose and both NaCl-stressed and non-stressed yeast cells grown in a medium containing 400 g/L glucose, there was not a direct correlation between β-glucan concentration and cell wall resistance as these cells appeared to be more susceptible to lytic enzyme and sonication treatment, probably due to the additional impact of ethanol toxicity [48] (Figures 7 and 8).

Our results were in accordance with the observations of Kim et al. (2006), according to which, during the yeast growth in a defined medium, the cell wall resistance to the action of β-glucanase increases until the yeast cells enter the stationary phase, while their resistance to the enzyme decreases at the end of the stationary phase [18]. Also, our results support the reports of Klis et al. (2002), which indicate that as cells enter the stationary phase, they become thicker [49]. This yeast cell wall resistance is a result of the cell wall increase during the exponential phase [18].

In non-continuous fermentation, during the transition from the exponential to the stationary phase, the action of β-1,3-glucan synthetase decreases while the action of glycogen synthetase increases [50]. During the exponential phase, the autolytic action of endo β-glucanases is higher compared to that in the stationary phase of yeast cells' growth, while a significant increase in endoglucanases' activity into the soluble fraction of β-glucan during the stationary phase in non-continuous fermentation of *Saccharomyces exiguus* has also been reported [51]. The stationary phase is characterised by carbon limitation, and as the cell culture enters this, more glucan is necessary for the maintenance of the cell's viability [18]. The higher β-glucan concentration at the stationary phase can also be explained by the decreased action of cell's glucanases, with the less active cell's growth at the exponential phase, which is accompanied by a non-severe carbon limitation, similar to the one at the end of the stationary phase [18]. During aerobic fermentation, when yeast cells enter the stationary growth phase, the cells become more resistant to the action of β-1,3-glucanase and are less permeable to macromolecules with the expression of Sed1p protein [52,53].

In addition to the above, our results show that the increase of glucose osmotic stress in the growth medium, but also the NaCl hyperosmotic stress in the pre-cultured yeast cells, impacts negatively on the β-glucan concentration in the yeast cell wall. This negative impact on the β-glucan concentration was even more intense from the stationary phase until the end of the fermentation (death phase) (Figure 6).

Our results differed from those of other researchers, who reported a β-glucan increase at the end of fermentation; however, these researchers incorporated NaCl in the fermented medium and did not pre-culture yeast cells under salt hyperosmotic stress, and also used a different yeast strain [22]. Also, in their work, cell viability and total cell count were not taken into consideration and an increase in β-glucan concentration may have resulted from cell lysis, cytoplasm efflux and consequently a higher cell wall:cell ratio [14]. Our results cannot be compared with the results of Aguilar-Uscanga et al. (2003), who studied the impact of growth conditions and the mode of cultivation on the yeast cell wall structure and not the impact of stress conditions, with the collection of cell samples (50 mL volume) only in the early exponential phase [21].

Our β-glucan concentration results differed from the reports of other researchers that suggest that the cell wall integrity (CWI) pathway, in cooperation with the high osmotic response (HOG) pathway, regulates the action of zymolyase, the enzyme hydrolysing the β-1,3 glucan network, and thus result in an increase in the amount of cell wall β-glucan. These researchers used different yeast strains and different fermented medium components [24], while the data concerning osmotic stress responses were based on a previous research work in which KCl and NaCl had been diluted in the fermented substrate [54,55]. Our results are in accordance with the observations of Ene et al. (2015), who reported that the cell wall is not rigid but elastic and that sudden decreases in cell volume due to hyperosmotic conditions result in rapid increases in cell wall thickness, and thus a decrease in β-glucan content [25].

Our results are in accordance with the results of other researchers who used Glucanex 200G for the estimation of yeast cell wall resistance to lytic enzyme during batch fermentation and reported a higher cell resistance in the stationary phase [18]. Also, our results differ from other researchers' results (97 ± 0.18% breaking rate, 10–15% cell concentration), but these differences may arise from the fact that these researchers used sonication for yeast cell lysis during a β-glucan extraction protocol in dry yeast first treated with hot water for mannoprotein removal and not in yeast from various fermentation phases [38].

The yeast cells appeared to be more susceptible to sonication treatment compared to lytic enzyme treatment. The higher cell wall resistance to lytic enzymes in the various fermentation phases could be attributed to the variation in cell number during the fermentation process (Figures 3 and 7) and thus to the enzyme dispersion to a larger volume of cells [56] (e.g., the stable resistance of cells of fermented media with 400 g/L glucose to the action of the lytic enzyme at the end of the death phase where they appeared the highest cell number), while cell disruption of yeast cells treated with ultrasounds is independent of cell concentration but mainly proportional to the acoustic power [57,58] (Figures 3 and 8).

5. Conclusions

In recent years, there has been increasing interest in industrial production of yeast β-glucan and its incorporation in functional foods and medicines due to its immunological properties in human and animal health systems. This study could comprise a quantitative indicator for the industrial production of yeast β-glucan from defined cell cultures but also from other yeast sources like breweries' [23] and wineries' spent yeast biomass [40].

The stationary growth phase appears to be optimal for β-glucan isolation from pure cell fermentation cultures. NaCl and glucose stress have a negative impact on β-glucan production. Yeast NaCl-stressed cells have a reduced β-glucan concentration compared to non-stressed for the same glucose concentrations, while this difference is more significant at the end of the fermentation for a fermented medium containing 200 g/L glucose, which is close to the wine must concentration intended for wine production. For the other two glucose concentrations (300 and 400 g/L), the differences at the end of the fermentation are not statistically significant. In the present study, it seems that the two different stresses act synergistically, with an additive negative impact on the cell wall β-glucan concentration; as for the same glucose concentration, the preconditioned NaCl cells had a lower β-glucan concentration compared to the non-preconditioned ones. Another significant observation was that for a yeast cell grown in a medium containing 400 g/L glucose, the cell wall β-glucan concentration remained almost the same from the end of the exponential phase until the end of the fermentation for both NaCl-stressed and non-stressed cells.

The study of cell wall physiology, which is still a poorly explored research field, and yeast cell growth under various stress conditions (glucose, ethanol, temperature, SO_2, etc.), could comprise a key tool for the biotechnological development of new products like functional foods and antifungal drugs for medical and agricultural applications. Further study on the impact of stress conditions on the immunological properties of the isolated β-glucan from the various fermentation phases, as well as the study of glycolysis–glycogenesis biochemical pathways in the yeast cell wall under stress conditions, is proposed as it could offer new perspectives on yeast β-glucan-based drugs, adding to a deeper understanding of yeast stress biology phenomena and a more comprehensive view of alcoholic fermentation biochemical pathways. Additionally, the study of the genes encoding β-glucan accumulation in the yeast cell wall during the fermentation process and under various environmental stresses is proposed as it could enlighten us as to the molecular mechanisms of the yeast cell.

Author Contributions: Vassileios Varelas conceived, designed and performed the experiments, analyzed the data and wrote the paper. Evangelia Sotiropoulou and Maria Liouni analyzed the data. Xara Karambini performed the experiments. Elias T. Nerantzis conceived and designed the experiments.

References

1. Reggiori, F.; Klionsky, D.J. Autophagic processes in yeast: Mechanism, machinery and regulation. *Genetics* **2013**, *194*, 341–361. [CrossRef] [PubMed]
2. Ribéreau-Gayon, P.; Dubourdieu, D.; Donèche, B.; Lonvaud, A. Biochemistry of Alcoholic Fermentation and Metabolic Pathways of Wine Yeasts. In *Handbook of Enology*; John Wiley & Sons, Ltd.: Somerset, NJ, USA, 2006; pp. 53–77.
3. Branduardi, P.; Porro, D. Yeasts in Biotechnology. In *Yeast*; Wiley-VCH Verlag GmbH & Co. KGaA: Weinheim, Germany, 2012; pp. 347–370.
4. Ballou, C.E. Genetics of Yeast Mannoprotein Biosynthesis. In *Fungal Polysaccharides*; American Chemical Society: Washington, DC, USA, 1980; Volume 126, pp. 1–14.
5. François, J.M.; Walther, T.; Parrou, J.L. Genetics and Regulation of Glycogen and Trehalose Metabolism in *Saccharomyces cerevisiae*. In *Microbial Stress Tolerance for Biofuels*; Liu, Z.L., Ed.; Springer: Berlin/Heidelberg, Germany, 2012; pp. 29–55.
6. Wilson, W.A.; Hughes, W.E.; Tomamichel, W.; Roach, P.J. Increased glycogen storage in yeast results in less branched glycogen. *Biochem. Biophys. Res. Commun.* **2004**, *320*, 416–423. [CrossRef] [PubMed]
7. Arvindekar, A.U.; Patil, N.B. Glycogen—A covalently linked component of the cell wall in *Saccharomyces cerevisiae*. *Yeast* **2002**, *19*, 131–139. [CrossRef] [PubMed]
8. Orlean, P. Architecture and biosynthesis of the *Saccharomyces cerevisiae* cell wall. *Genetics* **2012**, *192*, 775–818. [CrossRef] [PubMed]
9. Klis, F.M. Review: Cell wall assembly in yeast. *Yeast* **1994**, *10*, 851–869. [CrossRef] [PubMed]
10. Van der Rest, M.E.; Kamminga, A.H.; Nakano, A.; Anraku, Y.; Poolman, B.; Konings, W.N. The plasma membrane of *Saccharomyces cerevisiae*: Structure, function, and biogenesis. *Microbiol. Rev.* **1995**, *59*, 304–322. [PubMed]
11. Rank, G.H.; Robertson, A.J. Protein and lipid composition of the yeast plasma membrane. In *Yeast Genetics: Fundamental and Applied Aspects*; Spencer, J.F.T., Spencer, D.M., Smith, A.R.W., Eds.; Springer: New York, NY, USA, 1983; pp. 225–241.
12. Stier, H.; Ebbeskotte, V.; Gruenwald, J. Immune-modulatory effects of dietary yeast β-1,3/1,6-D-glucan. *Nutr. J.* **2014**, *13*. [CrossRef] [PubMed]
13. Ahmad, A.; Anjum, F.M.; Zahoor, T.; Nawaz, H.; Dilshad, S.M.R. β glucan: A valuable functional ingredient in foods. *Crit. Rev. Food Sci. Nutr.* **2012**, *52*, 201–212. [CrossRef] [PubMed]
14. Varelas, V.; Liouni, M.; Calokerinos, A.C.; Nerantzis, E.T. An evaluation study of different methods for the production of β-D-glucan from yeast biomass. *Drug Test. Anal.* **2016**, *8*, 46–55. [CrossRef] [PubMed]
15. Klis, F.M.; Boorsma, A.; de Groot, P.W.J. Cell wall construction in *Saccharomyces cerevisiae*. *Yeast* **2006**, *23*, 185–202. [CrossRef] [PubMed]
16. Levin, D.E. Regulation of cell wall biogenesis in *Saccharomyces cerevisiae*: The cell wall integrity signaling pathway. *Genetics* **2011**, *189*, 1145–1175. [CrossRef] [PubMed]
17. Hahn-Hägerdal, B.; Karhumaa, K.; Larsson, C.U.; Gorwa-Grauslund, M.; Görgens, J.; van Zyl, W.H. Role of cultivation media in the development of yeast strains for large scale industrial use. *Microb. Cell Fact.* **2005**, *4*. [CrossRef] [PubMed]
18. Kim, K.S.; Yun, H.S. Production of soluble β-glucan from the cell wall of *Saccharomyces cerevisiae*. *Enzym. Microb. Technol.* **2006**, *39*, 496–500. [CrossRef]
19. Catley, B.J. Isolation and analysis of cell walls. In *Yeast, A Practical Approach*; Campbell, I., Duffus, J.H., Eds.; Oxford University Press: London, UK, 1988; pp. 163–183.
20. McMurrough, I.; Rose, A.H. Effect of growth rate and substrate limitation on the composition and structure of the cell wall of *Saccharomyces cerevisiae*. *Biochem. J.* **1967**, *105*, 189–203. [CrossRef] [PubMed]
21. Aguilar-Uscanga, B.; François, J.M. A study of the yeast cell wall composition and structure in response to growth conditions and mode of cultivation. *Lett. Appl. Microbiol.* **2003**, *37*, 268–274. [CrossRef] [PubMed]
22. Naruemon, M.; Romanee., S.; Cheunjit, P.; Xiao, H.; McLandsborough, L. A.; Pawadee, M. Influence of additives on *Saccharomyces cerevisiae* β-glucan production. *Int. Food Res. J.* **2013**, *20*, 1953–1959.
23. Bastos, R.; Coelho, E.; Coimbra, M.A. Modifications of *Saccharomyces pastorianus* cell wall polysaccharides with brewing process. *Carbohydr. Polym.* **2015**, *124*, 322–330. [CrossRef] [PubMed]

24. García, R.; Rodríguez-Peña, J.M.; Bermejo, C.; Nombela, C.; Arroyo, J. The high osmotic response and cell wall integrity pathways cooperate to regulate transcriptional responses to zymolyase-induced cell wall stress in *Saccharomyces cerevisiae*. *J. Biol. Chem.* **2009**, *284*, 10901–10911. [CrossRef] [PubMed]
25. Ene, I.V.; Walker, L.A.; Schiavone, M.; Lee, K.K.; Martin-Yken, H.; Dague, E.; Gow, N.A.R.; Munro, C.A.; Brown, A.J.P. Cell wall remodeling enzymes modulate fungal cell wall elasticity and osmotic stress resistance. *MBio* **2015**. [CrossRef] [PubMed]
26. Erasmus, D.J.; van der Merwe, G.K.; van Vuuren, H.J.J. Genome-wide expression analyses: Metabolic adaptation of *Saccharomyces cerevisiae* to high sugar stress. *FEMS Yeast Res.* **2003**, *3*, 375–399. [CrossRef]
27. Logothetis, S.; Nerantzis., E.T.; Giouliotí, A.; Kanelis, T.; Tataridis, P.; Walker, G. Influence of sodium chloride on wine yeast fermentation performance. *Int. J. Wine Res.* **2010**, *2*, 35–42. [CrossRef]
28. Ishmayana, S.; Learmonth, R.P.; Kennedy, U.J. Fermentation performance of the yeast *Saccharomyces cerevisiae* in media with high sugar concentration. In Proceedings of the 2nd International Seminar on Chemistry, Jatinangor, Indonesia, 24–25 November 2011; pp. 379–385.
29. Lei, H.; Xu, H.; Feng, L.; Yu, Z.; Zhao, H.; Zhao, M. Fermentation performance of lager yeast in high gravity beer fermentations with different sugar supplementations. *J. Biosci. Bioeng.* **2016**, *122*, 583–588. [CrossRef] [PubMed]
30. Novo, M.; Gonzalez, R.; Bertran, E.; Martínez, M.; Yuste, M.; Morales, P. Improved fermentation kinetics by wine yeast strains evolved under ethanol stress. *LWT Food Sci. Technol.* **2014**, *58*, 166–172. [CrossRef]
31. Trainotti, N.; Stambuk, B.U. NaCl stress inhibits maltose fermentation by *Saccharomyces cerevisiae*. *Biotechnol. Lett.* **2001**, *23*, 1703–1707. [CrossRef]
32. Pratt, P.L.; Bryce, J.H.; Stewart, G.G. The effects of osmotic pressure and ethanol on yeast viability and morphology. *J. Inst. Brew.* **2003**, *109*, 218–228. [CrossRef]
33. Ren, H.; Wang., X.; Liu, D.; Wang, B. A glimpse of the yeast *Saccharomyces cerevisiae* responses to NaCl stress. *Afr. J. Microbiol. Res.* **2012**, *6*, 713–718.
34. Dhar, R.; Sägesser, R.; Weikert, C.; Yuan, J.; Wagner, A. Adaptation of *Saccharomyces cerevisiae* to saline stress through laboratory evolution. *J. Evol. Biol.* **2011**, *24*, 1135–1153. [CrossRef] [PubMed]
35. Tilloy, V.; Ortiz-Julien, A.; Dequin, S. Reduction of ethanol yield and improvement of glycerol formation by adaptive evolution of the wine yeast *Saccharomyces cerevisiae* under hyperosmotic conditions. *Appl. Environ. Microbiol.* **2014**, *80*, 2623–2632. [CrossRef] [PubMed]
36. Logothetis, S.; Nerantzis, E.T.; Tataridis, P.; Goulioti, A.; Kannelis, A.; Walker, G.M. Alleviation of stuck wine fermentations using salt-preconditioned yeast. *J. Inst. Brew.* **2014**, *120*, 174–182. [CrossRef]
37. Prieto, M.A.; Vázquez, J.A.; Murado, M.A. Comparison of several mathematical models for describing the joint effect of temperature and ph on glucanex activity. *Biotechnol. Prog.* **2012**, *28*, 372–381. [CrossRef] [PubMed]
38. Magnani, M.; Calliari, C.M.; de Macedo Jr, F.C.; Mori, M.P.; de Syllos Cólus, I.M.; Castro-Gomez, R.J.H. Optimized methodology for extraction of (1→3)(1→6)-*β*-D-glucan from *Saccharomyces cerevisiae* and in vitro evaluation of the cytotoxicity and genotoxicity of the corresponding carboxymethyl derivative. *Carbohydr. Polym.* **2009**, *78*, 658–665. [CrossRef]
39. Varelas, V.; Tataridis, P.; Liouni, M.; Nerantzis, E.T. Application of different methods for the extraction of yeast β-glucan. *e-J. Sci. Technol.* **2016**, *11*, 75–89.
40. Varelas, V.; Tataridis, P.; Liouni, M.; Nerantzis, E.T. Valorization of winery spent yeast waste biomass as a new source for the production of β-glucan. *Waste Biomass Valori.* **2016**, *7*, 807–817. [CrossRef]
41. Logothetis, S.; Tataridis, P.; Kanellis, A.; Nerantzis, E.T. The effect of preconditioning cells under osmotic stress on high alcohol production. *Zb. Matice Srp. za Prir. Nauke* **2013**, 405–414. [CrossRef]
42. Bauer, F.F.; Pretorius, I.S. Yeast stress response and fermentation efficiency: How to survive the making of wine—A Review. *S. Afr. J. Enol. Viticult.* **2000**, *21*, 27–51.
43. Patynowski, R.J.; Jiranek, V.; Markides, A.J. Yeast viability during fermentation and sur lie ageing of a defined medium and subsequent growth of *Oenococcus oeni*. *Aust. J. Grape Wine Res.* **2002**, *8*, 62–69. [CrossRef]
44. Morris, G.J.; Winters., L.; Coulson, G.E.; Clarke, K.J. Effect of osmotic stress on the ultrastructure and viability of the yeast *Saccharomyces cerevisiae*. *Microbiology* **1986**, *132*, 2023–2034. [CrossRef] [PubMed]
45. Naruemon, M.; Romanee., S.; Cheunjit, P.; Xiao, H.; McLandsborough, L.A.; Pawadee, M. Effect of three additives on the cell morphology and β-glucan production in *Saccharomyces cerevisiae*. *Res. J. Pharm. Biol. Chem. Sci.* **2011**, *2*, 283–295.

46. Gilliland, R.B. Determination of yeast viability. *J. Inst. Brew.* **1959**, *65*, 424–429. [CrossRef]
47. Logothetis, S.; Walker, G.; Nerantzis, E.T. Effect of salt hyperosmotic stress on yeast cell viability. *Zb. Matice Srp. za Prir. Nauke* **2007**, *113*, 271–284. [CrossRef]
48. Stanley, D.; Bandara, A.; Fraser, S.; Chambers, P.J.; Stanley, G.A. The ethanol stress response and ethanol tolerance of *Saccharomyces cerevisiae*. *J. Appl. Microbiol.* **2010**, *109*, 13–24. [CrossRef] [PubMed]
49. Klis, F.M.; Mol, P.; Hellingwerf, K.; Brul, S. Dynamics of cell wall structure in *Saccharomyces cerevisiae*. *FEMS Microbiol. Rev.* **2002**, *26*, 239–256. [CrossRef] [PubMed]
50. Fleet, G.H. Cell walls. In *The Yeasts*; Rose, A.H.H.J.D., Ed.; Academic Press: London, UK, 1991; Volume 4, pp. 199–277.
51. Inouhe, M.; Sugo, E.; Tohoyama, H.; Joho, M.; Nevins, D.J. Cell wall metabolism and autolytic activities of the yeast *Saccharomyces exiguus*. *Int. J. Biol. Macromol.* **1997**, *21*, 11–14. [CrossRef]
52. Shimoi, H.; Kitagaki, H.; Ohmori, H.; Iimura, Y.; Ito, K. Sed1p is a major cell wall protein of *Saccharomyces cerevisiae* in the stationary phase and is involved in lytic enzyme resistance. *J. Bacteriol.* **1998**, *180*, 3381–3387. [PubMed]
53. De Nobel, H.; Ruiz, C.; Martin, H.; Morris, W.; Brul, S.; Molina, M.; Klis, F.M. Cell wall perturbation in yeast results in dual phosphorylation of the slt2/mpk1 map kinase and in an slt2-mediated increase in fks2–lacz expression, glucanase resistance and thermotolerance. *Microbiology* **2000**, *146*, 2121–2132. [CrossRef] [PubMed]
54. Rep, M.; Krantz, M.; Thevelein, J.M.; Hohmann, S. The transcriptional response of *Saccharomyces cerevisiae* to osmotic shock: Hot1p and msn2p/msn4p are required for the induction of subsets of high osmolarity glycerol pathway-dependent genes. *J. Biol. Chem.* **2000**, *275*, 8290–8300. [CrossRef] [PubMed]
55. O'Rourke, S.M.; Herskowitz, I. Unique and redundant roles for HOG MAPK pathway components as revealed by whole-genome expression analysis. *Mol. Biol. Cell* **2004**, *15*, 532–542. [CrossRef] [PubMed]
56. Kim, K.S.; Chang, J.E.; Yun, H.S. Estimation of soluble-glucan content of yeast cell wall by the sensitivity to glucanex® 200g treatment. *Enzym. Microb. Technol.* **2004**, *35*, 672–677. [CrossRef]
57. Liu, D.; Zeng, X.A.; Sun, D.W.; Han, Z. Disruption and protein release by ultrasonication of yeast cells. *Innov. Food Sci. Emerg. Technol.* **2013**, *18*, 132–137. [CrossRef]
58. Apar, D.K.; Ozmek, B. Protein releasing kinetics of bakers' yeast cells by ultrasound. *Chem. Biochem. Eng. Q.* **2008**, *22*, 113–118.

12

A Bacterial Laccase for Enhancing Saccharification and Ethanol Fermentation of Steam-Pretreated Biomass

Antonio D. Moreno [1,2], David Ibarra [3,*], Antoine Mialon [4] and Mercedes Ballesteros [2]

1 IMDEA Energía, Biotechnological Processes for Energy Production Unit, Móstoles, Madrid 28935, Spain; david.moreno@ciemat.es

2 CIEMAT, Renewable Energy Division, Biofuels Unit, Avda. Complutense 40, Madrid 28040, Spain; m.ballesteros@ciemat.es

3 INIA-CIFOR, Forestry Products Department, Cellulose and Paper Laboratories, Ctra de La Coruña Km 7.5, Madrid 28040, Spain

4 MetGen Oy, Rakentajantie 26, Kaarina 20780, Finland; antoine@metgen.com

* Correspondence: ibarra.david@inia.es

Academic Editor: Ronnie G. Willaert

Abstract: Different biological approaches, highlighting the use of laccases, have been developed as environmentally friendly alternatives for improving the saccharification and fermentation stages of steam-pretreated lignocellulosic biomass. This work evaluates the use of a novel bacterial laccase (MetZyme) for enhancing the hydrolysability and fermentability of steam-exploded wheat straw. When the water insoluble solids (WIS) fraction was treated with laccase or alkali alone, a modest increase of about 5% in the sugar recovery yield (glucose and xylose) was observed in both treatments. Interestingly, the combination of alkali extraction and laccase treatment boosted enzymatic hydrolysis, increasing the glucose and xylose concentration in the hydrolysate by 21% and 30%, respectively. With regards to the fermentation stage, the whole pretreated slurry was subjected to laccase treatment, lowering the phenol content by up to 21%. This reduction allowed us to improve the fermentation performance of the thermotolerant yeast *Kluyveromyces marxianus* CECT 10875 during a simultaneous saccharification and fermentation (SSF) process. Hence, a shorter adaptation period and an increase in the cell viability—measured in terms of colony forming units (CFU/mL)—could be observed in laccase-treated slurries. These differences were even more evident when a presaccharification step was performed prior to SSF. Novel biocatalysts such as the bacterial laccase presented in this work could play a key role in the implementation of a cost-effective technology in future biorefineries.

Keywords: alkaline extraction; bacterial Metzyme laccase; lignocellulosic ethanol; simultaneous saccharification and fermentation; thermotolerant yeast

---

1. Introduction

The transition towards a post-petroleum society for mitigating global climate change is currently led by the development and implementation of biorefineries. Biorefineries will be competitive, innovative and sustainable local industries for the production of plant- and waste-derived fuels, materials and chemicals. Due to its low costs and wide distribution, lignocellulosic biomass is the most promising feedstock to be used in biorefineries, and lignocellulose-derived fuels, including ethanol, the most significant product.

Many different feedstocks, conversion methods, and process configurations have been studied for lignocellulosic ethanol production, with the biochemical route being the most promising option [1]. Lignocellulose is a complex matrix where a 'skeleton' polymer, cellulose, is coated by two 'protective' polymers, hemicellulose and lignin. Biochemical conversion of lignocellulosic biomass includes a pretreatment step to open up the structure and increase biomass digestibility. Subsequently, cellulose and hemicellulose polymers are subjected to an enzymatic saccharification process to obtain the fermentable sugars. The optimal performance of cellulolytic enzymes is therefore a crucial step that determines the overall process efficiency. Finally, the resulting sugars are converted into ethanol via microbial fermentation [1].

Pretreatment influences lignocellulose digestibility by an extensive modification of the structure. A large number of pretreatment technologies, mainly physical and/or chemical, have been developed and applied on a wide variety of feedstocks [2]. Among them, hydrothermal pretreatments, such as steam explosion, are considered the most effective methods and are commonly used for lignocellulose-to-ethanol conversion. The action mechanism of these pretreatment technologies lies in the solubilisation of hemicellulose fraction and the redistribution and/or modification of lignin, which increase outstandingly the hydrolysis of cellulose without the need of adding any catalyst [3]. These pretreatment technologies, however, still present several drawbacks that must be overcome. First, the residual lignin that is left in the pretreated materials represents an important limiting factor during the enzymatic hydrolysis of carbohydrates, promoting the non-specific adsorption of hydrolytic enzymes and, in turn, decreasing saccharification yields [4]. Second, these pretreatment methods generate some soluble compounds, derived from sugar degradation (furan derivatives and weak acids) and partial lignin solubilisation (aromatic acids, alcohols and aldehydes), which inhibit cellulolytic enzymes and fermentative microorganisms [5]. Performing a delignification step prior to the addition of hydrolytic enzymes may reduce the non-productive adsorption of these enzymes, enhancing the saccharification yields. In the same way, a detoxification process may reduce the amount of inhibitors produced after steam explosion pretreatment, boosting the saccharification and fermentation steps. Different physico/chemical technologies have been studied for delignification and detoxification of pretreated materials [2,6]. However, most of these methods require extra equipment and additional steps and have high energy demands, complicating the lignocellulose-to-ethanol process and increasing the production costs. As an alternative to physico/chemical methods, the use of ligninolytic enzymes such as laccases may provide further integration into the process and lower energy requirements [7].

Laccases are multicopper oxidases that catalyze the one-electron oxidation of phenols, anilines and aromatic thiols to their corresponding radicals with the concomitant reduction of molecular oxygen to water. Laccases are mainly produced by plants and fungi, including the white-rot basidiomycetes responsible for lignin degradation in nature [8]. Also, some bacterial laccases have been described and fully characterized, generally showing lower redox potential and more stable at high pH and temperatures compared to fungal laccases [9]. The role of laccases in lignin degradation makes them attractive biocatalysts for the pulp and paper industry as substitutes of chlorine-containing reagents in pulp bleaching [10,11]. Both fungal and bacterial laccases have been studied with beneficial results [12,13]. Moreover, they are used in wastewater treatment to detoxify industrial effluents with high phenolic content—such as the streams obtained during pulp and paper production—due to their ability to oxidize phenolic compounds [14,15].

The vast experience gained from the extensive use of laccases in the paper pulp industry has provided an excellent starting point for the application of laccases within a broader perspective. In this context, different fungal laccases have been widely studied for improving the conversion efficiency of lignocellulose into ethanol, and consequently increasing final product concentrations [7,16–25]. Nevertheless, little is known about the use of bacterial laccases for these purposes. The present work evaluates the commercial bacterial laccase MetZyme, exploring its potential for improving the hydrolysability and fermentability of steam-exploded wheat straw.

2. Materials and Methods

2.1. *Raw Material and Steam Explosion Pretreatment*

Wheat straw, supplied by CEDER-CIEMAT (Soria, Spain), was used as raw material. It presented the following composition (% dry weight (DW)): cellulose, 40.5 $\pm$ 2.1; hemicelluloses, 26.1 $\pm$ 1.1 (xylan, 22.7 $\pm$ 0.5; arabinan 2.1 $\pm$ 0.4; and galactan, 1.3 $\pm$ 0.2); Klason lignin, 18.1 $\pm$ 0.8; ashes, 5.1 $\pm$ 0.3; and extractives, 14.6 $\pm$ 0.4.

Prior to steam explosion, wheat straw was milled in a laboratory hammer mill to obtain a chip size between 2 and 10 mm. Then, the milled material was pretreated in a 10 L reactor at 200 °C for 2.5 min. The recovered slurry was handled differently depending on its further use. For analytical purposes, one portion was vacuum filtered with the aim of obtaining a liquid fraction or prehydrolysate and a solid fraction. Subsequently, the solid fraction was thoroughly washed with distilled water to obtain the water insoluble solids (WIS) fraction. Chemical composition of both raw and pretreated material (WIS fraction) was determined using the Laboratory Analytical Procedures (LAP) for biomass analysis, provided by the National Renewable Energies Laboratory [26]. Sugars and degradation compounds contained in the liquid fraction were also measured. Most of the sugars present in the liquid fraction were in oligomeric form, and therefore a mild acid hydrolysis (4% (*v*/*v*) H_2SO_4, 120 °C and 30 min) was required to determine the concentration of monomeric sugars. The obtained WIS fraction was also used for saccharification studies since the majority of the inhibitory compounds were removed. On the other hand, the remained slurry was used as substrate to evaluate its fermentability due to the higher inhibitor content. Both WIS and slurry were stored at 4 °C until their use.

2.2. *Enzymes*

An industrial thermostable bacterial laccase (pH range 3–8) was specifically selected from MetGen's products portfolio (MetZyme, Cat.-No: 10-101-UF, MetGen Oy, Kaarina, Finland), and used in both saccharification and fermentation assays. Laccase activity (284 IU/g of laccase activity) was measured by oxidation of 5 mM 2,2′-azino-bis(3-ethylbenzothiazoline-6-sulphonic acid) (ABTS) to its cation radical in 0.1 M sodium acetate (pH 5) at 24 °C. Formation of the ABTS cation radical was monitored at 436 nm (ε_{436} = 29,300 $M^{-1} \cdot cm^{-1}$).

A mixture of NS50013 and NS50010, both produced by Novozymes (Bagsvaerd, Denmark), was used for the saccharification of steam-pretreated what straw. NS50013 (60 FPU/mL of cellulase activity) is a cellulase preparation that presents low β-glucosidase activity and therefore it requires the supplementation with NS50010 (810 IU/mL of β-glucosidase activity), which mainly presents β-glucosidase activity. Overall cellulase activity was determined using filter paper (Whatman No. 1 filter paper strips), while β-glucosidase activity was measured using cellobiose as a substrate. The enzymatic activities were followed by the release of reducing sugars [27].

One unit of enzyme activity was defined as the amount of enzyme that transforms 1 μmol of substrate per minute.

2.3. *Microorganism and Growth Conditions*

Kluyveromyces marxianus CECT 10875, a thermotolerant strain selected by Ballesteros *et al.* [28], was employed as fermentative microorganism in this study. Active cultures for inoculation were obtained in 100-mL flasks with 50 mL of growth medium containing 30 g/L glucose, 5 g/L yeast extract, 2 g/L NH_4Cl, 1 g/L KH_2PO_4, and 0.3 g/L $MgSO_4 \cdot 7H_2O$. After 16 h on an orbital shaker at 150 rpm and 42 °C, the precultures were centrifuged at 9000 rpm for 10 min. Supernatant was discarded and cells were washed once with distilled water and diluted accordingly to obtain an inoculum level of 1 g/L DW.

2.4. *Laccase Treatment and Saccharification of the WIS Fraction*

The WIS fraction obtained after steam explosion (200 °C, 2.5 min) was subjected to a sequential laccase treatment and saccharification directly (Strategy 1) or after a mild alkaline extraction (Strategy 2).

Strategy 1, *sequential laccase treatment and saccharification*: 2.5 g DW of the corresponding WIS fraction were suspended in 50 mM sodium citrate buffer (pH 5.5) in 100-mL shake flasks to reach a final concentration of 5% (*w*/*v*) total solids (TS). This solution was treated with MetZyme laccase (10 IU/g DW substrate) for 24 h at 50 °C and 150 rpm in an orbital shaker. After 24 h of laccase treatment, solids were filtered through a Büchner funnel, washed with 1 L of water and dried at 60 °C. In a subsequent step, the laccase-treated WIS fraction was resuspended with 50 mM sodium citrate buffer (pH 5.5) in 100-mL flasks to reach a final concentration of 5% TS (*w*/*v*). Solids were subjected to saccharification at 50 °C for 72 h in an orbital shaker (150 rpm), with an enzyme loading of 5 FPU/g DW substrate of NS50013 and 5 IU/g DW substrate of NS50010.

Strategy 2, *mild alkaline extraction and sequential laccase treatment and saccharification*: 2.5 g DW of the corresponding WIS fraction was extracted with alkali (2.5% NaOH, for 1 h at 60 °C and 5% TS (*w*/*v*) substrate loading) followed by filtration and water washing. Then, the alkali-extracted WIS fraction was resuspended in 50 mM sodium citrate buffer (pH 5.5) in 100-mL flasks to reach a final concentration of 5% TS (*w*/*v*) and subjected to sequential laccase treatment and saccharification as explained above.

The effects of bacterial laccase treatments on both WIS fractions were evaluated in terms of (1) chemical composition and (2) saccharification yields. The chemical composition of laccase-treated WIS, subjected or not to a mild alkaline extraction, was determined using the NREL-LAP for biomass analysis [26]. On the other hand, the enzymatic hydrolysates obtained from laccase-treated WIS (with and without a previous mild alkaline extraction step) were centrifuged to remove the remaining solids, and the supernatants were analyzed to determine glucose and xylose concentration. For a better comparison between assays, relative glucose/xylose recoveries (RGR; RXR) were calculated as following Equation (1):

$$\text{RGR}\,(\%) = \text{g/L}\,\text{glucose}_{\text{assay}} \times 100/\text{g/L}\,\text{glucose}_{\text{control}} \quad (1)$$

For RXR (%), similar equation was used but with xylose concentration instead.

Control assays were performed under same conditions in Strategy 1 and Strategy 2 without the addition of MetZyme laccase. All the experiments were carried out in triplicate.

2.5. *Laccase Treatment and Fermentation of the Whole Slurry*

The whole slurry obtained after steam explosion (200 °C, 2.5 min) was subjected to laccase treatment and simultaneous saccharification and fermentation without (Strategy 3) and with (Strategy 4) a presaccharification step to evaluate its fermentability.

Strategy 3, *consecutive laccase treatment and simultaneous saccharification and fermentation (LSSF)*: 2.5 g DW of the corresponding slurry was suspended with 50 mM sodium citrate buffer (pH 5.5) in 100-mL flasks to reach a final concentration of 10% TS (*w*/*v*). Then, 10 IU/g DW substrate of MetZyme laccase were added and the mixture was incubated at 50 °C and 150 rpm in an orbital shaker for 24 h. After laccase treatment, the slurries were subsequently subjected to a simultaneous saccharification and fermentation (SSF) process at 42 °C for 72 h in an orbital shaker (150 rpm). Laccase-treated slurries were subjected to SSF after the supplementation with 15 FPU/g DW substrate of NS50013, 15 IU/g DW substrate of NS50010, nutrients (those described for cell propagation, except glucose) and 1 g/L DW of *K. marxianus*.

Strategy 4, *consecutive laccase treatment with presaccharification and simultaneous saccharification and fermentation (LPSSF)*: 2.5 g DW of the corresponding slurry were suspended in 50 mM sodium citrate buffer (pH 5.5) in 100-mL flasks to reach a final concentration of 10% TS (*w*/*v*). Then, 10 IU/g DW substrate of MetZyme laccase were added and the mixture was incubated at 50 °C and 150 rpm in an orbital shaker. After 16 h of laccase treatment, a presaccharification step was carried out for 8 h by supplementing the slurries with 15 FPU/g DW substrate of NS50013 and 15 IU/g DW substrate of NS50010. Afterwards, the temperature was reduced to 42 °C and nutrients and 1 g/L DW of *K. marxianus* were added, which turned the process into a SSF. The experiments were run for another 72 h.

The effect of MetZyme laccase on specific inhibitory compounds was evaluated before yeast addition, *i.e.*, right after laccase treatment or laccase treatment with presaccharification. For that, prior starting SSF processes samples were taken and centrifuged, and the supernatants were analyzed for the identification and quantification of inhibitory compounds. In the same way, samples were periodically withdrawn during SSF processes to determine cell viability and glucose and ethanol concentration (a centrifugation step was included prior to analyze glucose and ethanol concentration).

Control assays were performed under the same conditions without the addition of MetZyme laccase. All the experiments were carried out in triplicate.

2.6. Analytical Methods

Ethanol was analyzed by gas chromatography, using a 7890A GC System (Agilent, Waldbronn, Germany) equipped with an Agilent 7683B series injector, a flame ionization detector and a Carbowax 20 M column operating at 85 °C. Injector and detector temperature was maintained at 175 °C.

Sugar concentration was quantified by high-performance liquid chromatography (HPLC) in a Waters chromatograph equipped with a refractive index detector (Waters, Mildford, MA, USA). A CarboSep CHO-682 carbohydrate analysis column (Transgenomic, San Jose, CA, USA) operating at 80 °C with ultrapure water as a mobile-phase (0.5 mL/min) was used for the separation.

Furfural and 5-hydroxymethylfurfural (5-HMF) were analyzed and quantified by HPLC (Agilent, Waldbronn, Germany), using a Coregel 87H3 column (Transgenomic, San Jose, CA, USA) at 65 °C equipped with a 1050 photodiode-array detector (Agilent, Waldbronn, Germany). As mobile phase, 89% 5 mM H_2SO_4 and 11% acetonitrile at a flow rate of 0.7 mL/min were used.

Formic acid and acetic acid were also quantified by HPLC (Waters, Mildford, MA, USA) using a 2414 refractive index detector (Waters, Mildford, MA, USA) and a Bio-Rad Aminex HPX-87H (Bio-Rad Labs, Hercules, CA, USA) column maintained at 65 °C with a mobile phase (5 mmol/L H_2SO_4) at 0.6 mL/min of flow rate.

Total phenolic content was analyzed according to the Folin-Ciocalteu procedure [29]. 0.5 mL of sample and the serial standard solution (gallic acid) were introduced into test tubes with 2.5 mL of Folin-Ciocalteu's reagent (1:10 dilution in water) and 2 mL of sodium carbonate (7.5% w/v). The tubes were incubated for 5 min at 50 °C. After cooling down the temperature, the absorbance was measured at 760 nm using a Lambda 365 spectrophotometer (PerkinElmer, Boston, MA, USA).

Cell viability was measured by cell counting using agar plates (30 g/L glucose, 5 g/L yeast extract, 2 g/L NH_4Cl, 1 g/L KH_2PO_4, and 0.3 g/L $MgSO_4 \cdot 7H_2O$, 20 g/L agar). Plates were incubated at 42 °C for 24 h.

All analytical values were calculated from duplicates or triplicates. Statistical analyses were performed using IBM SPSS Statistics v22.0 for MacOs X Software (SPSS, Inc., Chicago, IL, USA). The mean and standard deviation were calculated for descriptive statistics. When appropriate, analysis of variance (ANOVA) with or without Bonferroni's post-test was used for comparisons between assays. The level of significance was set at $p < 0.05$, $p < 0.01$ or $p < 0.001$.

3. Results and Discussion

3.1. Pretreated Biomass Composition

Steam explosion pretreatment was performed at 200 °C and 2.5 min (Table 1). In comparison to the cellulose content of the untreated wheat straw (40.5%), steam explosion increased the cellulose proportion of the WIS fraction (53.5%) due to an extensive hemicellulose solubilization and degradation. This solubilization is evidenced by the lower proportion of the remaining hemicellulose (11.7%) fraction of the WIS residue and the high xylose content (32 g/L) in the liquid fraction. Also, different degradation products were recovered in the liquid fraction due to biomass degradation. The most predominant inhibitors were acetic acid, formic acid, furfural, 5-HMF and phenols (Table 1). Acetic acid is formed by the hydrolysis of acetyl groups contained in the hemicellulose structure. Formic acid derives from furfural and 5-HMF degradation, which in turn, results from pentoses (mainly xylose) and hexoses degradation, respectively. Finally, phenols are released during lignin partial solubilization and degradation of lignin [5,30]. A wide variety of phenolic substituted compounds such as 4-hydroxybenzaldehyde, vanillin, syringaldehyde, *p*-coumaric acid or ferulic acid, have been identified in steam-exploded wheat straw [3,31].

Table 1. Composition of steam-exploded wheat straw at 200 °C, 2.5 min.

WIS		**Liquid Fraction**				
Component	**% DW**	**Sugar Component**	**Monomeric Form (g/L)**	**Oligomeric Form (g/L)**	**Inhibitors**	**g/L**
Cellulose	53.5 ± 1.1	Glucan	2.3 ± 0.2	12.4 ± 0.3	Furfural	0.8 ± 0.0
Hemicellulose	11.7 ± 0.7	Xylan	2.8 ± 0.1	29.2 ± 0.7	5-HMF	0.3 ± 0.0
Lignin	30.4 ± 3.2	Arabinan	1.3 ± 0.2	1.1 ± 0.0	Acetic acid	6.9 ± 0.3
		Galactan	0.4 ± 0.1	1.4 ± 0.1	Formic acid	6.3 ± 0.2
		Mannan	n.d.	n.d.	Phenols	5.9 ± 0.8

5-HMF, 5-hydroxymethylfurfural; n.d., not determined.

3.2. *Laccase Treatment and Saccharification of the WIS Fraction*

The WIS fraction obtained after pretreatment of wheat straw at 200 °C, 2.5 min was subjected to laccase treatment and saccharification with and without a mild alkaline extraction: Strategy 1, sequential laccase treatment and saccharification; and Strategy 2, mild alkaline extraction and sequential laccase treatment and saccharification.

3.2.1. Effect of Bacterial Laccase Treatment on the Chemical Composition of WIS

The chemical composition of laccase-treated WIS, without and with a previous mild alkaline extraction step, was determined and compared with their respective controls (Table 2). In the case of those pretreated materials that were not subjected to an alkaline extraction, no relevant changes in the lignin content were observed after treatment with MetZyme laccase. Contradictory results have been described with fungal laccases on steam-pretreated materials. Moilanen *et al.* [21] obtained no substantial variation in the lignin content after laccase (*Cerrena unicolor*) treatment of steam-pretreated giant reed (*Arundo donax*). Similar results were obtained by Martín-Sampedro *et al.* [20,32] when steam-exploded eucalypt was treated with *Myceliophtora thermophila* laccase. In contrast, Oliva-Taravilla *et al.* [33] observed a slight increment in the lignin content of unwashed steam-exploded wheat straw after treatment with *Pycnoporus cinnabarinus* laccase. Likewise, Moilanen *et al.* [21] also described a lignin content increment in steam-pretreated spruce (*Picea abis*) treated with *C. unicolor* laccase.

Table 2. Composition of WIS samples treated with bacterial MetZyme laccase without or with a prior alkaline extraction.

Assay	**Composition (% DW, *w/w*) [a]**		
	Cellulose	**Hemicellulose**	**Lignin**
C	53.5 ± 1.1	11.7 ± 0.7	30.4 ± 3.2
L	54.1 ± 0.6	11.2 ± 0.7	30.5 ± 3.1
Alk + C	58.2 ± 2.9	12.3 ± 0.5	27.6 ± 1.6
Alk + L	57.9 ± 2.7	12.4 ± 0.6	27.1 ± 0.6

[a] The remaining percent (of the whole 100%) for biomass composition is represented by other components, including ashes and acid soluble lignin. C, control without alkaline extraction; ALK + C, control with alkaline extraction; L, laccase treatment, Alk + L, alkaline extraction and laccase treatment. Analysis of variance (ANOVA) with Bonferroni's post-test was performed to identify differences between C, L, Alk + C or Alk + L. Differences in means are not statistically significant.

It is known that alkaline treatment of steam-exploded materials decreases lignin content considerably [23,34,35]. In our study, the alkaline treatment was performed at mild conditions and caused 9% delignification (the mean difference is not significant at the 0.05 level) of steam-exploded wheat straw. When MetZyme laccase treatment was subsequently applied to the alkali-extracted WIS, no benefit were found by combining both treatments and similar values of delignification (11%) were observed (Table 2).

3.2.2. Effect of Bacterial Laccase Treatment on Saccharification Yields

RGRs and RXRs obtained after the saccharification of the WIS fractions treated with laccase are shown in Figure 1. In the case of Strategy 1 (sequential laccase treatment and saccharification), RGR of laccase-treated assays was increased by almost 5% (the mean difference is not significant at the 0.05 level) compared to control hydrolysates (Figure 1A). Similarly, an increment on RXR (3%, the mean difference is not significant at the 0.05 level) was also observed (Figure 1B). Even though no major changes were observed in the lignin content after treatment with this bacterial laccase, the slightly better saccharification yields could be attributed to the modification of the lignin structure on the WIS surface, which would affect the interaction of hydrolytic enzymes with the pretreated material. In this context, the action mechanism of laccases towards phenolic lignin units is altering the hydrophobicity of lignin and, consequently, lowering the non-specific adsorption of cellulases to this polymer. Palonen and Viikari [24] reported an increment of carboxyl groups of lignin from steam-pretreated spruce after treatment with the fungal *T. hirsuta* laccase, decreasing the hydrophobicity of lignin and increasing surface charge. These changes reduced the non-specific adsorption of hydrolytic enzymes on lignin, enhancing saccharification yields. Similar results were also obtained by Moilanen *et al.* [21] when steam-pretreated spruce was treated with *C. unicolor* laccase. Nevertheless, these authors also reported an increase in the non-specific adsorption of cellulases and lower glucose recovery yields when laccase treatment was performed on steam-pretreated giant reed. Oliva-Taravilla *et al.* [33] also described lower saccharification yields when steam-exploded wheat straw was treated with the fungal *P. cinnabarinus* laccase. In that work, the increment in Klason lignin observed in laccase-treated WIS was related to a grafting phenomenon of soluble phenols onto the lignin polymer, which hinders the accessibility of cellulolytic enzymes to cellulose and therefore reduces sugar recoveries.

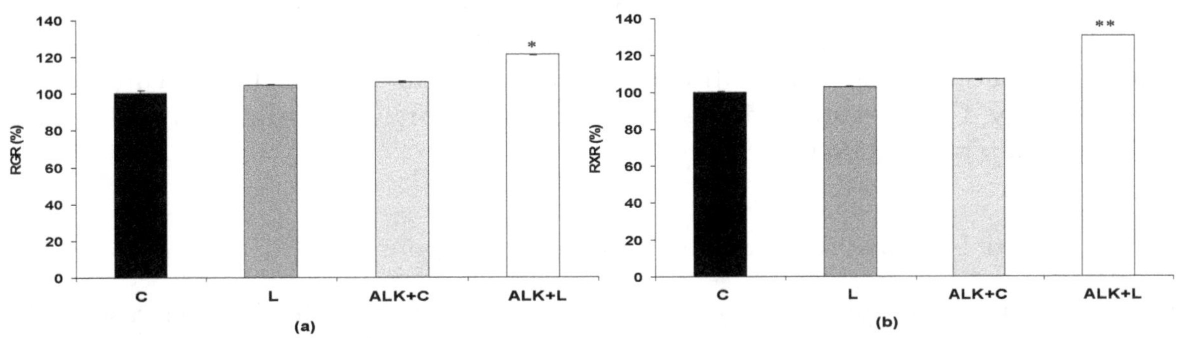

Figure 1. Relative glucose (RGR) (**a**) and xylose (RXR) (**b**) recoveries at 72 h of enzymatic hydrolysis of WIS samples resulting from the different MetZyme laccase treatment and saccharification strategies. Strategy 1, sequential laccase treatment and saccharification (C, control sample; L, laccase sample). Strategy 2, alkaline extraction and sequential laccase treatment and saccharification (ALK + C, control sample with alkaline extraction; ALK + L, laccase sample with alkaline extraction). Glucose concentration values after 72 h of saccharification of control samples were 13.1 and 13.9 g/L for strategies 1 and 2, respectively. Xylose concentration values after 72 h of saccharification of control samples were 2 and 2.2 g/L for strategies 1 and 2, respectively. Mean values and standard deviations were calculated from the triplicates to present the results. Analysis of variance (ANOVA) with Bonferroni's post-test was performed to identify differences between C, L, Alk + C or Alk + L. The mean difference is significant at the (*) 0.05 or (**) 0.01 level.

In the case of Strategy 2 (mild alkaline extraction and sequential laccase treatment and saccharification), the enzymatic hydrolysis of control assays extracted with alkali produced higher RGR (6%, the mean difference is not significant at the 0.05 level) and RXR (7%, the mean difference is not significant at the 0.05 level) values than the control assays not subjected to mild alkaline treatment (Figure 1). This enhancement in saccharification yields after the extraction with alkali is very well known [23,34,35]. Alkali extraction generates new irregular pores as a result of the removal of lignin and the disruption of lignin-carbohydrate complexes, contributing to an increase in the accessibility and susceptibility of cellulose and hemicellulose polymers to the action of hydrolytic enzymes. These advantages can be boosted by a subsequent laccase treatment due to the possibility of obtaining higher delignification ranges, increase the porosity and the available surface area, and decrease the non-specific adsorption of hydrolytic enzymes [19,24,25]. Thus, when alkali-treated WIS were subsequently subjected to laccase treatment, a synergistic effect was observed in the saccharification process, enhancing sugar recovery yields by 21% ($p < 0.05$) and 30% ($p < 0.01$) in RGR and RXR, respectively (Figure 1). The increase in both porosity and surface area promoted by the mild alkali extraction enables an easier penetration of laccase into the fibers, allowing a better accessibility to the lignin polymer. Similar results were found by Yang *et al.* [25] when using *Brassica campestris* straw as raw material. These authors observed by scanning electron microscopy (SEM) some irregular holes on the surface of *B. campestris* straw after alkali treatment, being increased not only in number and density but also in width and depth when the laccase extracted from the fungus *Ganoderma lucidum* was subsequently used. The same effect was described by Li *et al.* [19] in corn straw after combining pretreatment with NaOH and crude ligninolytic enzyme produced by the fungus *Trametes hirsuta*. These results strongly highlight the benefits of combining a mild alkali treatment with a bacterial laccase treatment for improving the hydrolysability of steam-exploded wheat straw.

3.3. *Laccase Treatment and Fermentation of the Whole Slurry*

In addition to offering the possibility of increasing the sugar content during the enzymatic hydrolysis, laccase can work as a detoxification agent to improve the fermentability of pretreated lignocellulosic materials [7]. With the aim of evaluating the effect of bacterial laccase treatment on the fermentability of steam-pretreated wheat straw, the whole slurry was subjected to laccase treatment and simultaneous saccharification and fermentation without and with a presaccharification step: Strategy 3, consecutive laccase treatment and simultaneous saccharification and fermentation (LSSF); and Strategy 4, consecutive laccase treatment with presaccharification and simultaneous saccharification and fermentation (LPSSF).

3.3.1. Effect of Bacterial Laccase Treatment on Inhibitory Compounds

The concentration of inhibitory compounds after treatment with MetZyme laccase, without and with an enzymatic presaccharification step, was determined and compared with their respective controls assays (Table 3). Inhibitory compounds can alter the growth of fermenting microorganisms and also inhibit/deactivate cellulolytic enzymes, decreasing final yields and productivities [5,36–38]. Furfural and 5-HMF have a direct inhibition effect on either the glycolytic or fermentative enzymes of the yeast, reducing equally biomass formation and ethanol yields. Acetic acid and formic acid reduce biomass formation by modifying the intracelular pH and promoting an imbalance in the ATP/ADP ratio. Finally, phenols alter biological membranes, affecting the growth rates and also inhibiting and deactivating hydrolytic enzymes.

Table 3. Inhibitory compounds concentration (g/L) of slurry samples treated with bacterial MetZyme laccase without or with enzymatic presaccharification.

Assay	Inhibitors (g/L)				
	Furfural	5-HMF	Acetic Acid	Formic Acid	Phenols
C	0.3 ± 0.0	0.1 ± 0.0	2.5 ± 0.0	2.2 ± 0.1	1.6 ± 0.0
L	0.3 ± 0.0	0.1 ± 0.0	2.4 ± 0.0	2.3 ± 0.4	1.3 ± 0.0 **
CP	0.3 ± 0.0	0.1 ± 0.0	3.3 ± 0.0	2.1 ± 0.0	1.8 ± 0.1
LP	0.3 ± 0.0	0.1 ± 0.0	3.3 ± 0.0	2.1 ± 0.0	1.4 ± 0.1 ***

C, control without presaccharification; L, laccase treatment without presaccharification; CP, control with presaccharification; LP, laccase treatment with presaccharification; 5-HMF, 5-hydroxymethylfurfural. Analysis of variance (ANOVA) was performed to identify differences between C and L or CP and LP. Difference in means is significant at the (**) 0.01 or (***) 0.001 level.

In general, laccases catalyze the oxidation of phenols, generating unstable phenoxy radicals. These newly formed radicals interact with each other and lead to polymerization into aromatic compounds with lower inhibitory capacity. Total depletion in phenolic content seems to be impossible due to the structural characteristics of phenols [39]. Laccases can easily convert certain compounds, such as syryngaldehyde or cinnamic acids, whilst other phenolic compounds are oxidized with lower rates (vanillin) or remain intact (hydroxybenzaldehyde) [31,39]. Several studies have reported an incomplete removal of phenolic compounds. Kalyani *et al.* [18] described a phenol reduction of 76% when the whole slurry from steam-exploded rice straw was treated with *Coltricia perennis* laccase. Moreno *et al.* [22] achieved higher phenol reductions (93%–94%) when *P. cinnabarinus* and *T. villosa* laccases were used on steam-exploded wheat straw. The same range was observed by Jurado *et al.* [17] with steam-exploded wheat straw and *Coriolopsis rigida* laccase and by Jönsson *et al.* [16] with SO_2 steam-pretreated willow and *Trametes versicolor* laccase. These mentioned studies have in common the use of fungal laccases, mainly from white-rot basidiomycetes. In our case, the bacterial MetZyme laccase decreased the total phenolic content by 20%–21% ($p < 0.01$), independently of whether a presaccharification step is included (Table 3). In comparison to fungal laccases, the lower efficiency in phenol removal by this particular bacterial laccase can be attributed to the lower redox potential of bacterial laccases in general. The redox potential of fungal laccases is estimated to be around +730 mV and +790 mV, while bacterial or plant laccases have a redox potential of about +450 mV. This higher redox potential of fungal laccases increases their capability to act towards a wider range of phenolic compounds. Nevertheless, the lower redox potential might not represent the only explanation for the reduction in the oxidation capacity of bacterial laccases, as other factors such as K_{cat}/K_M ratio (as a measure of enzyme efficiency) may also play an important role [40].

In contrast to the phenols reduction, furan derivatives and weak acids were altered by bacterial laccase in none of the strategies assayed (Table 3). The absence of laccase action on these type of inhibitory compounds has been already reported in previous studies with fungal laccases [16,17,22,31]. This substrate-specific reaction of laccases towards phenols offers some advantages over chemical and physical detoxification methods such as mild reaction conditions, the generation of fewer inhibitory sub-products and lower energy [41].

3.3.2. Effect of Bacterial Laccase Treatment on Cell Viability and Ethanol Production

Control and detoxified slurries, resulting from MetZyme laccase treatments without (L) and with the enzymatic presaccharification (LP), were subjected to SSF for 72 h at 42 °C. *K. marxianuns* CECT 10875 was used as the fermenting microorganism due to its ability to tolerate relatively high temperatures. Thermotolerant yeasts are gaining great significance due to the possibility of better integration between both saccharification and fermentation stages. Optimal temperatures for enzymatic hydrolysis are about 50 °C. In this context, the use of thermotolerant yeasts capable of growing and fermenting around those temperatures is beneficial for the performance of hydrolytic enzymes [31]. During fermentation assays, cell viability, glucose consumption and ethanol production were monitored (Figures 2 and 3).

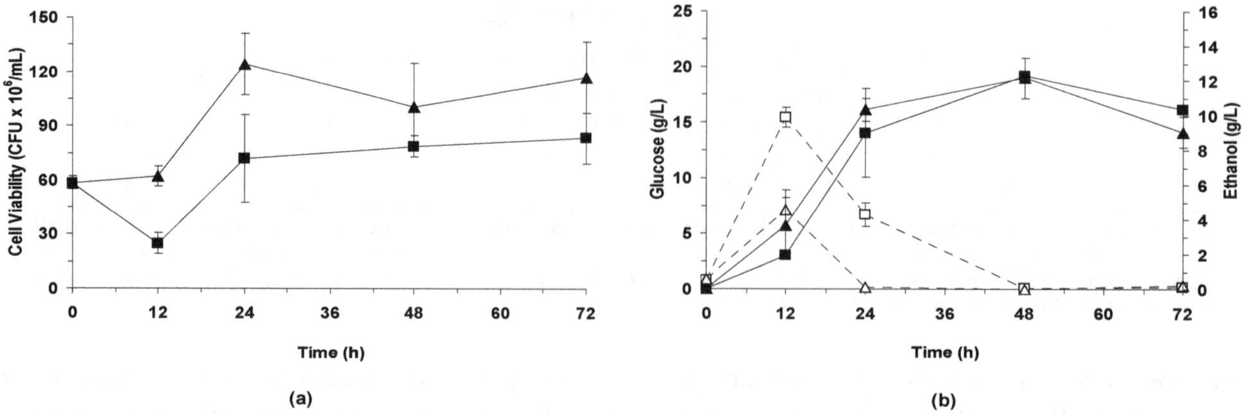

Figure 2. Consecutive laccase treatment and simultaneous saccharification and fermentation (LSSF, Strategy 3). (**a**) Viable cells during simultaneous saccharification and fermentation (SSF) assay with *K. marxianus* of slurry samples resulting from Metzyme laccase treatment. Symbols used: control (■) and laccase (▲) samples. (**b**) Time course for ethanol (filled symbols and continuous lines) and glucose (open symbols and discontinuous lines) during simultaneous saccharification and fermentation (SSF) assay with *K. marxianus* of slurry samples resulting from Metzyme laccase treatment. Symbols used: control (■, □) and laccase (▲, △) samples. Mean values and standard deviations were calculated from the triplicates to present the results.

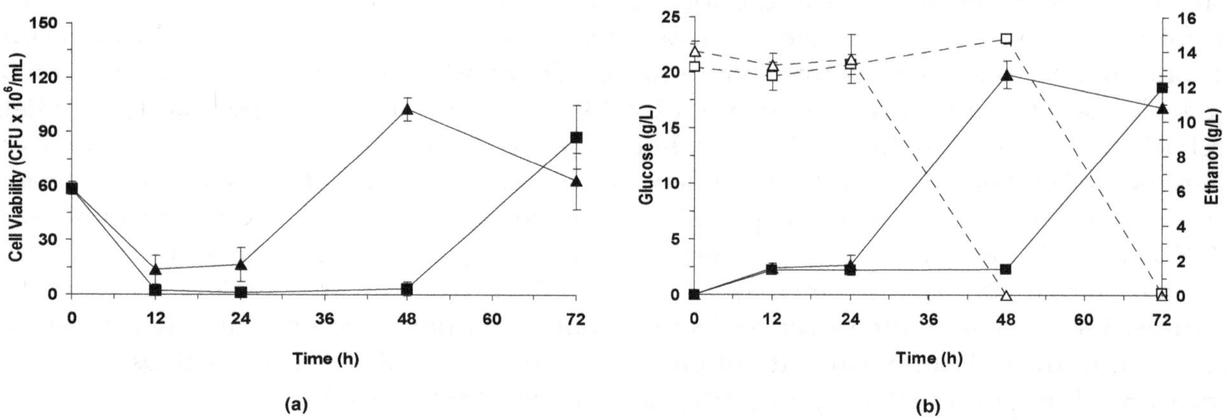

Figure 3. Consecutive laccase treatment with presaccharification and simultaneous saccharification and fermentation (LPSSF, Strategy 4). (**a**) Viable cells during simultaneous saccharification and fermentation (SSF) assay with *K. marxianus* of slurry samples resulting from Metzyme laccase treatment with presaccharification. Symbols used: control (■) and laccase (▲) samples. (**b**) Time course for ethanol (filled symbols and continuous lines) and glucose (open symbols and discontinuous lines) during simultaneous saccharification and fermentation (SSF) assay with *K. marxianus* of slurry samples resulting from Metzyme laccase treatment with presaccharification. Symbols used: control (■, □) and laccase (▲, △) samples. Mean values and standard deviations were calculated from the triplicates to present the results.

LSSF

During SSF of wheat straw slurry, cell viability—in the form of CFU/mL—decreased within the first 12 h in control assays (Figure 2A). This effect is correlated with the adaptation of the yeast to the different components in the fermentation medium, and usually promotes a delay in glucose consumption and ethanol production (Figure 2B). This adaptation period depends on different factors, such as the presence of inhibitory compounds, the nature and concentration of inhibitors and the synergistic effects between them [5,36].

The adaptation phase is overcome by *K. marxianus* after converting certain inhibitory compounds, including furfural and 5-HMF, to their less toxic forms. After being adapted, the yeast showed a remarkable increase in cell viability between 12 and 24 h of SSF, until reaching the value of 80 CFU/mL that remained constant for the rest of the process (Figure 2A). The increase in cell viability made it possible to obtain the maximum glucose consumption (values were not estimated due to the continuous release of glucose) and ethanol production rates (0.59 g/L· h between 12 and 24 h), lowering the glucose concentration to values below 0.1 g/L and reaching the highest ethanol concentration (12.3 g/L) within 48 h (Figure 2B, Table 4).

Table 4. Summary of simultaneous saccharification and fermentation (SSF) assays with *K. marxianus* of slurry samples treated with bacterial MetZyme laccase without or with enzymatic presaccharification.

Assay	Adaptation Phase (h)	$EtOH_{max}$ (g/L)	$Y_{E/G}$ (g/g)	$Y_{E/ET}$ (%)	Q_{Emax} (g/L· h)
C	12	12.3 ± 0.1 [a]	0.29 ± 0.00	65.1	0.58 ± 0.21 [c]
L	<12	12.2 ± 1.2 [a]	0.29 ± 0.03	64.4	0.56 ± 0.09 [c]
CP	48	12.0 ± 0.1 [b]	0.29 ± 0.1	63.4	0.43 ± 0.00 [d]
LP	24	12.7 ± 0.8 [a]	0.30 ± 0.1	67.1	0.44 ± 0.02 [e]

C, control without presaccharification; L, laccase treatment without presaccharification; CP, control with presaccharification; LP, laccase treatment with presaccharification; $EtOH_{max}$, maximum ethanol concentration reached at 48 h [a] and 72 h [b]; $Y_{E/G}$, ethanol yield based on total glucose content in the slurry. The ethanol yield is calculated considering that the liquid volume of the SSF system is constant [42]; $Y_{E/ET}$, percentage of ethanol produced from potential glucose, assuming maximum ethanol yields 0.45 g/g in *K. marxianus* [43]; Q_{Emax}, volumetric ethanol productivity within 12–24 h [c], 48–72 h [d] or 24–48 h [e] of SSF.

When laccase-treated slurries were subjected to SSF, the fermentation performance of *K. marxianus* was slightly improved due to the lower phenolic content. This improvement was more evident in cell viability, where no reduction in the number of CFU/mL was observed within the first 12 h and a significant increase to about 120 CFU/mL was obtained between 24 and 72 h of fermentation (Figure 2A). Similar values of maximum ethanol production rates and maximum ethanol concentrations were observed for both control and laccase-treated slurries (Figure 2B, Table 4). However, a shorter adaptation phase was observed in those slurries treated with Metzyme, which can aid in reducing the overall process time. Similar improvements on the fermentation performance of *K. marxianus* and other fermenting microorganisms have been also observed when using fungal laccases. Moreno *et al.* [22,31] reported higher cell viability, glucose consumption rates and ethanol productivity values when steam-exploded wheat straw was treated with *P. cinnabarinus* and *T. villosa* laccases. Jönsson *et al.* [16] reported higher glucose consumption rate, ethanol productivity and ethanol yield when the liquid fraction from acid steam-exploded willow was treated with *T. versicolor* laccase and fermented with *Saccharomyces cerevisiae*. In the same way, Jurado *et al.* [17] observed higher biomass concentration, sugar consumption and ethanol yield after treating steam-exploded wheat straw with *C. rigida* laccase and fermenting it with *S. cerevisiae*.

LPSSF

In comparison to assays without a presaccharification step, the enzymatic prehydrolysis (P) extended the adaptation period of yeast cells during the subsequent SSF process (Figure 3A,B). A remarkable drop in cell viability was measured in non-treated slurries within the first 12 h of SSF, followed by a long stationary phase where no cell growth was observed. After 48 h, a sudden increase in cell viability could be seen, reaching a value of about 95 CFU/mL (Figure 3A). Regarding glucose concentration, the prehydrolysis stage increased the glucose content before inoculation up to 20 g/L. After inoculation, the adaptation phase of *K. marxianus* allowed the continuous accumulation of this sugar during the first 48 h of SSF, reaching a maximum value of 23 g/L. Supported by the increase in cell viability, glucose started to be consumed after 48 h, and values below 0.1 g/L were observed at 72 h of SSF (Figure 3B). Maximum ethanol concentration (12 g/L) and yield (0.29 g/g) were similar to those obtained in SSF processes without presaccharification, but this value was reached with a delay of 24 h (Table 4). The extended lag phase of *K. marxianus* during PSSF processes in comparison to that observed during SSF can be justified by the presence of a higher concentration of inhibitory compounds after the presaccharification step, especially for acetic acid and phenols (Table 3). According to Thomsen *et al.* [44], acetic acid is produced by the hydrolysis of acetyl groups in hemicelluloses, which involves the synergistic action of both hemicellulase and acetyl esterase activities. In this sense, the cellulolytic NS50013 preparation used in this study—obtained from *Trichoderma* spp. strains—contains xylanase and acetyl esterase activities that can release the acetyl groups that remain in the fibers, increasing the acetic acid concentration [45]. Similarly, the increment in the phenol content can be explained by the release of *p*-coumaric acid and ferulic acid during the presaccharification step. Ferulic acid and *p*-coumaric acid are two typical lignin phenolic compounds present in wheat straw [3,31]. The hydrolysis of these cinnamic acids is attributed to the complementary action of xylanase and phenolic acid esterase activities [31]. The esterase activities, highlighting feruloyl esterase activity, are naturally produced by *Aspergillus niger*, which is the strain producing the β-glucosidase NS50010 preparation [46].

When laccase treatment was combined with the presaccharification step (LP), the adaptation phase was reduced from 48 h in control assays to 24 h in laccase-treated slurries. This reduction can be seen in Figure 3A, where an increase from about 15 CFU/mL to above 100 CFU/mL was observed within 24 and 48 h of SSF. In comparison to LSSF process without a presaccharification stage where the number of CFU/mL was kept constant after the adaptation phase, in LPSSF process a remarkable decrease in cell viability took place within 48 and 72 h, which is an indicator of the higher inhibitory content even after laccase treatment. In relation to glucose consumption and ethanol production, similar rates (0.43 g/L·h and 0.44 g/L·h in control and laccase-treated slurries, respectively) were observed after the adaptation phase either in control or laccase-treated assays, which resulted in a maximum ethanol concentration of about 12 g/L—similar to those values obtained in the SSF processes without a presaccharification step (Figure 3B, Table 4). It is important to notice that higher ethanol concentrations than those obtained in the present work are needed for the cost-effectiveness of a commercial lignocellulosic ethanol production. Working at higher substrate concentrations is therefore imperative and a presaccharification step is typically included in the process in order to avoid certain problems such as mixing. In this context, the use of laccases could play a crucial role to increase the fermentability of steam-pretreated lignocellulosic materials.

4. Conclusions

The present work shows the potential of the bacterial MetZyme laccase for improving both the hydrolysability and fermentability of steam-pretreated materials. The laccase treatment of the WIS fraction resulted in slightly higher glucose and xylose recoveries during a saccharification process. This improvement was increased synergistically by the action of a mild alkaline extraction performed prior to laccase treatment. MetZyme laccase also showed modest phenol removal when treating the whole pretreated slurries, reducing the inhibitory effects of steam-exploded wheat straw. The lower inhibitory content led to improve the fermentation performance of *K. marxianus* in SSF processes with or without a presaccharification step, shortening its adaptation period and the overall fermentation times. These results represent an interesting approach to improve the efficiency of the ethanol production process, which might contribute to making lignocellulosic ethanol production economically viable. Nevertheless, other parameters, including laccase dosages and production costs, need to be further studied and optimized for the cost-effectiveness of the process.

Acknowledgments: Authors wish to thank the Spanish MIMECO for funding this study via Project CTQ2013-47158-R. Antonio D. Moreno acknowledges a "Juan de la Cierva" contract (FJCI-2014-22385).

Author Contributions: Antonio D. Moreno, David Ibarra, Antoine Mialon, and Mercedes Ballesteros participated in the design of the study. Antonio D. Moreno and David Ibarra performed the experimental work and wrote the manuscript. Antonio D. Moreno, David Ibarra, Antoine Mialon, and Mercedes Ballesteros conceived the study and commented on the manuscript. All the authors read and approved the final manuscript.

Abbreviations

The following abbreviations are used in this manuscript:

WIS	Water Insoluble Solids fraction
NREL-LAP	National Renewable Energies Laboratory-Laboratory Analytical Procedures
SSF	Simultaneous Saccharification and Fermentation
CFU	Colony Forming Units
DW	Dry Weight
LAP	Laboratory Analytical Procedures
IU	International Units
FPU	Filter Paper Units
TS	Total Solids
RGR	Relative Glucose Recovery
RXR	Relative Xylose Recovery
LSSF	consecutive Laccase treatment and Simultaneous Saccharification and Fermentation
LPSSF	consecutive Laccase treatment with Presaccharification and Simultaneous Saccharification and Fermentation
5-HMF	5-hydroxymethylfurfural
C	Control treatment
CP	Control treatment with Presaccharification
L	Laccase treatment
LP	Laccase treatment with Presaccharification
Alk	Alkaline extraction
ATP/ADP	Adenosine-5′-triphosphate/Adenosine-5′-diphosphate

References

1. Ballesteros, M. Enzymatic hydrolysis of lignocellulosic biomass. In *Bioalcohol Production. Biochemical Conversion of Lignocellulosic Biomass*; Waldron, K., Ed.; Woodhead Publishing: Cambridge, UK, 2010; pp. 159–177.
2. Alvira, P.; Tomás-Pejó, E.; Ballesteros, M.; Negro, M.J. Pretreatment technologies for an efficient bioethanol production process based on enzymatic hydrolysis: A review. *Bioresour. Technol.* **2010**, *101*, 4851–4861. [CrossRef] [PubMed]
3. Ballesteros, I.; Negro, M.; Oliva, J.M.; Manzanares, P.; Ballesteros, M. Ethanol production from steam-explosion pretreated wheat straw. *Appl. Biochem. Biotechnol.* **2006**, *130*, 496–508. [CrossRef]
4. Berlin, A.; Balakshin, M.; Gilkes, N.; Kadla, J.; Maximenko, V.; Kubo, S. Inhibition of cellulase, xylanase and β-glucosidase activities by sofwood lignin preparations. *J. Biotechnol.* **2006**, *125*, 198–209. [CrossRef] [PubMed]
5. Palmqvist, E.; Hahn-Hägerdal, B. Fermentation of lignocellulosic hydrolysates. II: Inhibitors and mechanism of inhibition. *Bioresour. Technol.* **2000**, *74*, 25–33. [CrossRef]
6. Palmqvist, E.; Hahn-Hägerdal, B. Fermentation of lignocellulosic hydrolysates. I: Inhibition and detoxification. *Bioresour. Technol.* **2000**, *74*, 17–24. [CrossRef]
7. Moreno, A.D.; Ibarra, D.; Alvira, P.; Tomás-Pejó, E.; Ballesteros, M. A review of biological delignification and detoxification methods for lignocellulosic bioethanol production. *Crit. Rev. Biotechnol.* **2015**, *35*, 342–354. [CrossRef] [PubMed]
8. Thurston, C.F. The structure and function of fungal laccases. *Microbiology* **1994**, *140*, 19–26. [CrossRef]

9. Enguita, F.J.; Matias, P.M.; Martins, L.O.; Placido, D.; Henriques, A.O.; Carrondo, M.A. Spore-coat laccase CotA from *Bacillus subtilis*: Crystallization and preliminary X-ray characterization by the MAD method. *Acta Crystallogr. D Biol. Cristallogr.* **2002**, *58*, 1490–1493. [CrossRef]
10. Bajpai, P. Biological bleaching of chemical pulps. *Crit. Rev. Biotechnol.* **2004**, *24*, 1–58. [CrossRef] [PubMed]
11. Paice, M.G.; Bourbonnais, R.; Reid, I.D.; Archibald, F.S.; Jurasek, L. Oxidative bleaching enzymes: A review. *J. Pulp Pap. Sci.* **1995**, *21*, 280–284.
12. Ibarra, D.; Camarero, S.; Romero, J.; Martínez, M.J.; Martínez, A.T. Integrating laccase-mediator treatment into an industrial-type sequence for totally chlorine free bleaching of eucalipt kraft pulp. *J. Chem. Technol. Biotechnol.* **2006**, *81*, 1159–1165. [CrossRef]
13. Eugenio, M.E.; Hernández, M.; Moya, R.; Martín-Sampedro, R.; Villar, J.C.; Arias, M.E. Evaluation of a new laccase produced by Streptomyces ipomoea on biobleaching and ageing of kraft pulps. *BioResources* **2011**, *6*, 3231–3241.
14. Chandra, R.; Chowdhary, P. Properties of bacterial laccases and their application in biorremediation of industrial wastes. *Environ. Sci. Process. Impacts* **2015**, *17*, 326–342. [CrossRef] [PubMed]
15. Milstein, O.; Haars, A.; Majcherczyk, A.; Tautz, D.; Zanker, H.; Hüttermann, A. Removal of chlorophenols and chlorolignins from bleaching effluents by combined chemical and biological treatment. *Water Sci. Technol.* **1988**, *20*, 161–170.
16. Jönsson, L.J.; Palmqvist, E.; Nilvebrant, N.-O.; Hahn-Hägerdal, B. Detoxification of wood hydrolysates with laccase and peroxidase from the white-rot fungus *Trametes versicolor*. *Appl. Microbiol. Biotechnol.* **1998**, *49*, 691–697. [CrossRef]
17. Jurado, M.; Prieto, A.; Martínez-Alcalá, A.; Martínez, A.T.; Martínez, M.J. Laccase detoxification of steam-exploded wheat straw for second generation bioethanol. *Bioresour. Technol.* **2012**, *100*, 6378–6384. [CrossRef] [PubMed]
18. Kalyani, D.; Dhiman, S.S.; Kim, H.; Jeya, M.; Kim, I.-W.; Lee, J.-K. Characterization of a novel laccase from the isolated *Coltricia perennis* and its application to detoxification of biomass. *Process Biochem.* **2012**, *47*, 671–678. [CrossRef]
19. Li, J.; Sun, F.; Li, X.; Yan, Z.; Yuan, Y.; Liu, X.F. Enhanced saccharification of corn straw pretreated by alkali combining crude ligninolytic enzymes. *J. Chem. Technol. Biotecnhol.* **2012**, *87*, 1687–1693. [CrossRef]
20. Martín-Sampedro, R.; Eugenio, M.E.; García, J.C.; López, F.; Villar, J.C.; Díaz, M.J. Steam explosion and enzymatic pre-treatments as an approach to improve the enzymatic hydrolysis of *Eucalytpus globulus*. *Biomass Bioenergy* **2012**, *42*, 97–106. [CrossRef]
21. Moilanen, U.; Kellock, M.; Galkin, S.; Viikari, L. The laccase-catalyzed modification of lignin for enzymatic hydrolysis. *Enzym. Microb. Technol.* **2011**, *49*, 492–498. [CrossRef] [PubMed]
22. Moreno, A.D.; Ibarra, D.; Fernández, J.L.; Ballesteros, M. Different laccase detoxification strategies for ethanol production from lignocellulosic biomass by the thermotolerant yeast *Kluyveromyces marxianus* CECT 10875. *Bioresour. Technol.* **2012**, *106*, 101–109. [CrossRef] [PubMed]
23. Moreno, A.D.; Ibarra, D.; Alvira, P.; Tomás-Pejó, E.; Ballesteros, M. Exploring laccase and mediators behavior during saccharification and fermentation of steam-exploded wheat straw for bioethanol production. *J. Chem. Technol. Biotehcnol.* **2015**. [CrossRef]
24. Palonen, H.; Viikari, L. Role of oxidative enzymatic treatments on enzymatic hydrolysis of softwood. *Biotechnol. Bioeng.* **2004**, *86*, 550–557. [CrossRef] [PubMed]
25. Yang, P.; Jiang, S.; Zheng, Z.; Shuizhong, L.; Luo, S.; Pan, L. Effect of alkali and laccase pretreatment of Brassica campestris straw: Architecture, crystallisation, and saccharification. *Polym. Renew. Resour.* **2011**, *2*, 21–34.
26. National Renewable Energy Laboratory. *Chemical Analysis and Testing Laboratory Analytical Procedures*; National Renewable Energy Laboratory: Golden, CO, USA, 2007. Available online: www.1.ere.energy.gov/biomass/analytical_procedures.html (accessed on 15 March 2015).
27. Ghose, T.K. Measurement of cellulase activity. *Pure Appl. Chem.* **1987**, *59*, 257–268. [CrossRef]
28. Ballesteros, I.; Ballesteros, M.; Cabañas, A.; Carrasco, J.; Martín, C.; Negro, M.J.; Saez, F.; Saez, R. Selection of thermotolerant yeasts for simultaneous saccharification and fermentation (SSF) of cellulose to ethanol. *Appl. Biochem. Biotechnol.* **1991**, *28*, 307–315. [CrossRef] [PubMed]
29. Singleton, V.L.; Rossi, J.A. Colorimetry of total phenolic with phosphomolibdic-phosphotungstic acid reagent. *Am. J. Enol. Vitic.* **1965**, *16*, 144–158.

30. Oliva, J.M.; Saez, F.; Ballesteros, I.; González, A.; Negro, M.J.; Manzanares, P.; Ballesteros, M. Effect of lignocellulosic degradation compounds from steam explosion pretreatment on ethanol fermentation by thermotolerant yeast *Kluyveromyces marxianus*. *Appl. Biochem. Biotechnol.* **2003**, *105*, 141–154. [CrossRef]
31. Moreno, A.D.; Ibarra, D.; Ballesteros, I.; Fernández, J.L.; Ballesteros, M. Ethanol from laccase-detoxified lignocellulose by the thermotolerant yeast *Kluyveromyces marxianus*—Effects of steam pretreatment conditions, process configurations and substrate loadings. *Biochem. Eng. J.* **2013**, *79*, 94–103. [CrossRef]
32. Martín-Sampedro, R.; Capanema, E.A.; Hoeger, I.; Villar, J.C.; Rojas, O.J. Lignin changes after steam explosion and laccase-mediator treatment of eucalytpus wood chips. *J. Agric. Food Chem.* **2011**, *59*, 8761–8769. [CrossRef] [PubMed]
33. Oliva-Taravilla, A.; Moreno, A.D.; Demuez, M.; Ibarra, D.; Tomás-Pejó, E.; González-Fernández, C.; Ballesteros, M. Unraveling the effects of laccase treatment on enzymatic hydrolysis of steam-exploded wheat straw. *Bioresour. Technol.* **2015**, *175*, 209–215. [CrossRef] [PubMed]
34. Fernandes, M.C.; Ferro, M.D.; Paulino, A.F.C.; Mendes, J.A.S.; Gravitis, J.; Evtuguin, D.V.; Xavier, A.M.R.B. Enzymatic saccharification and bioethanol production from *Cynara cardunculus* pretreated by steam explosion. *Bioresour. Technol.* **2015**, *186*, 309–315. [CrossRef] [PubMed]
35. Yang, B.; Boussaid, A.; Mansfield, S.D.; Gregg, D.J.; Sadler, J.N. Fast and efficient alkaline peroxide treatment to enhance the enzymatic digestibility of steam-exploded softwood substrates. *Biotechnol. Bioeng.* **2002**, *77*, 678–684. [CrossRef] [PubMed]
36. Klinke, H.B.; Thomsen, A.B.; Ahring, B.K. Inhibition of ethanol-producing yeast and bacteria by degradation products produced during pretreatment of biomass. *Appl. Microbiol. Biotechnol.* **2004**, *66*, 10–26. [CrossRef] [PubMed]
37. Ximenes, E.; Kim, Y.; Mosier, N.; Dien, B.; Ladisch, M. Inhibition of cellulases by phenols. *Enzym. Microb. Technol.* **2010**, *46*, 170–176. [CrossRef]
38. Ximenes, E.; Kim, Y.; Mosier, N.; Dien, B.; Ladisch, M. Deactivation of cellulases by phenols. *Enzym. Microb. Technol.* **2011**, *48*, 54–60. [CrossRef] [PubMed]
39. Kolb, M.; Sieber, V.; Amann, M.; Faulstich, M.; Schieder, M. Removal of monomer delignification products by laccase from *Trametes versicolor*. *Bioresour. Technol.* **2012**, *104*, 298–304. [CrossRef] [PubMed]
40. Lahtinen, M.; Kruus, K.; Boer, H.; Kemell, M.; Andberg, M.; Viikari, L.; Sipilä, J. The effect of lignin model compound structure on the rate of oxidation catalyzed by two different fungal laccases. *J. Mol. Catal. B Enzym.* **2009**, *57*, 204–210. [CrossRef]
41. Parawira, W.; Tekere, M. Biotechnological strategies to overcome inhibitors in lignocellulose hydrolysates for ethanol production: Review. *Crit. Rev. Biotechnol.* **2011**, *31*, 20–31. [CrossRef] [PubMed]
42. Zhang, J.; Bao, J. A modified method for calculating practical ethanol yield at high lignocellulosic solids content and high ethanol titer. *Bioresour. Technol.* **2012**, *116*, 74–79. [CrossRef] [PubMed]
43. Tomás-Pejó, E.; Oliva, J.M.; González, A.; Ballesteros, I.; Ballesteros, M. Bioethanol production from wheat straw by the thermotolerant yeast *Kluyveromyces marxianus* CECT 10875 in a simultaneous saccharification and fermentation fed-batch process. *Fuel* **2009**, *88*, 2142–2147. [CrossRef]
44. Thomsen, M.T.; Thygesen, A.; Thomsen, A.B. Identification and characterization of fermentation inhibitors formed during hydrothermal treatment and following SSF of wheat straw. *Appl. Microbiol. Biotechnol.* **2009**, *83*, 447–455. [CrossRef] [PubMed]
45. Juhász, T.; Szengyel, K.; Réczey, K.; Siika-Aho, M.; Viikari, L. Characterization of cellulase and hemicellulases produced by *Trichoderma reesei* on various carbon sources. *Process Biochem.* **2005**, *40*, 3519–3525. [CrossRef]
46. Dien, B.S.; Ximenes, E.A.; O'Brien, P.J.; Moniruzzaman, M.; Li, X.L.; Balan, V.; Dale, B.; Cotta, M.A. Enzyme characterization for hydrolysis of AFEX and liquid hot water pretreated distillers'grains and their conversion to ethanol. *Bioresour. Technol.* **2008**, *99*, 5216–5225. [CrossRef] [PubMed]

13

Yeast Nanometric Scale Oscillations Highlights Fibronectin Induced Changes in *C. albicans*

Anne-Céline Kohler [1,*], Leonardo Venturelli [1], Abhilash Kannan [2], Dominique Sanglard [2], Giovanni Dietler [1,3], Ronnie Willaert [3,4,5] and Sandor Kasas [1,2,6]

1 Laboratoire de Physique de la Matière Vivante, EPFL, 1015 Lausanne, Switzerland; leonardo.venturelli@epfl.ch (L.V.); giovanni.dietler@epfl.ch (G.D.); sandor.kasas@epfl.ch (S.K.)
2 Institute of Microbiology, Lausanne University Hospital, CH-1011 Lausanne, Switzerland; Abhilash.Kannan@chuv.ch (A.K.); Dominique.Sanglard@chuv.ch (D.S.)
3 International Joint Research Group VUB-EPFL NanoBiotechnology & NanoMedicine (NANO), Vrije Universiteit Brussel, 1050 Brussels, Belgium; Ronnie.Willaert@vub.be
4 Research Group Structural Biology Brussels, Alliance Research Group VUB-UGent NanoMicrobiology (NAMI), 1050 Ixelles, Belgium
5 Department Bioscience Engineering, University Antwerp, 2000 Antwerp, Belgium; Ronnie.Willaert@uantwerpen.be
6 Plateforme de Morphologie UFAM, CUMRL, University of Lausanne, 1015 Lausanne, Switzerland
* Correspondence: anne-celine.kohler@epfl.ch

Abstract: Yeast resistance to antifungal drugs is a major public health issue. Fungal adhesion onto the host mucosal surface is still a partially unknown phenomenon that is modulated by several actors among which fibronectin plays an important role. Targeting the yeast adhesion onto the mucosal surface could lead to potentially highly efficient treatments. In this work, we explored the effect of fibronectin on the nanomotion pattern of different *Candida albicans* strains by atomic force microscopy (AFM)-based nanomotion detection and correlated the cellular oscillations to the yeast adhesion onto epithelial cells. Preliminary results demonstrate that strongly adhering strains reduce their nanomotion activity upon fibronectin exposure whereas low adhering Candida remain unaffected. These results open novel avenues to explore cellular reactions upon exposure to stimulating agents and possibly to monitor in a rapid and simple manner adhesive properties of *C. albicans*.

Keywords: *Candida albicans*; adhesion; fibronectin; nanomotion; atomic force microscope (AFM)

---

1. Introduction

Yeast biotechnology is a recent field where nanotechniques are used to manipulate and analyze yeast cells and cell constituents at the nanoscale [1]. Among the nanotechniques, AFM-related approaches played a major role in unveiling morphological, mechanical and biochemical properties of yeast [2–4]. Recently, our team demonstrated that living cells attached onto a soft cantilever induce nanometric scale oscillations (referred to as nanomotion) that stop as soon as the organism dies [5]. Commercially available atomic force microscopes (AFM) or dedicated devices easily detect these oscillations. Nanomotion detection has been applied to numerous biological samples such as proteins, single organelles, and a plethora of living cells such as prokaryotes (bacteria) and eukaryotes (fungal, vegetal and mammalian cells) [6]. The most straightforward application of the technique is the ultra-rapid antibiotic sensitivity test (AST). AST can be performed within an hour as compared to long-lasting traditional AST methods, which depend on the replication rate of the bacteria [7–9]. The test consists in attaching the organism of interest onto an AFM cantilever and monitoring its oscillations as a function of time upon addition of antibiotics in the analysis chamber. It is worth noting that the nanometric scale oscillations do not only reflect the living or death state of the organisms but also its activity [5,10].

Fungal infections are a major public health issue nowadays; it is estimated that every year fungi infect about 1.2 billion people [11]. *C. albicans* is a common fungal pathogen that belongs to the human microbiome of healthy individuals [12]. This commensal relationship is a complex interplay of candidial and human factors. However, impairment of the host immunity or the normal host microbiota can lead to *C. albicans* infection (candidiasis) [13]. *C. albicans* is the predominant cause of virtually all types of candidiasis [14]. The first step of the infection is the adhesion of *C. albicans* onto the host. This step is an essential determinant of pathogenesis, as it allows *C. albicans* to attach to host cells and to form biofilms or to disseminate in the host blood vessels. The biofilm increases yeast cell resistance to antifungal therapeutics and protects it from the host immune system [15]. *C. albicans* has developed multiple ways to colonize and infect host cells and tissues. One such mechanism is the specific ligand–receptor interaction through a whole range of adhesins displayed on the yeast cell wall [16–18]. These cell wall proteins are capable of recognizing protein ligands [16], glycolipids [19–22] and carbohydrates [23–29] on the host cells. Fibronectin is an important protein ligand of the host extracellular matrix (ECM) that plays an essential role in *C. albicans* adhesion [30]. Furthermore, targeting fibronectin has shown to alter *C. albicans* biofilm formation [31]. Therefore, a better understanding of the yeast–fibronectin interaction could lead to novel therapeutic options to fight candidiasis.

In this work, we applied nanomotion analysis to monitor the oscillatory activity of *C. albicans* upon exposure to fibronectin. We used an AFM-based nanomotion detector to follow the evolution of cellular oscillations in the absence and the presence of fibronectin on strongly and poorly adherent *C. albicans* cells. Interestingly, these two isolates reacted very differently to the interaction with fibronectin. These preliminary results demonstrate the potential of nanomotion analysis to monitor ligand–receptor interactions in a label free manner.

2. Materials and Methods

2.1. Yeast Strains

The *C. albicans* isolate 101 and CEC 3675 were kindly provided by Salomé Leibundgut and Christophe D'Enfert laboratories [32], respectively. The yeasts were cultured in yeast-extracted peptone-dextrose (YPD) medium (1% m/v yeast extract (Difco Laboratories, Fisher Scientific, Hampton, NH, USA), 2% m/v peptone (Difco Laboratories, Fisher Scientific, Hampton, NH, USA) and 2% m/v glucose (Sigma, St. Louis, MI, USA)) overnight at 30 °C with shaking (160 rpm).

2.2. Experimental Procedures

Rectangular tipless cantilevers (qp-CONT, NanoandMore GmbH, Wetzlar, Germany), with a nominal spring constant of 0.1 N/m and an average resonant peak in liquids of 8 kHz, were coated with 2 mg/mL of concanavalin A (Con A) (Sigma, St. Louis, MI, USA) for 30 min at room temperature. After removing the excess of Con A, the yeast cells were placed in contact with the cantilever for 1 h at room temperature to allow them to attach to its surface. Poorly attached *C. albicans* cells were removed by washing gently with YPD medium. Finally, the *C. albicans* covered cantilever was inserted into the analysis chamber containing 2 mL of filtered (0.2 μm syringe filter, Merck Millipore, Burlington, MA, USA) YPD medium. The measurements were performed at room temperature in YPD medium and in YPD medium containing 25 μg/mL of fibronectin (Sigma, USA). Fibronectin was directly added inside the chip reservoir. For the experiments performed with antifungals, caspofungin (Sigma, USA) was diluted in the YPD present in the analysis chamber to reach a final concentration of 100 μg/mL.

2.3. *Nanomotion Detector*

The cantilever oscillations were collected in real time using an in-house developed nanomotion detection device. The system relies on a laser-based signal transduction as typically used in commercial AFMs. A typical experiment lasted for 2 h. The control experiments were carried out for at least 4 h.

2.4. *Software and Nanomotion Analysis*

The cantilever oscillations were recorded and saved at 20 kHz using a USB-4431 DAQ card (National Instruments, Austin, TX, USA). The data acquisition program was developed in LabView. A dedicated Python program was used to process the recorded data and to display the deflection of the cantilever as a function of time. The software first removes the low frequency cantilever displacement signal by calculating a first order fit of the raw signal (deflection of the cantilever) by taking 20 seconds-long window frames. The obtained fit is then subtracted from the raw signal to remove thermally induced cantilever deflection. The thermal drift essentially occurs at the beginning of the experiment and during the fluid exchange procedures. The thermal drift free signal is further processed to obtain its variance in 10 seconds-window frames.

2.5. *Viability Assay*

Cells were placed inside a commercially available microfluidic chip (Ibidi, Planegg, Germany), and stained with calcofluor white (Sigma, USA), according to the manufacturer's instructions. To detect dead cells, propidium iodide (PI, Sigma, USA) was added to the YPD medium and the fluorescence of the yeast cells was recorded using an Axiovert microscope (Zeiss, Oberkochen, Germany).

2.6. *Adhesion Assay*

Adherence of *C. albicans* to TR146 cells was measured using the protocols previously described [33,34] with slight modifications (Figure S1). TR146 cells grown as monolayers in 6-well plates were incubated with 100 *C. albicans* cells for 20 min at 37 °C. The supernatant was carefully removed and spread on YPD agar plates to determine the number of non-adherent fungal cells. The adherent fungal cells that were left behind in the 6-well plates were rinsed with PBS and were overlaid with melted Wort agar at 40 °C. The plates were incubated at 30 °C for 36 h to count the colonies. Adherence was determined as the ratio of the number of colonies grown on Wort agar to the number of colonies grown on Wort agar and the number of colonies grown from the culture supernatant.

2.7. *Statistical Analysis*

Statistical analysis of nanomotion experiments were performed with the Python package Scipy. We performed the non-parametric Mann–Whitney U test for the three independent replicates. We used standard student t-test to process the adhesion assay on three independent replicates using the Graphpad Prism software.

3. Results

To assess a putative differential reaction of strongly and weakly interacting *C. albicans* to fibronectin, we quantified the adhesion of two different isolates, 101 and CEC3675, on oral keratinocytes (TR 146). As shown in Figure 1 isolate 101 was measured to have a significantly higher adhesion compared to isolate CEC3675.

To investigate the *C. albicans*–fibronectin interaction we used an in house nanomotion detector depicted in Figure 2A. The set up consists in an analysis chamber filled with liquid (in our case YPD) containing the cantilever to which yeast cells are attached (Figure 2B). The cantilever oscillations were recorded (Figure 2C) and processed to display the signal variance as a function of time (Figure 2D).

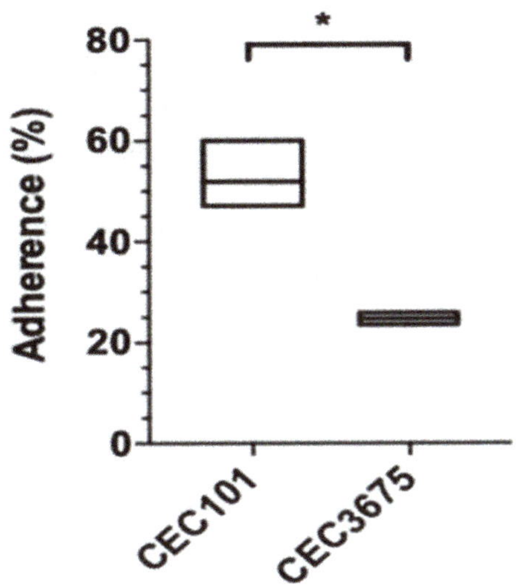

Figure 1. *C. albicans* isolates 101 and CEC3675 adhere differently to oral keratinocytes. Percentage of adherence of both isolates. Statistical analysis ($n = 3$) was done using standard t-test. The asterisk represents $p < 0.05$.

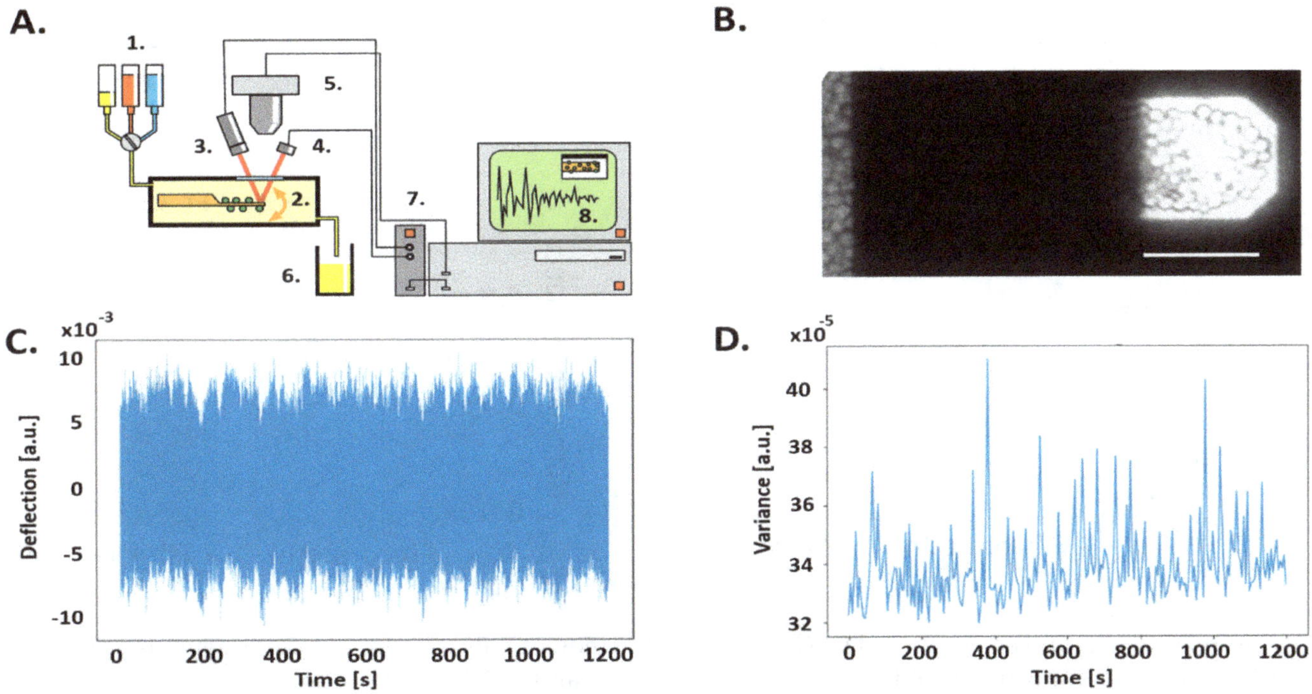

Figure 2. Nano-mechanical sensor system. (**A**) Representative image of a cantilever with attached *C albicans* cells. Scale bar 40 µm. (**B**) Schematic of the experimental system and data collection. (1) Liquids to be injected into the analysis chamber. In our case YPD, YPD containing fibronectin, and YPD containing caspofungin. (2) Analysis chamber with the AFM cantilever and *C. albicans* attached onto its surface (green circles). (3) Super luminescent diode. (4) Four-segment photodiode. (5) Optical microscopy with camera. (6) Liquid waste. (7) In-house dedicated electronics and National Instruments data acquisition card. (8) Desktop computer. (**C**). The collected raw data are processed; and (**D**). analyzed using the variance of the signal.

Using this system, we monitored the nanomotion pattern of *C. albicans* isolates 101 and CEC3675 in the absence and presence of fibronectin (Figure 3). Before addition of fibronectin, both isolates behaved similarly (Figure 3B). However, in the presence of fibronectin, nanomotion activity (variance) of isolate 101 drastically decreased (from 0.9 ± 0.5 to 0.3 ± 0.1) (Figure 3). In contrast, isolate CEC3675 did not present a significant decrease. To confirm that the drop of signal was not due to a change in the temperature, nor convective currents that can appear upon addition of a liquid in the analysis chamber, we performed control experiments, simultaneously, with another nanomotion detector. These experiments consisted in injecting the same quantity of medium, instead of fibronectin, into the analysis chamber. The obtained results showed no significant difference in the nanomotion pattern, for both isolates, upon addition of YPD media (Figure S2). Additionally, we assessed the number of cells present on the cantilever before and after the experiment to determine if the reduced signal was caused by cells being detached from the cantilever. The analysis of the images taken by the optical microscope located above the nanomotion detector (as depicted in the schematic in Figure 2A) confirmed that no cells detached from the cantilever throughout the experiments (Figure S3).

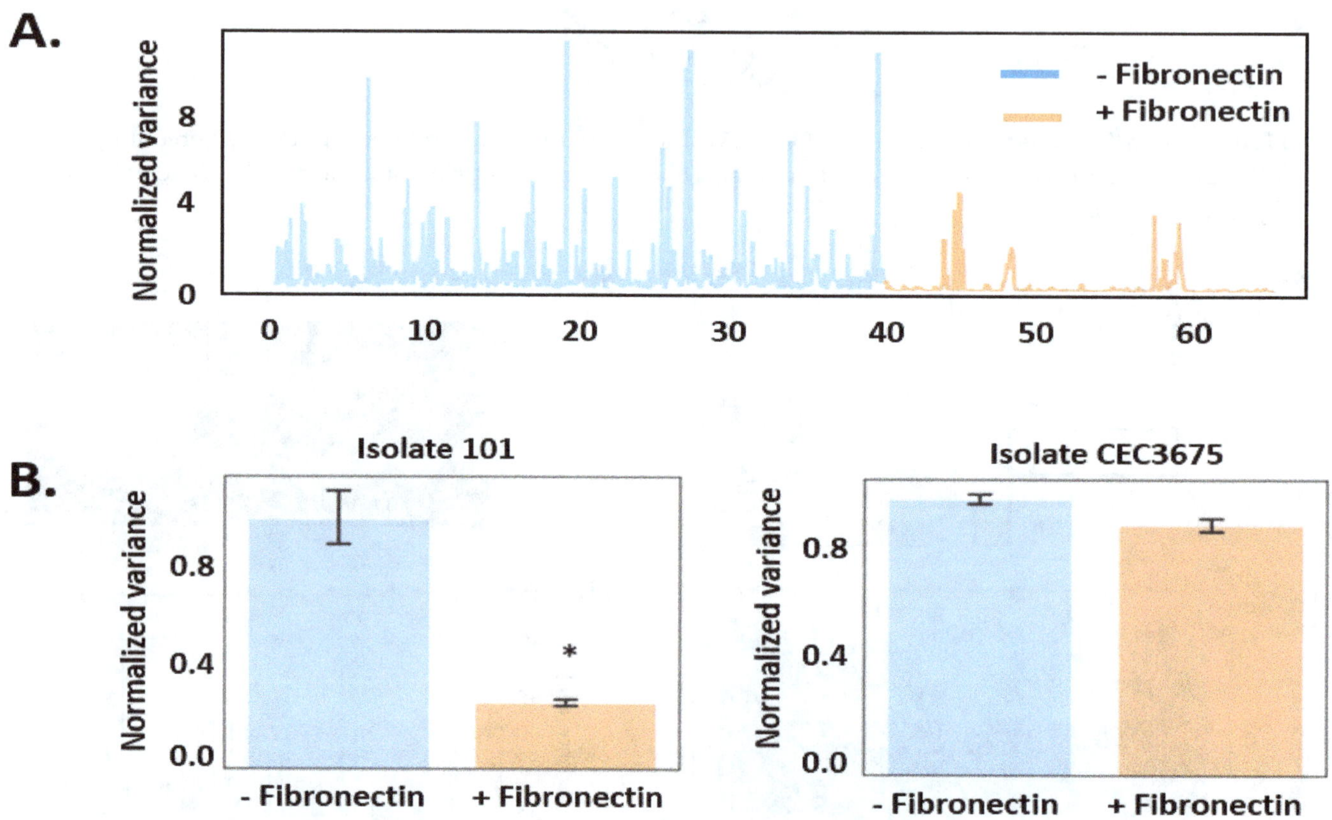

Figure 3. *C. albicans* isolate 101 and CEC 3675 react differently to fibronectin. (**A**). Representative graph of the normalized variance of isolate 101 in YPD (blue) and in YPD with fibronectin (orange). The decrease of the normalized variance is clearly visible between the two conditions. (**B**). The mean of the normalized variance (experiment in triplicate) represented as a bar plot for isolate 101 compared to isolate CEC 3675. Error bars are the confidence of intervals. Statistical analyses were done using Mann–Whitney U test, the asterisk represents $p < 0.05$.

To further exclude another cause of the decrease of the nanomotion signal for isolate 101, such as premature cell death, we monitored *C. albicans* viability by nanomotion and fluorescence microscopy in the absence and presence of fibronectin. Eventually the cells were killed by the antifungal caspofungin. As shown in Figure 3A, the variance of the nanomotion signal drastically dropped after the drug injection. The fluorescent viability test did not show any effect of fibronectin on the cellular viability as it can be noticed in Figure 4B. Similarly, fibronectin also did not have any effect on the viability of isolate CEC3675 (Figure S4).

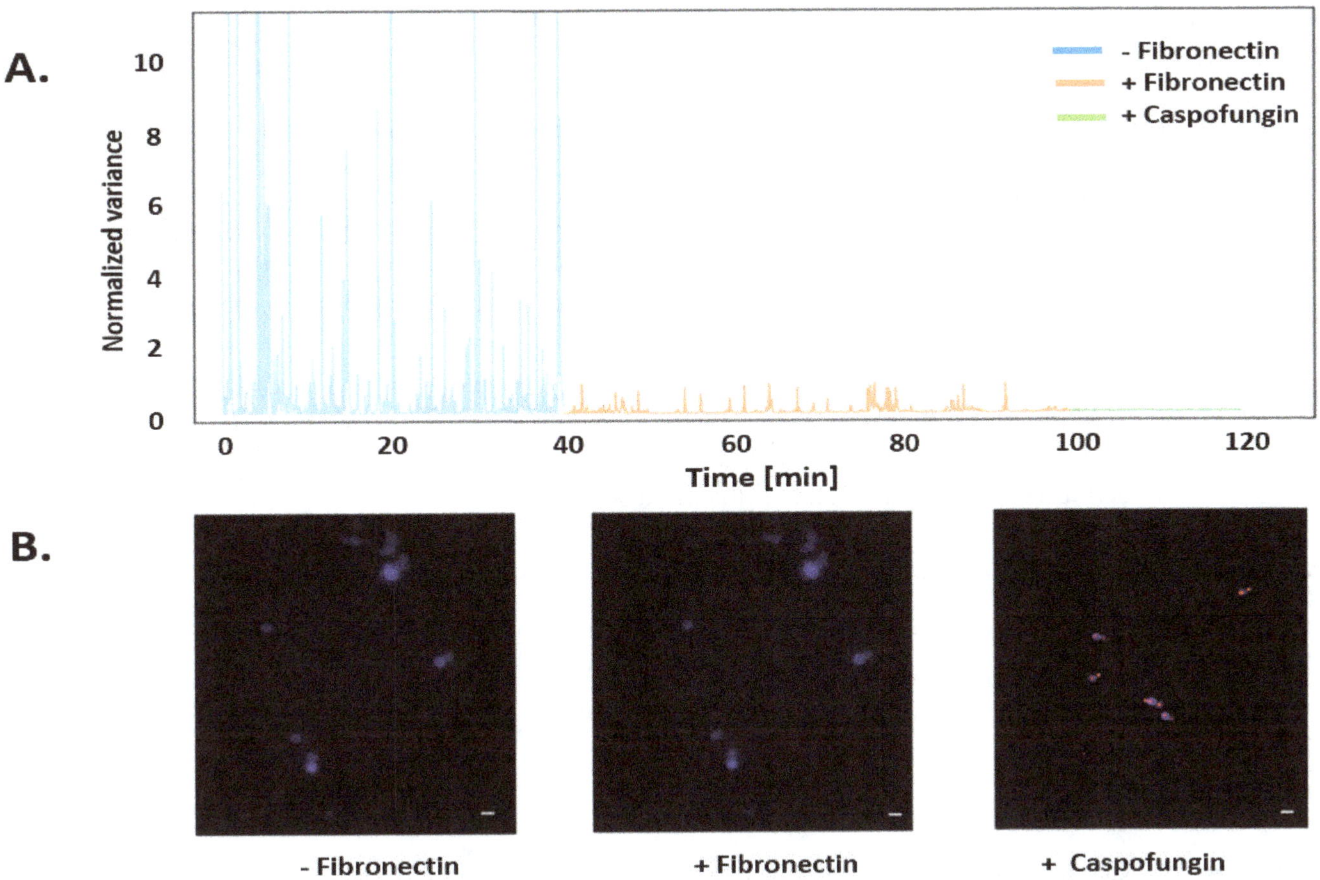

Figure 4. Viability assay of *C. albicans*. (**A**). Nanomotion signal of *C. albicans* isolate 101 in the absence (blue curve) and presence of fibronectin (orange curve), and after killing the cells by the antifungal caspofungin (green curve). (**B**). Representative fluorescence images of *C. albicans* isolate 101 in the absence (left panel) and presence of fibronectin (middle panel), and after killing (right panel). Scale bar 5 μm.

4. Discussion

C. albicans infection is a multistep process, consisting in the binding of *C. albicans* on epithelial cells. In a first adhesion step, the *C. albicans* adhesins of the agglutinin-like sequence (Als) family bind to ECM proteins of the host such as fibronectin [35,36], laminin and collagen. The attachment of the yeast cell to the host is followed by the penetration and transmigration of hypha into host cells, which then leads to vascular dissemination as soon the hypha reaches blood vessels. In this study we only explored the interaction of fibronectin with the yeast form. Adhesins playing a role in the planktonic *C. albicans* adhesion are the Als family members Als1 [27] and Als5 [37], Eap1 [38–40], Csh1 (cell surface hydrophobicity) [41,42], Ihd1 [43,44] and members of the SAP family [45–47]. It has been shown that Als1, Als5, and Csh1 interacts with fibronectin; Sap9 and Sap10 can interact with the ECM proteins collagen and vimentin. It has not yet been demonstrated that fibronectin is a ligand for Sap9/10, Eap1 and Ihd1.

Here, we used nanomotion detection to monitor the oscillation pattern of planktonic *C. albicans* cells upon exposure to fibronectin. Two different clinical isolates that showed a different adhesive phenotype, were used. The isolate 101 adhered significantly stronger to the host epithelial cells compared to isolate CEC3675. Nanomotion experiments showed that fibronectin affects isolate 101 significantly more than CEC3675. This drop of the nanomotion signal indicates a modification of the cellular activity upon fibronectin—*C. albicans* interaction. These results suggest that the initiation of adhesion related signaling in the yeast cell upon fibronectin attachment is mediated by the interaction with adhesins. Potential adhesion candidates that have been shown to interact with fibronectin are

Als1, Als3 and Csh1. The complete elucidation of the molecular mechanisms involved in the process are still unclear and deserve further research. We plan to investigate which specific adhesin(s) is (are) involved in the observed activity reduction. Additionally, the effect of other ligands such as laminin, collagen IV, fibrinogen and gelatin [28,48–50] should also be investigated.

This work demonstrated the ability of nanomotion detection to monitor in real time and in a label-free manner cellular activity changes induced by interacting ligands. Activity changes induced by increasing glucose concentration were observed for *Escherichia coli* in a previous study [5]. This technique opens novel avenues to detect cellular activation or inhibition induced by ligand–receptor interactions.

Author Contributions: Conceptualization, A.-C.K.; methodology, A.-C.K. and L.V.; software, A.-C.K.; validation, A.-C.K., L.V. and A.K.; formal analysis A.-C.K., investigation, A.-C.K., L.V., A.K.; resources, S.K., writing—original draft preparation, S.K. and A.-C.K.; writing—review and editing all the authors; visualization, A.-C.K., A.K.; supervision, A.-C.K., S.K.; project administration, S.K., G.D.; funding acquisition, S.K., G.D., D.S., R.W. All authors have read and agreed to the published version of the manuscript.

Acknowledgments: The authors thanks C. d'Enfert for providing the *Candida albicans* strains and S. Leibundgut for highly constructive discussions. The technical assistance of Danielle Brandalise is acknowledged. The Belgian Federal Science Policy Office (Belspo) and the European Space Agency (ESA) PRODEX program supported this work. The Research Council of the Vrije Universiteit Brussel (Belgium) and the University of Ghent (Belgium) are acknowledged to support the Alliance Research Group VUB-UGent NanoMicrobiology (NAMI), and the International Joint Research Group (IJRG) VUB-EPFL BioNanotechnology & NanoMedicine (NANO).

References

1. Willaert, R.; Kasas, S.; Devreese, B.; Dietler, G. Yeast Nanobiotechnology. *Fermentation* **2016**, *2*, 18. [CrossRef]
2. Formosa-Dague, C.; Duval, R.E.; Dague, E. Cell biology of microbes and pharmacology of antimicrobial drugs explored by Atomic Force Microscopy. *Semin. Cell Dev. Biol.* **2018**, *73*, 165–176. [CrossRef]
3. Alsteens, D.; Müller, D.J.; Dufrêne, Y.F. Multiparametric Atomic Force Microscopy Imaging of Biomolecular and Cellular Systems. *Acc. Chem. Res.* **2017**, *50*, 924–931. [CrossRef] [PubMed]
4. Kasas, S.; Stupar, P.; Dietler, G. AFM contribution to unveil pro- and eukaryotic cell mechanical properties. *Semin. Cell Dev. Biol.* **2018**, *73*, 177–187. [CrossRef] [PubMed]
5. Longo, G.; Alonso-Sarduy, L.; Rio, L.M.; Bizzini, A.; Trampuz, A.; Notz, J.; Dietler, G.; Kasas, S. Rapid detection of bacterial resistance to antibiotics using AFM cantilevers as nanomechanical sensors. *Nat. Nanotechnol.* **2013**, *8*, 522–526. [CrossRef]
6. Kohler, A.C.; Venturelli, L.; Longo, G.; Dietler, G.; Kasas, S. Nanomotion detection based on atomic force microscopy cantilevers. *Cell Surf.* **2019**, *5*, 100021. [CrossRef]
7. Villalba, M.I.; Stupar, P.; Chomicki, W.; Bertacchi, M.; Dietler, G.; Arnal, L.; Vela, M.E.; Yantorno, O.; Kasas, S. Nanomotion Detection Method for Testing Antibiotic Resistance and Susceptibility of Slow-Growing Bacteria. *Small* **2018**, *14*, 1702671. [CrossRef]
8. Stupar, P.; Opota, O.; Longo, G.; Prod'hom, G.; Dietler, G.; Greub, G.; Kasas, S. Nanomechanical sensor applied to blood culture pellets: A fast approach to determine the antibiotic susceptibility against agents of bloodstream infections. *Clin. Microbiol. Infect.* **2017**, *23*, 400–405. [CrossRef]
9. Mustazzolu, A.; Venturelli, L.; Dinarelli, S.; Brown, K.; Floto, R.A.; Dietler, G.; Fattorini, L.; Kasas, S.; Girasole, M.; Longo, G. A rapid unravelling of mycobacterial activity and of their susceptibility to antibiotics. *Antimicrob. Agents Chemother.* **2019**, *63*, 02194-18. [CrossRef]
10. Kasas, S.; Ruggeri, F.S.; Benadiba, C.; Maillard, C.; Stupar, P.; Tournu, H.; Dietler, G.; Longo, G. Detecting nanoscale vibrations as signature of life. *Proc. Natl. Acad. Sci. USA* **2015**, *112*, 378–381. [CrossRef]
11. Brown, G.D.; Denning, D.W.; Levitz, S.M. Tackling human fungal infections. *Science* **2012**, *336*, 647. [CrossRef] [PubMed]
12. Ghannoum, M.A.; Jurevic, R.J.; Mukherjee, P.K.; Cui, F.; Sikaroodi, M.; Naqvi, A.; Gillevet, P.M. Characterization of the Oral Fungal Microbiome (Mycobiome) in Healthy Individuals. *PLoS Pathog.* **2010** *6*, e1000713. [CrossRef] [PubMed]

13. Rupp, S. Interactions of the fungal pathogen Candida albicans with the host. *Future Microbiol.* **2007**, *2*, 141–151. [CrossRef] [PubMed]
14. Filler, S.G.; Sheppard, D.C. Fungal Invasion of Normally Non-Phagocytic Host Cells. *PLoS Pathog.* **2006**, *2*, e129. [CrossRef] [PubMed]
15. Nobile, C.J.; Johnson, A.D. *Candida albicans* Biofilms and Human Disease. *Annu. Rev. Microbiol.* **2015**, *69*, 71–92. [CrossRef] [PubMed]
16. Chaffin, W.L.; López-Ribot, J.L.; Casanova, M.; Gozalbo, D.; Martínez, J.P. Cell Wall and Secreted Proteins ofCandida albicans: Identification, Function, and Expression. *Microbiol. Mol. Biol. Rev.* **1998**, *62*, 130–180. [CrossRef]
17. Sundstrom, P. Adhesion in Candida spp. *Cell. Microbiol.* **2002**, *4*, 461–469. [CrossRef]
18. Filler, S.G. Candida-host cell receptor-ligand interactions. *Curr. Opin. Microbiol.* **2006**, *9*, 333–339. [CrossRef]
19. Ghannoum, M.A.; Burns, G.R.; Abu Elteen, K.; Radwan, S.S. Experimental evidence for the role of lipids in adherence of Candida spp. to human buccal epithelial cells. *Infect. Immun.* **1986**, *54*, 189–193. [CrossRef]
20. Jimenez-Lucho, V.; Ginsburg, V.; Krivan, H.C. Cryptococcus neoformans, Candida albicans, and other fungi bind specifically to the glycosphingolipid lactosylceramide (Gal beta 1-4Glc beta 1-1Cer), a possible adhesion receptor for yeasts. *Infect. Immun.* **1990**, *58*, 2085–2090. [CrossRef]
21. Yu, L.; Lee, K.K.; Sheth, H.B.; Lane-Bell, P.; Srivastava, G.; Hindsgaul, O.; Paranchych, W.; Hodges, R.S.; Irvin, R.T. Fimbria-mediated adherence of Candida albicans to glycosphingolipid receptors on human buccal epithelial cells. *Infect. Immun.* **1994**, *62*, 2843–2848. [CrossRef] [PubMed]
22. Cameron, B.J.; Douglas, L.J. Blood group glycolipids as epithelial cell receptors for Candida albicans. *Infect. Immun.* **1996**, *64*, 891–896. [CrossRef] [PubMed]
23. Sandin, R.L.; Rogers, A.L.; Patterson, R.J.; Beneke, E.S. Evidence for mannose-mediated adherence of Candida albicans to human buccal cells in vitro. *Infect. Immun.* **1982**, *35*, 79–85. [CrossRef] [PubMed]
24. Critchley, I.A.; Douglas, L.J. Role of glycosides as epithelial cell receptors for Candida albicans. *J. Gen. Microbiol.* **1987**, *133*, 637–643. [CrossRef]
25. Macura, A.B.; Tondyra, E. Influence of some carbohydrates and concanavalin A on the adherence of Candida albicans in vitro to buccal epithelial cells. *Zentralbl. Bakteriol.* **1989**, *272*, 196–201. [CrossRef]
26. Brassart, D.; Woltz, A.; Golliard, M.; Neeser, J.R. In vitro inhibition of adhesion of Candida albicans clinical isolates to human buccal epithelial cells by Fuc alpha 1—-2Gal beta-bearing complex carbohydrates. *Infect. Immun.* **1991**, *59*, 1605–1613. [CrossRef]
27. Donohue, D.S.; Ielasi, F.S.; Goossens, K.V.Y.; Willaert, R.G. The N-terminal part of Als1 protein from Candida albicans specifically binds fucose-containing glycans. *Mol. Microbiol.* **2011**, *80*, 1667–1679. [CrossRef]
28. Ielasi, F.S.; Alioscha-Perez, M.; Donohue, D.; Claes, S.; Sahli, H.; Schols, D.; Willaert, R.G. Lectin-glycan interaction network-based identification of host receptors of microbial pathogenic adhesins. *MBio* **2016**, *7*. [CrossRef]
29. Everest-Dass, A.V.; Kolarich, D.; Pascovici, D.; Packer, N.H. Blood group antigen expression is involved in C. albicans interaction with buccal epithelial cells. *Glycoconj. J.* **2017**, *34*, 31–50. [CrossRef] [PubMed]
30. Calderone, R.A.; Scheld, W.M. Role of fibronectin in the pathogenesis of candidal infections. *Rev. Infect. Dis.* **1987**, *9*, S400–S403. [CrossRef]
31. Nett, J.E.; Cabezas-Olcoz, J.; Marchillo, K.; Mosher, D.F.; Andes, D.R. Targeting fibronectin to disrupt in vivo Candida albicans biofilms. *Antimicrob. Agents Chemother.* **2016**, *60*, 3152–3155. [CrossRef] [PubMed]
32. Ropars, J.; Maufrais, C.; Diogo, D.; Marcet-Houben, M.; Perin, A.; Sertour, N.; Mosca, K.; Permal, E.; Laval, G.; Bouchier, C.; et al. Gene flow contributes to diversification of the major fungal pathogen Candida albicans. *Nat. Commun.* **2018**, *9*, 2253. [CrossRef] [PubMed]
33. Murciano, C.; Moyes, D.L.; Runglall, M.; Tobouti, P.; Islam, A.; Hoyer, L.L.; Naglik, J.R. Evaluation of the Role of Candida albicans Agglutinin-Like Sequence (Als) Proteins in Human Oral Epithelial Cell Interactions. *PLoS ONE* **2012**, *7*, e33362. [CrossRef] [PubMed]
34. Schönherr, F.A.; Sparber, F.; Kirchner, F.R.; Guiducci, E.; Trautwein-Weidner, K.; Gladiator, A.; Sertour, N.; Hetzel, U.; Le, G.T.T.; Pavelka, N.; et al. The intraspecies diversity of C. albicans triggers qualitatively and temporally distinct host responses that determine the balance between commensalism and pathogenicity. *Mucosal Immunol.* **2017**, *10*, 1335–1350. [CrossRef] [PubMed]
35. Skerl, K.G.; Calderone, R.A.; Segal, E.; Sreevalsan, T.; Scheld, W.M. In vitro binding of Candida albicans yeast cells to human fibronectin. *Can. J. Microbiol.* **1984**, *30*, 221–227. [CrossRef] [PubMed]

36. Douglas, L.J. Adhesin—Receptor interactions in the attachment of Candida albicans to host epithelial cells. *Can. J. Bot.* **1995**, *73*, 1147–1153. [CrossRef]
37. Alsteens, D.; Beaussart, A.; Derclaye, S.; El-Kirat-Chatel, S.; Park, H.R.; Lipke, P.N.; Dufrêne, Y.F. Single-cell force spectroscopy of Als-mediated fungal adhesion. *Anal. Methods* **2013**, *5*, 3657–3662. [CrossRef] [PubMed]
38. Li, F.; Palecek, S.P. EAP1, a Candida albicans Gene Involved in Binding Human Epithelial Cells. *Eukaryot. Cell* **2003**, *2*, 1266–1273. [CrossRef] [PubMed]
39. Li, F.; Svarovsky, M.J.; Karlsson, A.J.; Wagner, J.P.; Marchillo, K.; Oshel, P.; Andes, D.; Palecek, S.P. Eap1p, an adhesin that mediates Candida albicans biofilm formation in vitro and in vivo. *Eukaryot. Cell* **2007**, *6*, 931–939. [CrossRef] [PubMed]
40. Li, F.; Palecek, S.P. Distinct domains of the Candida albicans adhesin EAP1 p mediate cell-cell and cell-substrate interactions. *Microbiology* **2008**, *154*, 1193–1203. [CrossRef]
41. Hooshdaran, M.Z.; Barker, K.S.; Hilliard, G.M.; Kusch, H.; Morschhäuser, J.; Rogers, P.D. Proteomic analysis of azole resistance in Candida albicans clinical isolates. *Antimicrob. Agents Chemother.* **2004**, *48*, 2733–2735. [CrossRef] [PubMed]
42. Singleton, D.R.; Fidel, P.L.; Wozniak, K.L.; Hazen, K.C. Contribution of cell surface hydrophobicity protein 1 (Csh1p) to virulence of hydrophobic *Candida albicans* serotype A cells. *FEMS Microbiol. Lett.* **2005**, *244*, 373–377. [CrossRef] [PubMed]
43. de Groot, P.W.J.; Hellingwerf, K.J.; Klis, F.M. Genome-wide identification of fungal GPI proteins. *Yeast* **2003**, *20*, 781–796. [CrossRef] [PubMed]
44. McCall, A.D.; Pathirana, R.U.; Prabhakar, A.; Cullen, P.J.; Edgerton, M. Candida albicans biofilm development is governed by cooperative attachment and adhesion maintenance proteins. *NPJ Biofilms Microbiomes* **2019**, *5*, 21. [CrossRef]
45. Watts, H.; Cheah, F.S.; Hube, B.; Sanglard, D.; Gow, N.A. Altered adherence in strains of *Candida albicans* harbouring null mutations in secreted aspartic proteinase genes. *FEMS Microbiol. Lett.* **1998**, *159*, 129–135. [CrossRef]
46. Naglik, J.R.; Challacombe, S.J.; Hube, B. Candida albicans Secreted Aspartyl Proteinases in Virulence and Pathogenesis. *Microbiol. Mol. Biol. Rev.* **2003**, *67*, 400–428. [CrossRef]
47. Kumar, R.; Breindel, C.; Saraswat, D.; Cullen, P.J.; Edgerton, M. Candida albicans Sap6 amyloid regions function in cellular aggregation and zinc binding, and contribute to zinc acquisition. *Sci. Rep.* **2017**, *7*, 1–15. [CrossRef]
48. Klotz, S.A.; Gaur, N.K.; Lake, D.F.; Chan, V.; Rauceo, J.; Lipke, P.N. Degenerate peptide recognition by Candida albicans adhesins Als5p and Als1p. *Infect. Immun.* **2004**, *72*, 2029–2034. [CrossRef]
49. Sheppard, D.C.; Yeaman, M.R.; Welch, W.H.; Phan, Q.T.; Fu, Y.; Ibrahim, A.S.; Filler, S.G.; Zhang, M.; Waring, A.J.; Edwards, J.E. Functional and structural diversity in the Als protein family of Candida albicans. *J. Biol. Chem.* **2004**, *279*, 30480–30489. [CrossRef]
50. Argimón, S.; Wishart, J.A.; Leng, R.; Macaskill, S.; Mavor, A.; Alexandris, T.; Nicholls, S.; Knight, A.W.; Enjalbert, B.; Walmsley, R.; et al. Developmental regulation of an adhesin gene during cellular morphogenesis in the fungal pathogen Candida albicans. *Eukaryot. Cell* **2007**, *6*, 682–692. [CrossRef]

14

Yeast Nanobiotechnology

Ronnie Willaert [1,*], Sandor Kasas [2], Bart Devreese [3] and Giovanni Dietler [2]

1 IJRG VUB-EPFL, NanoBiotechnology & NanoMedicine (NANO), Alliance Research Group VUB-UGent NanoMicrobiology (NAMI), Vrije Universiteit Brussel, Brussels 1050, Belgium
2 IJRG VUB-EPFL, NanoBiotechnology & NanoMedicine (NANO), Laboratory of the Physics of Living Matter, Ecole Polytechnique de Lausanne, Lausanne 1015, Switzerland; Sandor.Kasas@epfl.ch (S.K.); Giovanni.Dietler@epfl.ch (G.D.)
3 IJRG VUB-EPFL, NanoBiotechnology & NanoMedicine (NANO), Alliance Research Group VUB-UGent NanoMicrobiology (NAMI), Laboratory for Protein Biochemistry and Biomolecular Engineering (L-ProBE), Ghent University, Ghent 9000, Belgium; Bart.Devreese@UGent.be
* Correspondence: Ronnie.Willaert@vub.ac.be

Academic Editor: Badal C. Saha

Abstract: Yeast nanobiotechnology is a recent field where nanotechniques are used to manipulate and analyse yeast cells and cell constituents at the nanoscale. The aim of this review is to give an overview and discuss nanobiotechnological analysis and manipulation techniques that have been particularly applied to yeast cells. These techniques have mostly been applied to the model yeasts *Saccharomyces cerevisiae* and *Schizosaccaromyces pombe*, and the pathogenic model yeast *Candida albicans*. Nanoscale imaging techniques, such as Atomic Force Microscopy (AFM), super-resolution fluorescence microscopy, and electron microscopy (scanning electron microscopy (SEM), transmission electron microscopy (TEM), including electron tomography) are reviewed and discussed. Other nano-analysis methods include single-molecule and single-cell force spectroscopy and the AFM-cantilever-based nanomotion analysis of living cells. Next, an overview is given on nano/microtechniques to pattern and manipulate yeast cells. Finally, direct contact cell manipulation methods, such as AFM-based single cell manipulation and micropipette manipulation of yeast cells, as well as non-contact cell manipulation techniques, such as optical, electrical, and magnetic cells manipulation methods are reviewed.

Keywords: yeasts; Atomic Force Microscopy (AFM); super-resolution fluorescence microscopy; electron microscopy; force spectroscopy; nanomotion analysis; yeast cell patterning; non- and direct-contact cell manipulation; optical/magnetic tweezer; nanoscale imaging

---

1. Introduction

Nanotechnology is the ability to work at the atomic, molecular, and supramolecular levels (on the scale of ~1–100 nm) to understand, create, and use material structures, devices, and systems with fundamentally new properties and functions resulting from their small structure [1]. Nanobiotechnology is defined as a field that applies nanoscale principles and techniques to understand and transform biosystems (living or nonliving) and that uses biological principles and materials to create new devices and systems integrated from the nanoscale [2]. The biological and physical sciences share a common interest in small structures (the definition of "small" depends on the application, but can range from 1 nm to 1 mm) [3]. A bacterial cell is approximately 1 μm, a yeast cell 5 μm, and a mammalian cell is 10 μm when rounded and 50 μm when fully spread in attached culture. A vigorous trade across the borders of these areas of science is developing around new materials and tools (largely from the physical sciences) and new phenomena (largely from the biological sciences). The physical sciences offer tools for the synthesis and fabrication of devices for measuring the characteristics of cells and sub-cellular components and of materials useful in cell and molecular biology. Biology offers a window into the most sophisticated collection of functional nanostructures that exist. Nanobiotechnology offers new solutions for the transformation of biosystems, and provides a broad technological platform for applications in several areas—including bioprocessing in industry, molecular medicine, investigating the health effects of nanostructures in the environment, improving food products (food conservation), and improving human performance [2].

This review discusses nanobiotechnological analysis and manipulation techniques that have been especially applied to yeast cells. Nanoscale imaging methods that allow imaging at nanometer resolution are reviewed: atomic force microscopy (AFM), super-resolution fluorescence microscopy, and electron microscopy (including scanning electron microscopy (SEM), transmission electron microscopy (TEM), and electron tomography). Force spectroscopy for the analysis of single biomolecule interactions or unfolding on live yeast cells, as well as single-cell force spectroscopy, and the recently developed AFM-based nanomotion analysis of cells is reviewed and discussed. Since single-cell analysis has increasingly been recognised as the key technology for the elucidation of cellular functions which are not accessible from bulk measurements of the population level, nano/micro single-yeast cell manipulation techniques are reviewed. Yeast cell patterning techniques, such as microcontact printing, mechanical cell patterning, and the use of robotic cell printing are discussed. Finally, direct-contact (such as AFM-based and micropipette-based) and non-contact (such as optical, electrical, and magnetic cell) yeast cell manipulation techniques are reviewed.

2. Yeast Nanobiotechnological Analyses

2.1. Nanoscale Imaging

We use microscopy in order to see objects in more detail. The best distance that one can resolve with optical instruments (disregarding all aberrations) is about 0.5 times the wavelength of light, or the order of 250 nm with visible radiation. High-resolution microscopy techniques that are used for nanoimaging and nanoscale characterisation have been developed in the last 20 years. They can be divided into three categories: optical microscopes, scanning probe microscopes (SPMs), and electron microscopes. Recently-developed microscopy-based technologies can also be used to control and manipulate objects at the nanoscale—i.e., single-cell as well as single-molecule manipulation and analysis.

2.1.1. Atomic Force Microscopy

Scanning probe microscopes (SPMs) are a family of instruments that are used to measure surface properties, and include atomic force microscopes (AFMs) and scanning tunneling microscopes (STMs). The main feature that all SPMs have in common is that the measurements are performed with a sharp probe operating in the near field; that is, scanning over the surface while maintaining a very close spacing to the surface. The STM—invented in the early nineteen-eighties by Binnig and Rohrer [4]—was the first to produce real-space images of atomic arrangements on flat surfaces. The development of the STM arose from an interest in the study of the electrical properties of thin insulating layers. This led to an apparatus in which the probe–surface separation was monitored by measuring electron tunneling between a conducting surface and a conducting probe. A few years later, Binnig and colleagues [5] announced the birth of the second member of the SPM family, the atomic force microscope (also known as the scanning force microscope, SFM). Numerous variations of these techniques have been developed since.

AFM is extensively used for imaging surfaces ranging from micro- to nanometer scales, with the objective of visualising and characterising surface textures and shapes [6]. It has evolved into an imaging method that yields structural details of biological samples, such as proteins, nucleic acids, membranes, and cells in their native environment. AFM is a unique technique for providing sub-nanometer resolution at a reasonable signal-to-noise ratio under physiological conditions.

It complements electron microscopy (EM) by allowing the visualisation of biological samples in buffers that preserve their native structure over extended time periods. Unlike EM, AFM yields 3D maps with an exceptionally good vertical resolution (less than a nanometer). Additionally, the measurement of mechanical forces at the molecular level provides detailed insights into the function and structure of biomolecular systems. Inter- and intramolecular interactions can be studied directly at the molecular level. Recently, improvements in the temporal resolution were made by the development of high-speed AFM [7]. This technique is capable of observing structure dynamics and dynamic processes at the sub-second to sub-100 ms temporal resolution and 2 nm lateral and 0.1 nm vertical resolution.

Since AFM imaging can be performed in physiological conditions, high resolution imaging of the yeast cell surface can be performed on living cells. Therefore, an appropriate cell immobilisation method has to be used that avoids cell detachment by the scanning probe. Several methods have been developed and used to perform high-resolution live yeast cell imaging and analysis (i.e., force spectroscopy). Yeast cells can be trapped in the pores of a filter membrane [8] (Figure 1A), in a hydrogel [9,10], or in microfabricated microwells [11,12]. Especially the model yeast *Saccharomyces cerevisiae* has been imaged at the nanoscale (Table 1). The cell walls of other yeasts have also been visualised: the model pathogenic yeast *Candida albicans*, and the other model yeast *Schizosaccharomyces pombe* (Table 1). Imaging can be easily combined with nanoindentation experiments to map the elasticity of the cell surface [13].

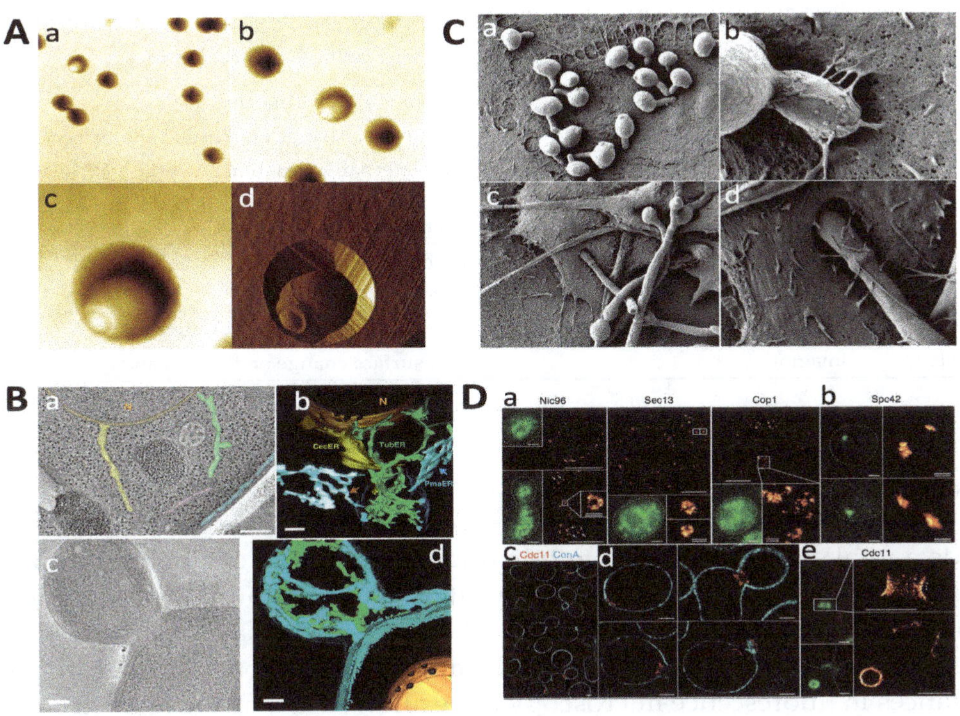

Figure 1. (**A**) Entrapment of a single *Saccharomyces cerevisiae* cell in the pore of a filter membrane. Atomic force microscopy (AFM) height (**a–c**) and amplitude (**d**) images. Courtesy of Dr. Ronnie Willaert, Vrije Universiteit Brussel, Belgium. (**B**) Electron tomography imaging of *S. cerevisiae*: 3D structural analysis of endoplasmic reticulum (ER) morphology. (**a,b**) 2D tomograph derived from a 200-nm-thick section shows the nuclear envelope (NE) (orange), plasma membrane ER (pmaER), central cisternal ER, tubular ER, and Golgi (pink; a) and the corresponding 3D model of (a) shows all ER domains in a WT yeast cell. The blue shade is the plasma membrane (PM); N is the nucleus; black holes on the NE are nuclear pores; (**c**) 2D tomograph of a mutant cell with a bud; (**d**) 3D model of ER domain organisation (the cytoplasmic face (cyto) of pmaER in blue and PM face of pmaER in red). Reprinted from ref. [14]. (**C**) Scanning electron microscopy (SEM) images of *Candida albicans* cells interacting with pharyngeal FaDu cells. (**a,b**) After 30 min of contact, the formation of germ tubes is visible. The cells attach to the FaDu cells through microvilli structures; (**c,d**) After 3 h of contact, the *C. albicans* cells produce long filamentous cells (hyphae and pseudohyphae), which penetrate FaDu cells. Magnification (a) ×4000,

(b) ×25,000, (c) ×4000, (d) ×16,000. Courtesy of Dr. Ronnie Willaert, Vrije Universiteit Brussel, Belgium. (**D**) Super-resolution photoactivatable localization microscopy (PALM) imaging of proteins in budding yeast green fluorescent protein (GFP)-fusion construct library. (**a**) Reconstructed super-resolution images of Nic96-, Sec13-, and Cop1-GFP; (**b**) Images of the spindle pole body protein Spc42-GFP; (**c**) Dual-colour reconstructed images of yeast cells expressing Cdc11-GFP (red) and the cell wall (cyan); (**d**) Reconstructed dual-colour images visualising the different organisational stages of Cdc11-GFP structures during the cell cycle; (**e**) Reconstructed images of septin Cdc11-GFP, which localises to a characteristic hourglass-shaped and later ring-like structure around the mother–bud neck. Reprinted with permission from ref. [15].

Table 1. Examples of AFM imaging of yeast cell surfaces.

Yeast Type	AFM Analysis	Objective	Refs
C. albicans	Imaging, cell surface elasticity	Effect of antifungal caspofungin	[16]
	Imaging, cell elasticity	Imaging mode evaluation	[17]
	Imaging, force spectroscopy using concanavalin A-functionalised tips	Mapping of adhesive properties	[18]
Candida parapsilosis	Imaging, adhesion force	Surface morphological characterisation	[19]
S. cerevisiae	Imaging	Immobilisation method	[8]
	Imaging	Immobilisation method	[9]
	Imaging, force spectroscopy using concanavalin A-functionalised tips	Mapping cell wall polysaccharides	[20]
	Imaging, cell elasticity	Mapping of cell elasticity	[21]
	Imaging	Cell surface change on thermal and osmotic stress	[22]
	Imaging, motion analysis	Nanomechanical motion analysis	[23]
	Imaging	Effect of electromagnetic field and antifungal nystatin on the cell wall	[24]
	Imaging, cell elasticity	Immobilisation method	[10]
	Imaging	Immobilisation method	[11]
	Imaging, cell surface elasticity	Effect of antifungal caspofungin	[16]
Sc. pombe	Imaging	Cell surface change on thermal and osmotic stress	[22]

2.1.2. Light Microscopy

Since the earliest examination of cellular structures, observing cells using a light microscope has fascinated biologists. Being able to observe processes as they happen with the use of light microscopy adds a vital extra dimension to our understanding of cell behaviour and function [25]. Microscopy has evolved to provide not only quantitative images but also a significant capability to perturb structure–function relationships in cells. These advances have been especially useful in the study of a wide range of biological processes, including cell adhesion and migration [26].

Recent advances in fluorescence microscopy have allowed the imaging of structures at extremely high resolutions [27]. The past decade witnessed an explosion of fluorescence microscopy-based approaches to image protein dynamics and interactions [28]. For example, fluorescence recovery after photobleaching (FRAP) or photo-activation using photo-convertible fluorescent proteins to assay protein mobility and maturation in cells [29]; and Förster resonance energy transfer (FRET) to monitor physical intra- or intermolecular associations in space and time [30,31].

Despite the advantages of standard fluorescence microscopy, ultra-structural imaging is not possible, owing to a resolution limit set by the diffraction of light (Rayleigh criterion) [32]. Therefore, the maximal spatial resolution of standard optical microscopy is around 200 nm. This limit is one to two orders of magnitude above the typical molecular length scales in cells. Several approaches have been used to break this diffraction limit (Table 2). The diffraction limit can be overcome by exploiting the distribution of fluorescence intensity from a single molecule. When imaged, a fluorophore behaves as a point source with an Airy disc point spread function. The center of mass of the function (and therefore the position of the molecule) can be obtained by performing a least-squares fit of an appropriate function (such as a Gaussian distribution) to the measured fluorescence intensity profile of the spot [33]. With a sufficient number of photons, these methods can provide a localisation of 1–2 nm (15 to 70 nm on intact cells), allowing the measurement of distances on the scale of individual proteins. Single-molecule detection offers new possibilities for obtaining sub-diffraction-limit spatial resolution [34–36].

Table 2. Super-resolution optical microscopy techniques (adapted from [37]).

<table>
<tr><th>Technique</th><th>Description</th><th>Spatial Resolution</th><th>Timescale</th></tr>
<tr><td>Fluorescence imaging with one-nanometer accuracy (FIONA)</td><td>Localises and tracks single-molecule emitters by finding the centre of their diffraction-limited point-spread function (PSF).</td><td>~1.5 nm</td><td>~0.3 ms</td></tr>
<tr><td>Single-molecule high-resolution colocalisation (SHREC)</td><td>Two-colour version of FIONA. Two fluorescent probes with different spectra are imaged separately and then localised and mapped onto the plane of the microscope.</td><td><10 nm</td><td>~1 s per frame</td></tr>
<tr><td>Single-molecule high-resolution imaging with photobleaching (SHRImP)</td><td>Uses the strategy wherein, upon photobleaching of two or more closely-spaced identical fluorophores, their position is sequentially determined by FIONA, starting from the last bleached fluorophore.</td><td>~5 nm</td><td>~0.5 s per frame</td></tr>
<tr><td>Nanometer-localised multiple single-molecules (NALMS)</td><td>Uses a similar principle to single-molecule high-resolution imaging with photobleaching to measure distances between identical fluorescent probes that overlap within a diffraction-limited spot.</td><td>~8 nm</td><td>~1 s per frame</td></tr>
<tr><td>Photoactivatable localization microscopy (PALM)</td><td>Serially photoactivates and photodeactivates many sparse subsets of photoactivatable fluorophores to produce a sequence of images that are combined into a super-resolution composite.</td><td>~2 nm</td><td>~1 min</td></tr>
<tr><td>PALM with independently running acquisition (PALMIRA)</td><td>Records non-triggered spontaneous off–on–off cycles of photoswitchable fluorophores without synchronising the detector to reach faster acquisition.</td><td rowspan="2">~50 nm</td><td rowspan="2">~2.5 min</td></tr>
<tr><td>Single particle tracking PALM</td><td>Combines PALM with live-cell single fluorescent particle tracking.</td></tr>
<tr><td>Stimulated emission depletion (STED)</td><td>Reduces the excitation volume below that dictated by the diffraction limit by coaligning one beam of light capable of fluorophore excitation with another that induces de-excitation by stimulated emission.</td><td>~16 nm</td><td>~10 min</td></tr>
<tr><td>Stochastic optical reconstruction microscopy (STORM)</td><td>Small sub-populations of photoswitchable fluorophores are turned on and off using light of different colours, permitting the localisation of single molecules. Repeated activation cycles produce a composite image of the entire sample.</td><td><20 nm</td><td>~mins</td></tr>
</table>

Photoactivatable or "optical highlighter" fluorescent proteins (FPs) have emerged as powerful new tools for cellular imaging [38–47]. The fluorescent properties of these proteins can be altered upon illumination at specific wavelengths. They either switch between a fluorescent and non-fluorescent state (photoswitching) [48–54], or they change their fluorescence emission from one wavelength to another (photoconversion) [39,44,55,56]. The controlled photoconversion/switching of these proteins provides unique opportunities to mark and track selected molecules in cells in space and time [42,48,50,57–59]. High-density mapping of single-molecule motions can be obtained using photoactivated localisation microscopy (PALM) [60–62].

Another promising application of photoswitchable proteins is their use in super-resolution microscopy. This technique relies on the stochastic photoactivation and localisation of single molecules, in which a fluorescence image is constructed from high-accuracy localisation of individual fluorescent molecules that are switched on and off optically [63–68]. Microscope techniques that are based on this principle are called RESOLFT (reversible saturable optical fluorescence transitions) microscopy. RESOLFT microscopy concepts are photoactivated localisation microscopy (PALM) [63,65], fluorescence photoactivation localisation microscopy (FPALM) [69], stochastic optical reconstruction microscopy (STORM) [64,69–72], and PALM with independently running acquisition

(PALMIRA) [73,74] (Table 2). Image resolution well below the Abbe diffraction limit is achieved. Labelled proteins can be localised with a precision down to about 2–10 nm. Stunning images have been obtained based on photoactivatable FPs [63,75] (Figure 1D). Recently, super-resolution microscopy has also been extended to dual-colour imaging [76]. Another recently developed super-resolution technique is STED (stimulated emission depletion) [72,77]. In a STED microscope, the focal spot of excitation light is overlapped with a doughnut-shaped spot of light of lower photon energy, quenching excited molecules in the excitation spot periphery by stimulated emission. A resolution of 15 to 70 nm has been realised to map, for example, the nanoscale distribution of proteins inside cells [78], on the plasma membrane [79], and the movement of synaptic vesicles inside the axons of cultured cells [80].

2.1.3. Electron Microscopy

Microscopes consist of an illumination source, a condenser lens to converge the beam on the sample, an objective lens to magnify the image, and a projector lens to project the image onto an image plane, which can then be photographed or stored. In electron microscopes, the wave nature of the electron is used to obtain an image. There are two important forms of electron microscopy: scanning electron microscopy (SEM) and transmission electron microscopy (TEM). Both use electrons as the source for sample illumination. The lenses used in electron microscopes are electromagnetic lenses. For high-resolution surface investigations, two commonly used techniques are scanning electron microscopy (SEM) (Figure 1C) and AFM. The operation of the SEM consists of applying a voltage between a conductive sample and filament, resulting in electron emission from the filament to the sample. This occurs in a vacuum environment. The electrons are guided to the sample by a series of electromagnetic lenses in the electron column. The resolution and depth of field of the image are determined by the beam current and the final spot size. The electrons interact with the sample within a few nanometers to several microns of the surface, depending on the beam parameters and sample type. Along with the secondary electron emission (which is used to form a morphological image of the surface in the SEM), several other signals are emitted as a result of the electron beam impinging on the surface. Each of these signals carries information about the sample that provides clues to its composition. Two of the most commonly used signals for investigating composition are X-rays and backscattered electrons. X-ray signals are commonly used to provide elemental analysis. The percentage of beam electrons that become backscattered electrons has been found to be dependent on the atomic number of the material, which makes it a useful signal for analysing the material composition.

Since electron microscopy is conducted in a vacuum environment, it is at a disadvantage for the study of hydrated samples. To image poorly-conductive surfaces without sample charging may require conductive coatings or staining (which may alter or obscure the features of interest), or it may require low-voltage operation or an environmental chamber, which may sacrifice resolution. Recently, an electron microscopy technique was described for imaging whole cells in liquid that offers nanometer spatial resolution and a high imaging speed using a scanning transmission electron microscope (STEM) [81,82]. The cells were placed in buffer solution in a microfluidic device with electron-transparent windows inside of the vacuum of the electron microscope.

In TEM, the transmitted electrons are used to create an image of the sample. Scattering occurs when the electron beam interacts with matter. Scattering can be elastic (no energy change) or inelastic (energy change). Elastic scattering can be coherent and incoherent (with and without phase relationship). TEMs with resolving powers in the vicinity of 1 Å are now common. A relatively recent electron microscopy technique that can be used to study cells at the nanoscale is electron tomography. Electron tomography (ET) is the most widely applicable method for obtaining three-dimensional information by electron microscopy [83–85]. A tomogram is a three-dimensional volume computed from a series of projection images that are recorded as the object in question is tilted at different orientations. ET has the potential to fill the gap between global cellular localisation and the detailed three-dimensional molecular structure, because it can reveal the localisation within the cellular context at true molecular resolution and the shapes and three-dimensional architecture of large molecular machines. It can also reveal the interaction of individual proteins and protein complexes with other cellular components, such as DNA and membranes. A recent development is cryo-electron tomography (cryo-ET), which allows the visualisation of cellular structures under close-to-life conditions [86–88] (see Figure 1B as an example). Rapid freezing followed by the investigation of the frozen-hydrated samples avoids artifacts notorious to chemical fixation and dehydration procedures. Furthermore, the biological material is observed directly, without heavy metal staining, avoiding problems in interpretation caused by unpredictable accumulation of staining material. Consequently, cryo-ET of whole cells has the advantage that the supramolecular architecture can be studied in unperturbed cellular environments.

The ultrastructure of yeast cells (the model yeasts *S. cerevisiae* and *Sc. pombe*) was first studied by TEM using thin sections in 1957 [14], and the freeze-etching replica method was introduced in 1969 to obtain the fine structure of yeast cells [89]. During the next 50 years, techniques for the analysis of the ultrastructure of yeasts advanced greatly [90]. Initially, yeast cells were fixed solely with potassium permanganate ($KMnO_4$), and not by the widely used osmium tetroxide (OsO_4), since the thick cell wall is a barrier for the penetration of OsO_4. Finer EM images were obtained by using a double fixation with glutaraldehyde (GA) and $KMnO_4$ [91]. Important landmark studies have used conventional chemical fixation using GA and OsO_4—after enzymatic removal of the cell wall—to describe the cellular features of *S. cerevisiae* and to compare ultrastructural defects that result from mutations in key genes [92–96]. Next, methods using cryo-immobilisation followed by freeze substitution have been developed to provide excellent preservation of intact yeast cells [97–99]. These approaches involve rapid freezing of the sample with subsequent substitution treatment to replace frozen water in the sample with an organic solvent and fixatives [100]. Currently, high pressure freezing followed by freeze substitution (HPF/FS) is the method of choice for preparing cells for ET. Yeast prepared with these methods are used in 3D electron tomography studies for which sampling of the cell is performed at unprecedented resolution [88,101] (Figure 1B).

2.2. Force Microscopy

AFM techniques have turned out to be a suitable and versatile tool for single-molecule interactions (Table 3) and for probing the physical properties of microbial cell surfaces [102]. Especially, it has been used to study yeast surfaces: to determine nanomechanical properties of the cell wall, map cell wall proteins (Figure 2A), molecular recognition forces (receptor–ligand interaction), and characterise biomolecules by single-molecule unfolding (Table 4). For these types of analyses, the force sensing capabilities of the AFM are used. AFM-based force spectroscopy exerts pulling forces on a single attached molecule by retraction of the tip in the z direction (perpendicular to the x–y scanning plane). Cantilever bending is detected by the deflection of a laser beam onto a position-sensitive detector, such as a quadrant photodiode. A piezoelectric actuator stage is used to control the positioning of the sample relative to the tip. AFM-based force spectroscopy is also used to study single cell interactions (cell–cell and cell–substrate adhesion).

Table 3. Examples of yeast molecule interaction studies using single-molecule force spectroscopy (SMFS).

Cell Type	Interacting Molecule 1	Interacting Molecule 2	Rupture Force (pN)	Refs
C. albicans	Als5p	Fibronectin	2800 ± 600	[103]
S. cerevisiae	Prion protein Sup35 hexapeptide	Prion protein Sup35 hexapeptide	—	[104]
	Prion protein Sup35 hexapeptide antiparallel hairpin structure	Prion protein Sup35 hexapeptide antiparallel hairpin structure	32–134	[105]
	Nucleoporin	Nucleoporin	—	[106]
	Nucleoporin	Importin	—	
	Flo1p	Flo1p	300 (100–600)	[107]

Table 4. Examples of yeast receptor–ligand interaction studies using single-molecule force spectroscopy (SMFS) on cell surfaces.

Cell Type	Cell Receptor	Ligand	Rupture Force (pN)	Refs
C. albicans	Cell surface β-mannan	Anti-β-1,2-mannoside antibodies	41 ± 14	[108]
	Cell surface β-glucans	Anti-β-1,3-glucan antibodies	38 ± 10	[108]
	Cell wall chitin	WGA [1] lectin	65 ± 19	[108]
	Cell wall	*Streptococcus mutans* exoenzyme glycosyltransferase B	1000–2000	[109]
C. glabrata	Epa6p	Hydrophobic surface	-	[110]
	Cell surface β-mannan	Anti-β-1,2-mannoside antibodies	54 ± 9	[108]
	Cell surface β-glucans	Anti-β-1,3-glucan antibodies	41 ± 8	[108]
	Cell wall chitin	WGA [1] lectin	41 ± 8	[108]
S. cerevisiae	Cell surface α-mannan	Con A [2] lectin	75-200	[111]
	Cell surface α-mannan	Con A [2] lectin	92 ± 35	[108]
	Cell surface β-glucans	Anti-β-1,3-glucan antibodies	42 ± 7	[108]
	Cell wall chitin	WGA [1] lectin	54 ± 19	[108]
	Wsc1p-His-tagged	NTA-Ni^{2+}	-	[112–114]
	HA [3]-tagged Ccw12p	Anti-HA antibody	69.3 ± 31.4	[115]
	Ste2p	α-factor	250	[116]
S. pastorianus	Flo protein	Glucose	121 ± 53	[111]
	Flo protein	Con A [2]	117 ± 41	[111]

[1] WGA: wheat germ agglutinin; [2] Con A: concanavalin A; [3] HA: human influenza hemagglutinin.

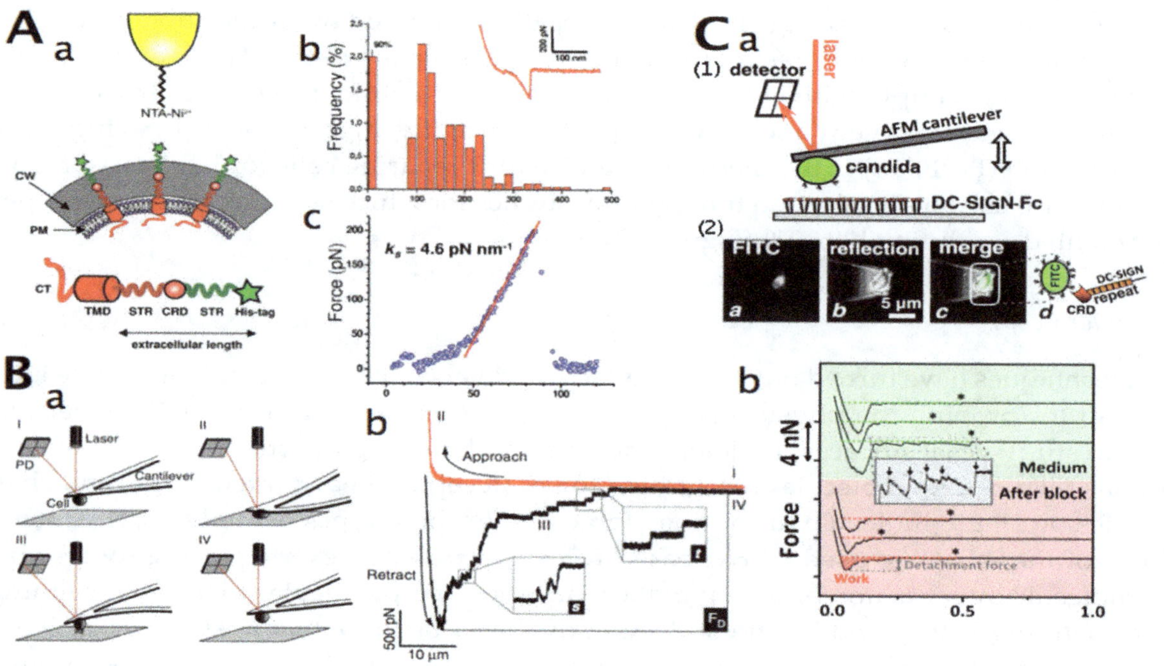

Figure 2. (**A**) (**a**) His-tagged modified Wsc1 membrane sensors were detected using AFM tips functionalised with Ni^{2+}-nitrilotriacetic acid (NTA) groups on live cells. The drawing shows a His-tagged elongated Wsc1 sensor with the cytoplasmic tail (CT), the transmembrane domain (TMD), the cysteine-rich domain (CRD), the serine/threonine-rich (STR) region, and the terminal His tag (in green) (CW = cell wall, PM = plasma membrane); (**b**) Adhesion force histograms and representative force curves recorded with a Ni^{2+}-NTA tip for *S. cerevisiae* cells expressing His-tagged elongated Wsc1 sensors; (**c**) Representative force extension curves obtained upon stretching a single Wsc1p. The curve displays a linear region, where force is directly proportional to extension. Reprinted with permission from [113]. (**B**) Experimental single-cell force spectroscopy (SCFS) setup. (**a**) A cell is attached to a coated cantilever. To measure the force acting on the cantilever, cantilever deflection is determined using a laser beam reflected by the cantilever onto a photodiode (PD). The cantilever-bound cell is lowered toward the substrate (I) until a preset force is reached (II). After a given contact time, the cantilever is retracted from the substrate (III) until cell and substrate are completely separated (IV);

(**b**) Force–distance (*F–D*) curve showing steps (I), (II), (III), and (IV), corresponding to those outlined in (a). Several unbinding events can be observed (*s*, force steps; *t*, unbinding of membrane tethers; F_D, maximal detachment force). Reprinted with permission from ref. [117]. (**C**) (**a**) (1) Interaction forces between the *C. albicans* cell and a dendritic cell-specific intercellular cell adhesion molecule-3 (ICAM-3)-grabbing non-integrin (DC-SIGN)-Fc-coated substrate, (2) a single fluorescein isothiocyanate (FITC)-labelled *C. albicans* cell immobilised on the apex of a tipless cantilever visualised by confocal microscopy. Single channels (*a–b*) and an overlay (*c–d*) show the FITC-labelled *C. albicans*; (**b**) Probing specific DC-SIGN–*C. albicans* interactions with atomic force microscope dynamic force spectroscopy. Examples of *F–D* curves of the interaction of DC-SIGN with *C. albicans*; single bond ruptures are visible as discrete steps (arrows in inset). The area enclosed by the curve and the zero-force line (no contact regime; dotted line) is a read-out for the adhesion between the cell and the substrate; the work needed to detach *C. albicans* from the DC-SIGN-Fc-coating. Next, three examples of *F–D* curves are shown after an in situ block with soluble *C. albicans* (CA)-mannan. The work and detachment force (indicated maximum force F_{max}) are smaller than before this block. The asterisk (*) indicates the distance at which the final bond detaches. The arrows in the inset indicate discrete rupture steps. Reprinted with permission from ref. [118].

Single-cell force spectroscopy (SCFS) assays on living cells have been applied to measure the strength of cell adhesion down to the contribution of single molecules [119–121] (Figure 2B,C). AFM-based SCFS is currently the most versatile method for the study of adhesive interactions of cells with other cells, proteins, and surfaces, since SCFS offers a large range of detectable forces (from 10 pN to 100 nN), and offers precise spatial (1 nm to 100 µm) and temporal (0.1 to >10 min) control over the adhesion experiment and experimental parameters [120]. A living cell can be attached to a tipless AFM cantilever and the interacting partner (molecule or cell) on a substrate-coated surface. Alternatively, the living cell can be fixed on a surface, and the tip functionalised with the interacting molecule. AFM force spectroscopy with a single cantilever-bound cell can be used to investigate cell–cell and cell–matrix interactions. The approach and withdrawal of this cell to and from its surface can be precisely controlled by parameters such as applied force, contact time, and pulling speed, benefiting from the AFM's high-force sensitivity and spatial resolution. The data collected in these experiments include information on repulsive forces before contact, cell deformability, maximum unbinding forces, individual unbinding events, and the total work required to remove a cell from the surface (Table 5, Figure 2B,C). Force spectroscopy can identify cell subpopulations and characterise the regulation of cell adhesion events with single-molecule resolution [122].

Table 5. Examples of yeast using single-cell (SCFS) force spectroscopy studies.

Cell Type	Interaction Partner	Variables	Refs
C. albicans	*Staphylococcus aureus*	Deletion of *ALS3* (adhesion gene)	[123]
	Hydrophobic DDP [1] coated surface	Surface hydrophilicity, hydrophobicity, deletion of *HGC1*, compared to *S. cerevisiae*	[124]
	C. albicans hyphae	Deletion of *ALS3* and *ALS1* (adhesion genes)	[125]
	DC-SIGN [2]	Differences in the N-mannan structure of the cell wall	[118]
C. glabrata	Adhesin Epa6p	Surface hydrophilicity, hydrophobicity, expressed and deleted *EPA6*	[110]
S. cerevisiae	Abiotic surface	Surface hydrophilicity, hydrophobicity BSA coating, life cycle stage, glutaraldehyde-treated cells	[126]
	Silica surface	Different silica with defined roughness	[127]
	Methacrylate polymers surface	Polymer imprinted and non-imprinted surface	[128]
	Bare and polydopamine-coated glass	Polydopamine coating	[129]

[1] DDP: dodecyl phosphate; [2] DC-SIGN: dendritic cell-specific intercellular cell adhesion molecule-3 (ICAM-3)-grabbing non-integrin.

2.3. Nanomotion Analysis

New sensor technologies based on microcantilevers have recently been developed [130,131]. Nanomechanical oscillators are increasingly being used for the detection of very small masses [132] or for nanostress sensing in molecular biology [133,134]. Cantilever resonators have been shown to possess a mass resolution in the pico- to femtogram ranges in both air [135] and liquid [136,137]. Many of the available systems are limited by the need to perform the measurements in air or in a humid environment, and most rely on the detection of the replication of the cells on the surface of the sensor. Thanks to the many advantages they offer, microcantilevers have recently been explored as nanosensors for cell studies; they are highly sensitive, selective, label-free, real-time, and provide in situ detection capabilities [138]. Single cell detection and monitoring on the cantilever sensor has been reported for *S. cerevisiae* cells [139,140], *E. coli* and *Bacillus subtilis* [140,141], HeLa cells [142], mouse lymphoblasts [140], and human lung carcinoma and mouse lymphocytic leukemia cells [143]. Cell growth detection has been demonstrated by monitoring resonance frequency changes of cantilevers as the mass increases from immobilized *S. cerevisiae* and fungal *Aspergillus niger* spores on the surface of the cantilevers in humid air [144]. *S. cerevisiae* cells were deposited onto the cantilever surface, and its bending as a function of time corresponded to the yeast growth behaviour [138].

Recently, the metabolic state of living organisms that are immobilized on the cantilever surface could be detected by cantilever nanomotion analysis in physiological conditions [145–147]. In nanomotion analysis mode, the sample is directly deposited onto the cantilever, and the analysis is performed with the functionalised cantilever in liquid. This differs from nanomechanical resonators, where the liquid sample is flowed through a capillary in the cantilever (Figure 3A). If the sample is alive, its nanometric-scale motions are transmitted to the cantilever, causing it to oscillate. These oscillations are detected by monitoring the cantilever displacements with the traditional laser–photodiode system; a typical set-up is depicted in Figure 3Ba. The cantilever and the sample of interest are immersed in an analysis chamber equipped with an inlet and an outlet that permits measurement in liquids, and, importantly, the exchange of liquids during measurements. It has been observed that any type of organism induces oscillations of the cantilever that only last while the organism is alive [147]. Once an efficient killing agent is applied, the cantilever oscillations stop. The exact origin of these vibrations is still under investigation. In the case of motile organisms, such as mammalian cells or flagella-equipped bacteria such as *E. coli*, the answer is straightforward. However, in the case of immotile microorganisms such as yeast or *Staphylococcus aureus*, the explanation is more challenging. Probably, a direct momentum transfer between the sample's surface proteins that undergo conformational changes and the cantilever plays an important role [146].

Figure 3Bb shows a typical nanomotion experiment with *C. albicans*. The AFM cantilever was pre-treated with glutaraldehyde and incubated in a solution containing the cells. Some *Candida* cells attached onto its surface. The cantilever was eventually inserted into the growth medium-filled analysis chamber, and its oscillations were recorded. After the injection of a buffer solution containing 10 μg/mL of caspofungin (an antifungal drug to which *Candida* is sensitive) in the analysis chamber, the cantilever oscillations dramatically decreased. This drop became noticeable after only 10 min post-caspofungin exposure. Such an application can be very efficient (in a timeframe of minutes) for the detection of chemicals to which living organisms are sensitive, or for simple assessment of the presence of living organisms in extreme environments.

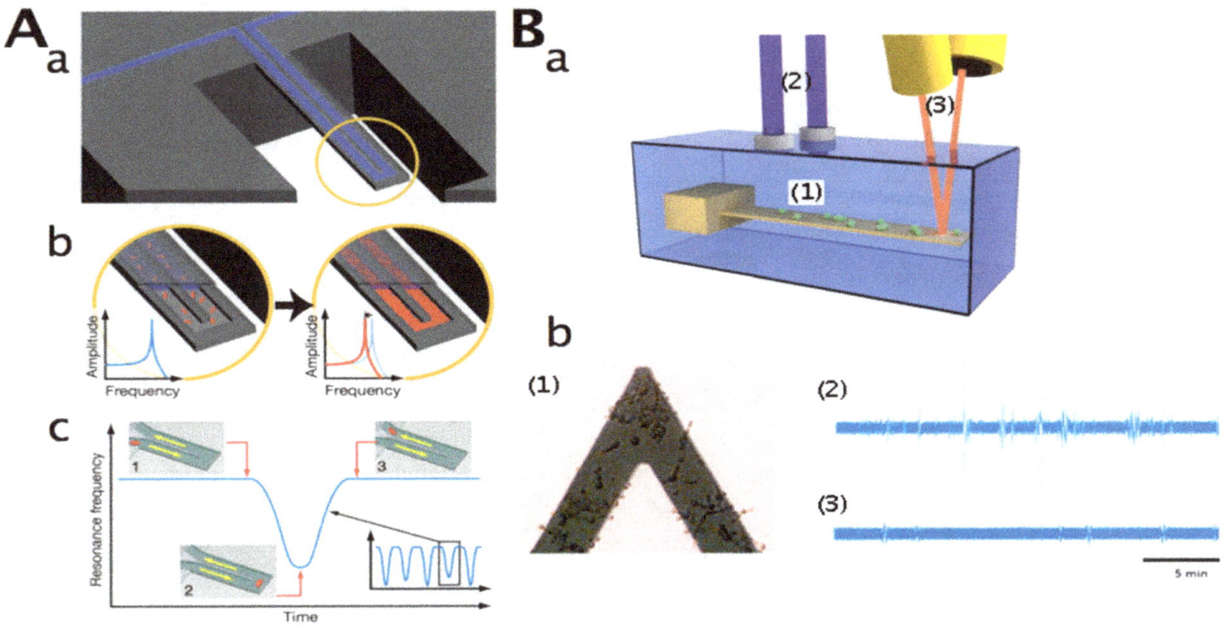

Figure 3. (**A**) Nanomechanical resonators enable the measurement of mass with extraordinary sensitivity. Illustration of two mass measurement modes enabled by a fluid-filled microcantilever. (**a**) A suspended microchannel translates mass changes into changes in resonance frequency. Fluid continuously flows through the channel and delivers biomolecules, cells, or synthetic particles. Sub-femtogram mass resolution is attained by shrinking the wall and fluid layer thickness to the micrometre scale and by packaging the cantilever under high vacuum; (**b**) While bound and unbound molecules both increase the mass of the channel, species that bind to the channel wall accumulate inside the device, and, as a result, their number can greatly exceed the number of free molecules in solution. This enables specific detection by way of immobilised receptors; (**c**) In another measurement mode, particles flow through the cantilever without binding to the surface, and the observed signal depends on the position of particles along the channel (insets 1–3). The exact mass excess of a particle can be quantified by the peak frequency shift induced at the apex. Reprinted with permission from [141]. (**B**) (**a**) Typical setup for the detection of the nanomotion of living organisms suspended in liquid medium: (1) analysis chamber with microbial cells (green) attached to the cantilever, (2) inlet and outlet of the fluid chamber, (3) laser and photodetector; (**b**) (1) *C. albicans* deposited onto a cantilever. Courtesy of Dr. Sandor Kasas, Ecole Polytechnique Fédéral de Lausanne, Switzerland, (2) oscillations of the cantilever in nourishing medium-filled analysis chamber, (3) oscillations of the cantilever after the replacement of the growth medium with caspofungin (antifungal agent)-containing buffer. The amplitude of cantilever oscillations is in the range of 1–8 nm (unpublished data).

3. Yeast Cell Patterning and Manipulation

3.1. Yeast Cell Patterning

Manipulating the physical location of cells is useful both to organize cells in vitro and to separate cells during screening and analysis [148,149]. The quest to manipulate cells on length scales commensurate with their size has led to a host of technologies exploiting chemical, mechanical, optical, electrical, and other phenomena. The major cell-patterning methods include patterning on adhesive micropatterns, mechanical cell patterning, and robotic cell patterning [150]. Cell-adherence methods have been especially developed for the adhesion of mammalian cells, but have also been developed for yeast cell patterning (Table 6). A variety of different patterning techniques have been developed to present adhesive ligands at a range of scales to investigate biological events, pushing the envelope on the minimum feature down to the nanometer scale [151–156].

Table 6. Applications of yeast cell patterning.

Yeast Type	Cell Patterning Method	Issue Addressed	Refs
S. cerevisiae	Patterning on adhesive micropatterns	Microcontact printing of concanavalin A	[157]
	Mechanical cell patterning in microfluidic microchambers	Monitoring dynamics of single-cell gene expression	[158]
	Robotic cell printing	Systematic profiling of cellular phenotypes	[159]
	Mechanical cell patterning using trap barriers	Single cell gene expression analysis	[160]
	Robotic cell printing	Localisation of the yeast proteome during polarised growth	[161]
	Mechanical cell patterning in microfluidic microchambers	Quantitative analysis of the yeast pheromone signalling response	[162]
	Patterning on adhesive micropatterns	Microcontact printing of biotinylated bovine serum albumin	[163]
	Mechanical cell patterning using trap barriers	Whole lifespan microscopic observation	[164]
	Mechanical cell patterning in microfluidic single-cell microwells	Real-time cellular responses of the mating MAPK pathway	[165]
	Mechanical and chemical patterning in microchambers	Molecular phenotyping of aging in single cells	[166]
	Mechanical cell patterning using trap barriers	Single cell analysis of yeast replicative aging	[167]
	Mechanical cell patterning in elongated cavities	Monitoring the dynamics of cell division	[168]
	Mechanical cell patterning using trap barriers	Studying ageing and dynamic single-cell responses	[169]
	Patterning in microcavity array by negative pressure, and embedded in agarose gel layer	Long-term single cell growth observation	[170]
	Mechanical cell patterning using trap barriers	Automated measurements of single-cell aging	[171]
	Mechanical cell patterning using trap barriers	High-throughput analysis of yeast replicative aging	[172]
Sc. pombe	Mechanical patterning in culture microchambers	Mechanical mechanisms redirecting cell polarity and cell shape in fission yeast	[173]
	Mechanical patterning in single-cell microwells	Determination of the mechanical forces involved in cell growth	[174]
	Mechanical patterning in microchambers	Time-lapse fluorescence observation of the effect of a microtubule-inhibiting drug	[175]
	Mechanical trapping in single-cell cavities	Fission yeast synchronisation	[176]
	Mechanical barrier single-cell trapping	Lon-term observation using super-resolution fluorescence microscopy	[177]
	Mechanical patterning in chemostat microchambers	Long-term single-cell analysis	[178]
	Mechanical patterning in culture microchambers	Studies of cellular aging	[179]

MAPK: Mitogen-activated protein kinases.

Microcontact printing has become the most popular technique [180]. A polydimethylsiloxane (PDMS) stamp with desired microfeatures is fabricated using soft lithography methods, and is used to print adhesive biomolecules onto the culture substrate [157,181]. For yeast cell adhesion, the lectin concanavalin A (which binds to cell wall mannose and glucose aminoglycans) can be used as an adhesive molecule. *S. cerevisiae* was also immobilised on cholesterol-modified microcontact-printed spots [163]. Despite its popularity, microcontact printing has several drawbacks for cell biology labs, such as the requirement of a clean room to microfabricate the stamp, and variations in the quality of the protein transfer [182].

In mechanical cell patterning, mechanical barriers capture the cells at specified spots. Cells can be trapped in microchambers (Figure 4A), microwells (Figure 4C), or by cell trap barriers (Figure 4D) (Table 6). Various microfabrication techniques have been used to fabricate microwell substrates for cell cultivation [150]. The microwell can have a diameter from several hundred micrometers up to the dimensions of a single cell [174]. Single-cell microwell arrays allow large numbers of cells to be stimulated and analysed (usually by fluorescence microscopy) in a massively parallel fashion [165,183]. Single-cell analysis has increasingly been recognised as the key technology for the elucidation of cellular functions, which are not accessible from bulk measurements on the population level [184,185]. Yeast cells have been trapped in microfluidic microchambers by using inlet and outlet valves [162,186] (Figure 4A). Culture chambers that are open on both sides [173,179,187] or on one side [176] have been constructed. These chambers fit single-cell dimensions and confine the cells. These culture chambers are suitable for non-adherent cells, such as yeast and bacteria [188].

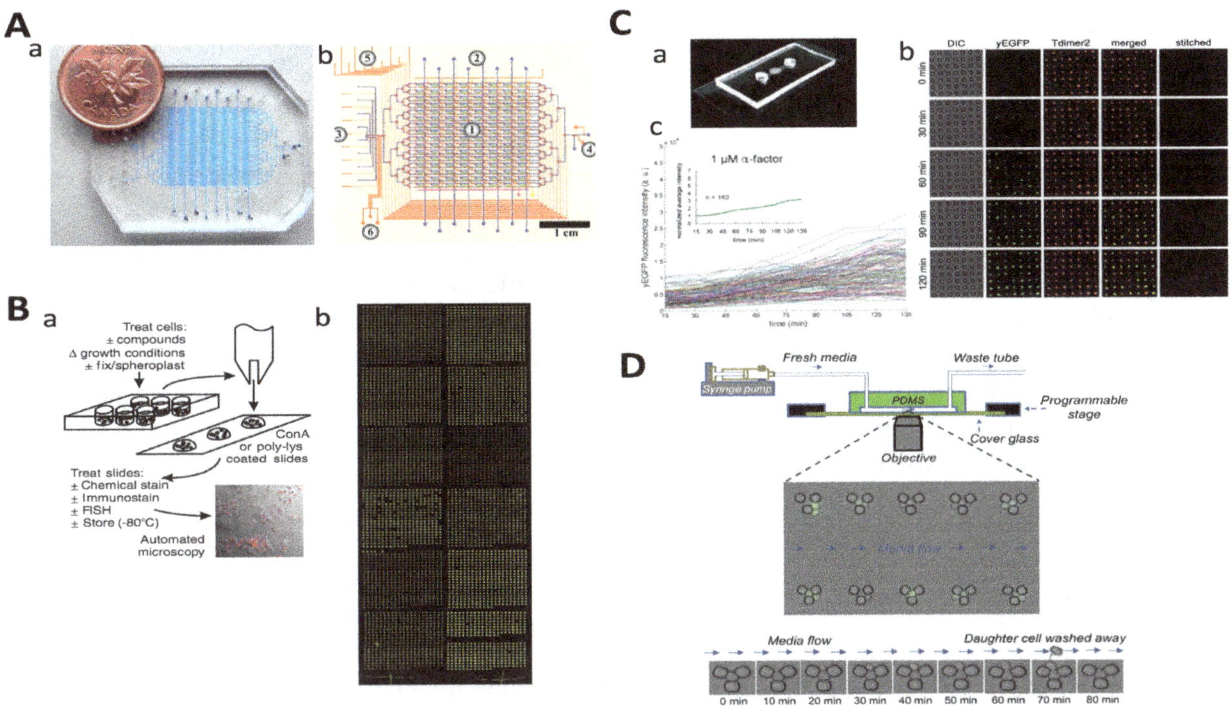

Figure 4. (**A**) Cell trapping in microfluidic chambers. (**a**) Image of the microfluidic device; (**b**) Working area of microfluidic device showing (1) array of 128 imaging chambers, (2) column inlets for loading different strains, (3) eight chemical inlets controlled by independent valves, (4) outlet ports, (5) fluidic multiplexer to deliver reagents to specified rows, (6) integrated peristaltic pump for on-chip formulation of stock reagents. Reprinted with permission from [162]. (**B**) (**a**) Spotted cell microarrays using contact cell printing. Cell chips are constructed using slotted steel pins to print cells robotically from multi-well plates onto glass slides; (**b**) Wide-field light scattering image of a cell microarray containing around 4800 viable haploid yeast deletion strains. From ref. [159]. (**C**) Single-yeast cell microwell array in a microfluidic chip. (**a**) An overview of the cell chip. The cell chip has one simple straight microfluidic channel and two punched reservoirs; (**b**) Representative microscopic images for 0, 30, 60, 90, and 120 min time points in the case of α-factor treatment (DIC: bright field, yEGFP: green fluorescent, and Tdimer2: red fluorescent images). The merged and stitched images show diverse colours from a mixture of green and red fluorescence; (**c**) Typical time-course measurements of mating responses of individual cells. The inset shows the normalised time-course average of yEGFP fluorescence intensity. Reprinted with permission from [165]. (**D**) A microfluidics platform that facilitates simultaneous lifespan and gene expression measurements of aging yeast cells. Schematics of the experimental setup (*upper panel*). The growth of a single cell that is trapped in a replicator as a function of time is shown (lower panel). From ref. [171].

Mechanical cell trap barriers have also been used to capture cells from suspensions in fluidic devices [158,189,190]. Fluid flow pushes the cells into the traps, and, therefore, these cell traps are also designated as hydrodynamic cell traps [191]. Barriers have been designed with a small fluidic leak that allowed single-cell trapping [160,177,192] (Figure 4D).

To create cellular microarrays, cells can be spotted or "printed" using a fluid-dispensing device ("cell printer") [150]. It is essential to obtain a highly reproducible number of living cells per spot and an optimised printing process that is qualified for the reproducible production of microarrays with cells that keep their vitality and function for analysis. Spot formation techniques are categorised as "contact printing" and "non-contact" printing [193,194]. Robotic yeast cell contact printing was initially used to print cells on an agar growth medium by using fluid-dispensing devices or pads [195], or cells were grown in multiwell culture plates and printed on a glass slide for high-throughput imaging [159] (Figure 3B), or only short-time analyses on living cells were performed. More often, non-contact-based devices are used to produce cellular arrays, such as modified inject printers or piezo-driven tips [196–199]. In non-contact printing techniques, the liquid metering is not determined by the complex interplay of the pin, the liquid, and substrate, but is separated from the substrate, because no contact between the printing tool and the substrate occurs. The fluid is ejected as a flying droplet or jet towards the surface from a certain distance, which makes metering more precise. One concept of non-contact printing is based on syringe–solenoid-driven printers, where a reservoir and a high-speed microsolenoid valve are connected to a high-resolution syringe (e.g., the M2-Automation, synQUAD, or Genomic Solutions system). Further non-contact microarrayers are piezoelectrically driven, where a technology similar to the one used in an ink-jet printer is used (e.g., M2-Automation, MicroDrop, PerkinElmer, Scienion, GeSim) [200,201]. A piezo-actuator is fixed at the top of the dispenser tip. The squeezing of the tip forced by the piezo-actuation induces droplet ejection out of the capillary. The fast response time of the piezoelectric crystal permits fast dispensing rates (kHz range), and the small deflection of the crystal generates droplets from tens of picoliters to a few nanoliters.

3.2. Direct Contact Cell Manipulation

3.2.1. AFM-Based Cell Manipulation

The desire to actively deliver precise amounts of biomolecules through nanosized probes initiated the development of novel microfluidic probes. Microfabrication processes have been introduced for the production of AFM cantilevers with embedded microchannels [202–205]. Microchannel cantilevers were connected to a pressure controller for active liquid handling in fluidic force microscopy (FluidFM) [206,207]. The ability to apply a pressure allows for negative pressure experiments involving suction for applications such as cell adhesion, or positive pressure experiments resulting in cell deposition on a specified spot or in controlled dispensing for applications such as the accurate delivery of bioactive compounds to a single targeted cell in physiological medium or even cell injection. FluidFM was used for the spatial manipulation of single *S. cerevisiae* cells [208]. Therefore, the hollow cantilever was positioned over a yeast cell and approached in AFM contact mode. An underpressure of ~50 mbar was applied to suck the cell against the channel aperture. After displacement, the cell was deposited onto the substrate with an AFM approach in contact mode, and the cell was released by applying a short overpressure pulse while retracting the probe. The underpressure single-cell immobilisation of cells on the cantilever also allows accelerating the pace of SCFS, since the conventional cell trapping cantilever chemistry can be avoided [124]. Single-cell *C. albicans* adhesion forces to a hydrophobic (dodecyl phosphate coated) surface were compared to adhesion to a hydrophilic (hydroxyl-dodecyl phosphate coated) surface, the *C. albicans* mutant Δ*hgc1* (which reduces the cell surface hydrophobicity), and to *S. cerevisiae* adhesion to the hydrophobic and hydrophilic substrate (Table 5). Force adherence measurements of *S. cerevisiae* cells on bare glass and polydopamine-coated glass substrates have been performed using a microfabricated hollow cantilever made entirely from SU-8 [129] (Table 5). Highly flexible SU-8 cantilevers with integrated microchannels have been fabricated for both additive and subtractive patterning of *S. cerevisiae* cells [209].

3.2.2. Micropipette Manipulation of Single Yeast Cells

The oldest and most commonly used approach for single-cell manipulation uses glass capillary micropipettes [210]. A negative pressure applied to growth media-filled capillary immersed in a cell culture dish controls the aspiration of a desired cell. A positive pressure dispenses the cell. Motion stages with multiple degrees-of-freedom were used to manually manipulate the micropipette and accurately control its tip position to perform either micromanipulation or microinjection [211]. Micromanipulators enable the controlled separation of selected living cells from suspension and even allow for isolation of prokaryotic cells [212]. They can also be used in adhesion studies, such as the interaction of a single *C. albicans* cell that is sucked to a micropipette with a diameter that is smaller than the cell, with a salivary pellicle-coated bead that is manipulated with a second micropipette [213].

Single cell manipulation systems that are based on capillaries are commercially available; for example: TransferMan (Eppendorf, Hamburg, Germany), PicoPipet (Bulldog Bio, Portsmouth, NH, USA), Stoelting Micromanipulators (Wood Dale, IL, USA), and miBot™ manipulator (Imina Technologies, Lausanne, Switzerland). These manipulation systems are manual, although the miBot micromanipulator is a mobile micro-robot that moves directly over the surface of the microscope base, has a nm spatial resolution, and can be remotely controlled. Micropipette cell manipulation systems that allow automatic selection and placement of a single yeast cell using vision-based feedback control have been developed [211]. A robotic micromanipulation system based on a general-purpose micromanipulator and a traditional glass micropipette was developed for pick-and-place positioning of single cells [214]. By integrating computer vision and motion control algorithms, the system visually tracks a cell in real time and controls multiple positioning devices simultaneously to accurately pick up a single cell, transfer it to a desired substrate, and deposit it at a specified location. A computer-controlled micropipette installed on an inverted fluorescence microscope was used to automatically recognise by computer vision, and both fluorescently labelled and unlabelled live cells in a Petri dish were picked up [215]. A recent developed computer vision-based automated single-cell isolation system allowed the isolation of single live cells from a very dense culture without immobilising cells on a surface [216].

Microchanneled AFM micropipettes have also been developed and used for cell adhesion and spatial cell manipulation applications (see previous section "AFM-based cell manipulation"). These AFM micropipettes are also designated as versatile nanodispensing (NADIS) systems [217]. Compared to conventional glass pipettes, this tool is particularly suitable when using substances of high cost or limited amounts, because significantly less volume is required for an experiment [218,219]. Another advantage over glass pipettes is the precise control wielded in the manipulation of sensitive targets, due to concurrent measurements of cantilever deflections without significant target damage [206]. Targets—such as functionalised surfaces or surface immobilised cells—can be precisely and gently manipulated physically, biologically, and chemically [129,207,208,220].

3.3. Non-Contact Cell Manipulation

3.3.1. Optical Manipulation of Single Yeast Cells

In the last decade, optical manipulation has evolved from a field of interest for physicists to a versatile tool widely used within life sciences [221]. Optical trapping and manipulation is a spin-off from research where lasers were used to study the effect of linear and angular momentum of light on small neutral particles. Arthur Ashkin first demonstrated that radiation pressure from a focused laser beam significantly affected the dynamics of micrometer-sized transparent and neutral particles, and two basic light-pressure forces were discovered: a scattering force in the direction of the incident light beam, and a gradient force in the direction of the intensity gradient of the beam [222]. The scattering component of the force works as a photonic "fire hose" pushing the particle in the direction of light propagation. The gradient force can be explained by a dipole in an inhomogeneous electric field that experiences a force in the direction of the intensity field gradient of the laser beam [223]. Using these forces, small particles (such as cells) can be accelerated, decelerated, and trapped in three dimensions.

Optical tweezers use light to levitate a particle (cell) of distinct refractive index [224]. The trapped cell is suspended at the waist of the focused (typically infrared) laser beam. The displacement of the cell from the focal center results in a proportional restoring force, and can be measured by interferometry or back-focal plane detection. Optical tweezers use a high gradient of optical pressure to guide cells by focusing a laser beam through a high numerical aperture (N.A.) lens on the cells. High optical intensity of about 10^{10} mW/cm^2 may cause damage on the cells, and are not suitable for long-term cell manipulation [225]. Micro-meter-sized homogeneous particles (or cells) can be trapped with forces ranging from a few pN to several tens of pN, depending on the optical properties of the particles and of the medium [226]. It is possible to track the position of the trapped particle with sub-nanometer accuracy at high (several MHz) repetition rates [227]. Due to the rapid advances in laser technology, optical manipulation setups have been developed that have become relatively uncomplicated. Optical manipulation is easily integrable with various microscopy setups, including confocal, super-resolution, or multiphoton microscopes. It allows for high spatial and temporal resolution, and interaction forces can be minimised.

Optical tweezers have been used to manipulate yeast cells, such as in cell trapping, cell positioning, and cell sorting (Table 7, Figure 5). Optical trapping was used to isolate single yeast cells from a mixture of two strains that were distinguishable in fluorescence microscopy [228]. An optical tweezer was used for the rapid separation and immobilisation of a single yeast cell by concomitant laser manipulation and locally thermosensitive hydrogelation [229]. Optical tweezers can be used to trap single yeast cells for further analysis, such as Raman microspectroscopy [230], time-lapse fluorescence microscopy to determine single-cell internal pH [231] (Figure 5C), or to study single cell dynamics by monitoring GFP-tagged proteins [232] (Figure 5D). Yeast cells are conducive to direct optical tweezing and can be used for single-cell force studies [233]. The effect of various factors (such as the ionic strength and the nature of the counter-ion in the solutions) on the adhesion and detachment force of yeast cells on glass was assessed [234]. Compared to AFM, magnetic tweezers, and more conventional ways of studying cell adhesion (such as shear-flow cells), optical tweezers present several advantages: direct measurements in physiological conditions, clear criterion to evaluate the proportion of adhering cells, and ease of examining the heterogeneity of cell behaviours in the population. However, the optical tweezer method is limited to low adherence forces (~1 to 100 pN) owing to the low refractive index of cells, and is sensitive to the cell optical heterogeneity.

Optical gradient forces generated by fast steerable optical tweezers are highly effective for sorting small populations of cells in a lab-on-a-chip environment (Figure 5). Reliable sorting of yeast cells in a microfluidic chamber by both morphological criteria and by fluorescence emission was demonstrated [235]. More than 200 yeast cells could be contact-free immobilised into a high-density array of optical traps in a microfluidic chip [236] (Figure 5B). The cell array could be moved to specific locations on the chip, enabling the controlled exposure of cells to reagents and the analysis of the responses of individual cells in a highly parallel format using fluorescence microscopy. Additionally, single cells were sorted within the microfluidic device using an additional steerable optical trap. Optical tweezers were used to spatially and temporally control pathogenic *C. albicans* and *Aspergillus fumigatus* and place them in proximity to host cells, which were subsequently phagocytosed [237,238].

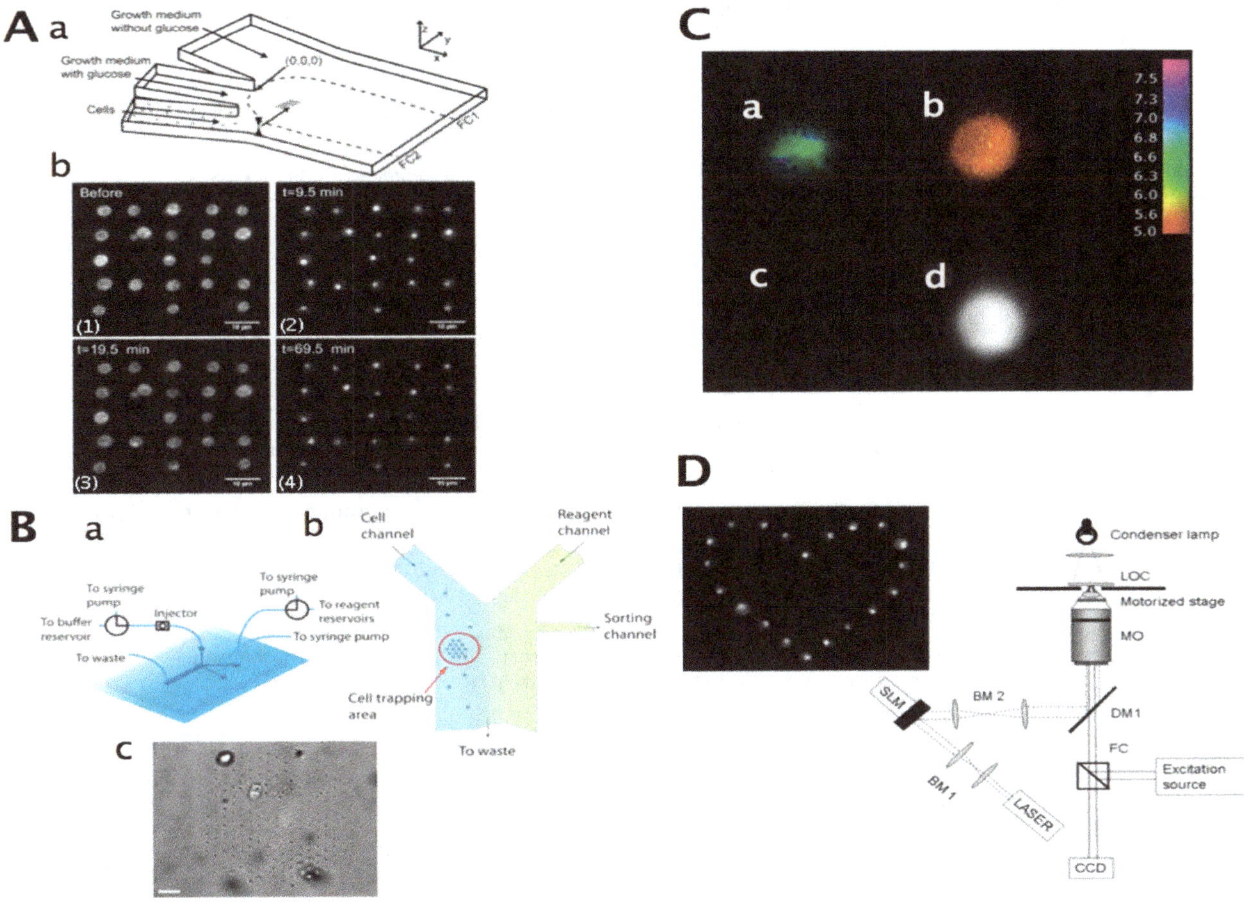

Figure 5. (**A**) (**a**) Schematic of the microfluidic device for subjecting single cells to environmental changes. The cells are collected in the flow of cells from the lower channel using optical tweezers and positioned within the measurement region. By changing the relative flow rates at the inlets (from FC1 to FC2 and back), the environment around the cells can be changed reversibly; (**b**) Images showing yeast cells expressing Msn2-GFP: (1) to (4) show the cellular response (i.e., the shuttling of Msn2-GFP proteins in and out of the nucleus) when four cycles of changing the medium back and forth between 4% and 0% glucose was performed in the microfluidic chip. Reprinted with permission from ref. [239]. (**B**) (**a**) Schematic description of the microfluidic chip; and (**b**) the fluidic circuit; (**c**) Transmission micrograph of more than 200 optically trapped yeast cells; scale bar: 30 μm. Reprinted with permission from ref. [236]. (**C**) Optically trapped single *S. cerevisiae* cell. All four images are of the same cell. (**a**,**b**) show the pH distribution at t = 0 min (30 °C) and t = 12 min (70 °C); (**c**,**d**) are propidium iodide images at t = 0 min (30 °C) and t = 12 min (70 °C). The colour bar represents pH values. Reprinted with permission from ref. [231]. (**D**) Typical optical tweezer setup. The trapping laser light is guided onto the spatial light modulator (SLM) via a beam expander (BM1). The laser beam with the imposed phase pattern then passes a second beam expander (BM2) to be imaged onto the back focal plane of the microscope objective (MO). The schematic figure of the setup also shows the lab-on-a-chip (LOC), a fluorescent excitation light source, a filter cube (FC), a dichroic mirror (DM1), a motorised microscope stage, and a condenser lamp. The inset image shows holographically-trapped *S. cerevisiae* cells that were stressed with sorbitol to induce localisation of Hog1-GFP to the cell nuclei. Reprinted with permission from ref. [221].

Table 7. Examples of yeast cell manipulation using optical manipulation.

Yeast Type	Issue Addressed	Refs
C. albicans	Control and manipulation of pathogenic yeast for live cell imaging and interaction with host cells	[237,238]
Hanseniaspora uvarum and *S. cerevisiae*	Confinement of an individual *H. uvarum* cell by *S. cerevisiae* cells increases the average generation time	[240]
S. bayanus	Study of growth pattern of cells under line optical tweezers generated by time-shared multiple optical traps	[241]
S. cerevisiae	On-chip single-cell separation and immobilisation using optical manipulation and thermosensitive hydrogel	[229]
	Real-time detection of hyperosmotic stress response in optically trapped single yeast cells using Raman microspectroscopy	[230]
	Optical manipulation of cells to microscopically observe environmentally-induced size modulations and spatial localisation of GFP-tagged proteins to elucidate various signalling pathways	[232]
	Optical trapping and surgery of living cells using two operational modes of a single laser	[242]
	Selection and positioning of single cells combined with microscopy analysis in a microfluidic channel; cycling of GFP-tagged Mig1p and Msn1p between the cytosol and nucleus	[239]
	Optical trapping and fluorescence microscopy investigation of the internal pH response and membrane integrity with increasing temperature	[232]
	Automated transportation of single cells	[243]
	Development of a microfluidic array cytometer based on refractive optical tweezers for parallel trapping, imaging, and sorting of individual cells	[236]
	Microfluidic sorting of arbitrary cells with dynamic optical tweezers	[235]
	Development of graded-index optical fibre tweezers with long manipulation length	[244]
	Position yeast cells in a microfluidic chamber to study glycolytic oscillations	[245,246]
	Tomographic phase microscopy with live cell rotation using holographic optical tweezers	[247]
	Development of a photonic crystal optical tweezer to trap an array of yeast cells	[248]
Sc. pombe	Displacement of the lipid granules	[249]
	Displacement of the nucleus	[250,251]
	Laser ablation of microtubules in vivo	[226]
	In vivo anomalous diffusion and weak ergodicity breaking of lipid granules	[252]
	Quantitative determination of optical trapping strength and viscoelastic moduli inside living cells	[253]

Optical tweezers can also be used to investigate the complex system of mechanical interactions taking place inside a living cell [226,249,250]. The viscoelastic properties of living *Sc. pombe* were investigated by studying the diffusion of lipid granules naturally occurring in the cytoplasm [251,252]. Optical manipulation techniques, such as optical tweezing, mechanical stress probing, or nano-ablation allow handling of probes and sub-cellular elements (such as organelles and individual molecules) with nanometric and millisecond resolution [254]. A near-infrared optical tweezer was used for yeast cell manipulation and micro-ablation [255]. This micro-nanosurgery system is based on a pulsed ultraviolet laser that induces plasma formation for intracellular surgery in live culture cells with submicron precision. Optical tweezers allow force probing of organelles and single molecules in vivo [256,257]. PicoNewton forces—such as those involved in cell motility or intracellular activity—can be measured with femtoNewton sensitivity, while controlling the biochemical environment. A method to perform a correct force calibration inside a living yeast cell (*Sc. pombe*) was developed [257]. This method takes the viscoelastic properties of the cytoplasm into account, and relies on a combination of active and passive recordings of the motion of the cytoplasmic object of interest. Absolute values for the in vivo viscoelastic moduli of the cytoplasm as well as the force constant describing the optical trap were determined.

3.3.2. Electrical and Magnetic Manipulation of Yeast Cells

Instead of using surface chemistry to prevent or allow cells to attach to certain regions, electromagnetic forces can be used to control cell positioning [149] and adhesion [258,259]. Electrical fields are very suitable for cell and bioparticle manipulation, with the advantages of strong controllability, easy operation, high efficiency, and minimal damage to targets [260]. Electrokinetic motion of cells refers to the migration of electrically charged or uncharged particles in a liquid medium or suspension in the presence of an electric field [261]. Electrical forces for manipulating cells at the microscale include electrophoresis and dielectrophoresis (DEP) [148,149]. Electrophoretic forces arise from the interaction of a cell's charge and a uniform or non-uniform electric field, whereas dielectrophoresis refers to the motion of polarised (uncharged) particles in only a non-uniform electric field [262]. Based on the applied electric field, DEP can be broadly divided into AC (AC DEP, classical DEP), DC (DC DEP), insulator-based DEP (iDEP, DC-iDEP), combined AC/DC (AC-iDEP), and travelling wave DEP (twDEP). In AC DEP, an array of metal electrodes is embedded inside a microdevice (such as a microfluidic chip) to generate a spatially non-uniform electric field, and can be used to separate particles by changing the medium property and frequency of the applied electric field [263,264]. It can also eliminate any electrophoretic (EP) and electroosmotic (EO) effect [265]. In DC DEP, the spatially non-uniform electric field is created by specially-designed insulators, such as electrically non-conducting obstructions or hurdles in a microdevice, and electrodes that are positioned at the ends of the microfluidic channels [263]. The DC electric field results in an EO force and eliminates the need for an external pump, which is required in the case of AC DEP. The combination of AC/DC DEP can transport cells by using electrokinetic effects (such as electroosmosis), and AC DEP can be used to separate cells. In the travelling wave DEP, transport and separation of cells can be performed with the AC electric field [266]. In this case, the spatial nonuniformity of the phase of the electric field is used to transport the particle, and the nonuniformity in magnitude of the field is used to separate the particles.

DEP and electrophoretic forces have been used to create microsystems that separate cell mixtures into their component cell types or act as electrical "handles" to transport cells or place them at specific locations [267] (Figure 6C). DEP has also been applied for cell sorting [268], focussing, filtration [269], and assembly [262]. DEP has been used to characterise cells, for example, to monitor cell viability changes (including morphology and internal structure) and isolate viable cells with minimal or no damage [270,271]. Electrophoretic and/or electroosmotic pumping can also be used to control and drive cell transport in microfluidic chip channels [272]. DEP tweezers have been developed that allow the positioning of a single cell in three dimensions or transfer a single cell to any designated area [260]. A DEP tweezer consisting of a sharp-tip glass needle with a pair of electrodes and using a pDEP force could hold a single yeast cell at the end of the micromanipulator [267]. The design of the tweezers was not adequately optimised for one-by-one manipulation; therefore, a round-tip shape for the DEP-based tweezers that shifts the electric field to the centre of the tweezers' tip due to the smooth geometry at the tip is most suitable for single-cell manipulation [273].

DEP traps for single-cell patterning in physiological solutions have been developed [274] (Table 8). DEP manipulation and trapping of yeast cells has been included in microdevices such as microfluidic chips [272,275–277]. Live and dead yeast cell separation was achieved with the "headlands and bays" electrodes [278]. Live yeast cells are attracted to the regions of the maximum field, while the dead ones are repelled to the regions of the minimum field, resulting in the separation of live and dead yeast cells. The separation of live and dead yeast cells by DEP could be enhanced by using the cross-linking agent glutaraldehyde (since glutaraldehyde selectively cross-links nonviable cells to a much greater extent than viable cells due to the higher cell wall permeability of nonviable cells) [279]. Live and dead *S. cerevisiae* cells were sorted by using AC DEP [280], multifrequency DEP [281], or AC/DC DEP using a quadrupole electrode array [282] (Figure 6A). Yeast cells could be pre-concentrated by trapping using DEP, and separated depending on their vitality by using hydrodynamic, DC electrophoretic, and DC electroosmotic forces [271]. Live and dead yeast cells were characterised based on dielectric properties [283,284]. By combining DEP and image processing, dielectrophoretic spectra of cells can be acquired [285]. From this, the membrane properties of the cells can be obtained. DEP was used to control the rotation and vibration of patterned yeast cell clusters [286] (Figure 6B). This strategy is based on the cellular spin resonance mechanism, but it utilises coating agents to create consistent rotation and vibration of individual cells.

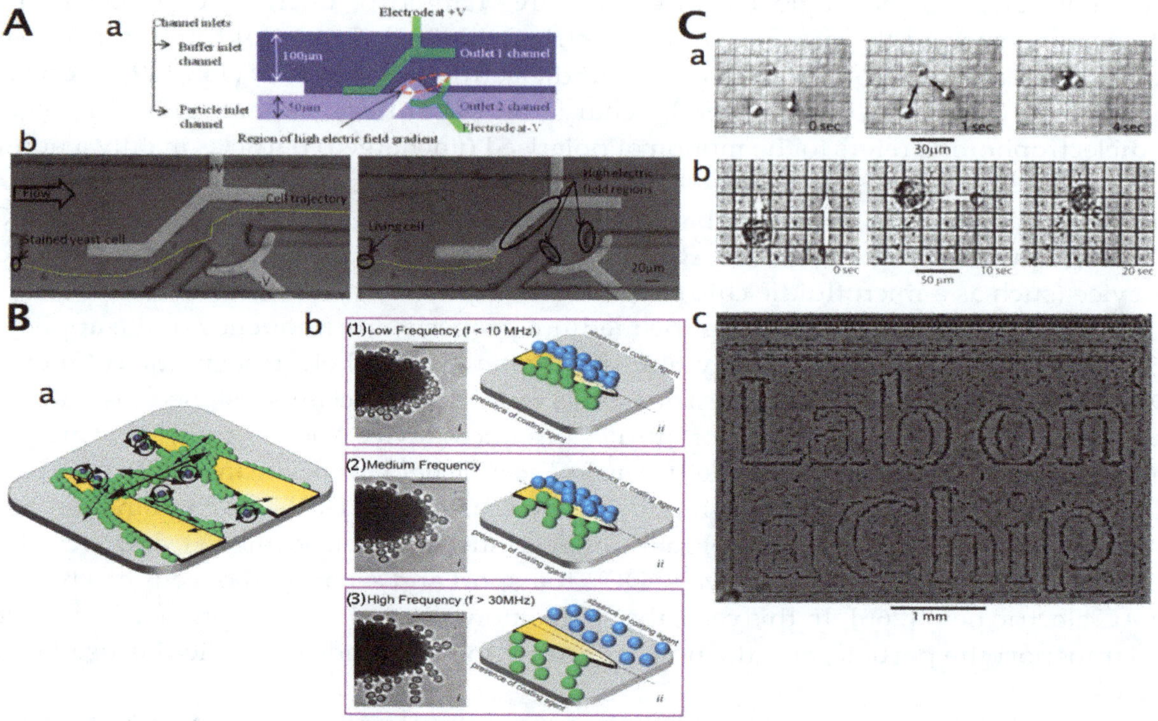

Figure 6. (**A**) (**a**) Schematic of a dielectrophoresis (DEP) sorter with buffer and particle flows being properly balanced; (**b**) Trajectory of a dead yeast cell experiencing negative DEP at the constriction (**left** panel); living yeast cells trapped under positive DEP (**right** panel). Reprinted with permission from ref. [282]. (**B**) (**a**) Three-dimensional schematic of the localised motion of patterned cell clusters under the influence of dielectrophoresis. Cells can exhibit rotational and vibrational movements according to their location within the cluster. Cells located at the free ends of pearl chains have the highest occurrence of rotation, which are indicated as blue cells. Alternatively, the cells packed along the long chains bridged between the microelectrodes have the highest occurrence of vibration. The arrows indicate the direction and strength of the vibrational movement of the cells; (**b**) Response of BSA-treated yeast cell clusters patterned onto a finger-shaped microelectrode array when operated with a 5 Vpk−pk AC sinusoid signal. (1-i, 2-i, 3-i) Response of BSA-treated cell clusters at 5, 20, and 40 MHz, respectively. (1-ii, 2-ii, 3-ii) Schematic representation of the response of patterned cell clusters at different frequencies, with BSA-treated cells (green) being compared to untreated cells (blue). The BSA-treated cells exhibit a positive DEP response, as the distance between the two adjacent cells increases at high frequencies. Alternatively, the untreated cells exhibit a negative DEP response at frequencies higher than 30 MHz. Reprinted with permission from ref. [286]. Copyright (2015) American Chemical Society. (**C**) (**a**) Time sequence of the DEP manipulation of yeast cells. Pixels are energised in sequence to move first one cell alone and then all three together; (**b**) Time sequence of yeast and rat alveolar macrophages manipulated with DEP. Pixels on the chip were energised to independently move the two cells and then bring them together; (**c**) Complex pattern of thousands of yeast cells patterned by DEP. Pixels across the array were energised to spell out "Lab on a Chip", attracting cells toward the local maxima of the electric field. Reprinted with permission from ref. [287].

Table 8. Examples of yeast cell electrical and magnetic manipulation.

Yeast Type	Manipulation Method	Issue Addressed	Refs
S. cerevisiae	Electrophoresis and electroosmosis	Cell transport in microfluidic channels	[272]
	Dielectrophoresis	Live and dead cell separation	[278]
	Magnetic patterning	Demonstration of magnetic micromanipulation of magnetically labelled cells	[276]
	Electroosmosis	Cell transport via electromigration in polymer-based microfluidic devices	[288]
	Dielectrophoresis (AC DEP)	Sorting live and dead cells	[280]
	Dielectrophoresis (AC DEP)	DEP tweezer for single cell manipulation	[267]
	Dielectrophoresis	Multiple frequency DEP separation and trapping of live and dead cells	[281]
	Diamagnetic trapping	Cell magnetic trapping in an array using a CoPt micromagnet array	[289]
	Magnetophoresis	Contactless diamagnetic trapping of cells onto a micromaget array	[289]
	Electrophoresis	Electrophoretic cell manipulation in a microfluidic device	[290]
	Dielectrophoresis (AC DEP)	Separation of yeast cells from blood cells in a microfluidic chip	[268]
	Dielectrophoresis	Live and dead cell separation	[279]
	Dielectrophoresis	Microfluidic chip for guiding cells by AC electrothermal effect and capturing by nDEP trap	[277]
	Dielectrophoresis (DC DEP)	Separation of a mixture of *S. cerevisiae* and *Escherichia coli* cells	[291]
	Dielectrophoresis (AC/DC DEP)	Sorting live and dead cells	[282]
	Dielectrophoresis, electroosmosis, electrophoresis	High-throughput trapping of cells, separation of live and dead cells	[271]
	Dielectrophoresis	Cell manipulation and immobilisation using photo-crosslinkable resin inside microfluidic devices	[292]
	Dielectrophoresis	Controlled rotation and vibration of cell clusters	[286]
	Magnetic manipulation	Magnetic manipulation of Fe_3O_4-doped hydrogel-coated cells	[293]

Magnetophoresis was applied to pattern yeast cells using a micromagnetic array [289]. Therefore, the diamagnetic *S. cerevisiae* were placed in an aqueous solution enriched in paramagnetic ions, and micromagnets that produce high magnetic field gradients were used. *S. cerevisiae* were coated with a single-layer of Fe_3O_4 nanoparticle-doped alginate hydrogel, which allowed their manipulation by a magnetic field [293]. Magnetic and electric manipulation of single or multiple yeast cells in a microfluidic channel was demonstrated using a microelectromagnet matrix and a micropost matrix [276]. The yeast cells labelled with magnetic beads were trapped by the microelectromagnet matrix, whereas the unlabelled cells were trapped by micropost matrix-generating electrical fields. The setup is suitable for the efficient sorting of yeast cells in a microfluidic chip. Yeast cells were trapped in a three-dimensional magnetic trap in an aqueous solution of paramagnetic ions [294].

Magnetic tweezers are similar in concept to optical tweezers; a magnetic particle in an external magnetic field experiences a force proportional to the gradient of the square of the magnetic field [295]. High forces can be achieved with relatively small magnetic field strengths, provided a very steep field gradient can be generated. The fields generated by sharp electromagnetic tips [296] or small permanent magnets [297] have been used to apply forces in excess of 200 pN on micron-sized magnetic particles. Magnetic tweezers are capable of exerting forces in excess of one nN (electromagnetic tweezers), and can be used to manipulate—and importantly, rotate—magnetic particles ranging in size from 0.5 to 5 μm. Magnetic tweezers are unique in that they afford passive, infinite bandwidth, force clamping over large displacements.

4. Conclusions

In recent years, single-molecule and single-cell analysis and manipulation techniques have been developed and applied to the study of yeast cells. Single-cell analysis has increasingly been recognised as the key technology for the elucidation of cellular functions, which are not accessible from bulk measurements at the population level. Various techniques are now available for the analysis of a single cell; with the aid of these techniques, many biological questions can be answered. A microfluidic device is now a suitable technique for single-cell analysis, because a microfluidic system can be manipulated with high throughput, and the amount of sample from a single cell is limited. As it became obvious from this review, the newly developed nanotechniques have been largely applied to the model yeasts *S. cerevisiae* and *Sc. pombe* for fundamental eukaryotic cell biology research, and the pathogenic model yeast *C. albicans* for elucidating the molecular basis of pathogen–host interactions.

High-resolution imaging techniques can provide up to single-biomolecule resolution. The most widely used imaging methods are scanning probe microscopy (i.e., AFM), super-resolution fluorescence microscopy, and electron microscopy. Their characteristics, advantages, and limitations are compared in Table 9. As can be noticed, nanoscale imaging methods are complementary, and they are therefore combined in recently developed imaging platforms, such as bio-AFM and super-resolution fluorescence microscopy [298,299], or the integration of EM and super-resolution microscopy in correlative light and electron microscopy (CLEM) [300–302]. The nanoscale exploration of surfaces of microbes such as yeast cells using AFM has expanded rapidly in the past years. Using AFM topographic imaging, the surface structure of live cells under physiological conditions is achieved with unprecedented resolution. Real-time imaging allows dynamic events to be followed. Chemical force microscopy (CFM)—in which AFM tips are functionalized with specific functional groups—can be used to measure interaction forces on the surface of live yeast cells. Molecular recognition imaging using spatially resolved force spectroscopy, dynamic recognition imaging, or immunogold detection can be used to localize specific receptors, such as yeast adhesins. Quantitative analysis of cell–cell or cell–substrate interactions can be performed with a number of techniques, where AFM single-cell force microscopy, optical tweezers, magnetic tweezers, and micropipette manipulation are the most popular. Understanding the fundamental forces involved in the adhesion of yeast cells is important not only in microbiology, to elucidate cellular functions (such as ligand-binding or biofilm formation), but also in medicine (host-pathogen interactions) and biotechnology (cell aggregation). These force spectroscopy techniques are compared in Table 10. These techniques are complementary, since each technique is most suitable for a specific force range.

Table 9. Comparison of high-resolution techniques for imaging yeast cells (adapted from [303]).

Characteristic	AFM	Electron Microscopy (SEM, TEM)	Super-Resolution Fluorescence Microscopy (PALM, STORM, SIM)
Resolution	~10 nm [1]	~1–10 nm	~5–50 nm
Live cell	Yes	No	Yes
Sample preparation requirement	Little	Little to substantial	Little to moderate
Sample preparation time	10 min–1 d	2 h–5 d	30 min
Image acquisition time	~5 min	5–10 min	Up to 24 h
Equipment cost	€150,000–350,000	€500,000	€250,000–500,000
Operational costs	Low	High	Moderate
Advantages	Localisation (and force spectroscopy) of single proteins; observation of dynamic processes; various environments (temperature, liquid, air, etc.)	Imaging of the cell ultrastructure at very high resolution	Time resolution.
Disadvantages	Only the cell surface is analysed; only one single cell at a time; slow temporal resolution; various sources of artifacts, such as cell or tip alteration	Fixation artifacts; no dynamics; no information on physical properties of proteins	Labelling is required

[1] Depends on the flatness of the surface; the provided value refers to the resolution for observing cells.

Table 10. Comparison of force spectroscopy techniques [295,304].

Characteristic	Optical Tweezers	Magnetic Tweezers	AFM	Micropipette
Type	Point Non-contact	Global/point Non-contact	Point Contact	Point Contact
Spatial resolution (nm)	0.1–2	5–10	0.5–1	-
Temporal resolution (s)	10^{-4}	10^{-1}–10^{-2}	10^{-3}	-
Stiffness (pN nm^{-1})	0.005–1	10^{-3}–10^{-6}	10–10^{5}	0.01–1000
Force range (pN)	0.1–100	10^{-3}–10^{2}	10–10^{4}	1–1000
Probe size (μm)	0.25–5	0.5–5	100–250	
Energy dissipation	Yes	No	No	No
Surface considerations	No	No	Yes	Yes
Features	Low noise and drift dumbbell geometry; access inside a cell	Force clamp, bead rotation, specific interactions; access inside a cell	High-resolution imaging	Controlled deposition/transfer of selected cells
Limitations	Photodamage, sample heating, non specific	No manipulation (force hysteresis)	Large high-stiffness probe, large minimal force, non specific	Low throughput

Several micro-nanomanipulation tools for cells have been developed. The methods can be based on direct-contact mechanical cell manipulation (such as AFM-based or micropipette-based manipulation), or based on non-contact cell manipulation (such as optical, electrical, and magnetic cell manipulation). Examples of these tools are the microchannel-embedded AFM microcantilevers that can be used to suck up one selected yeast cell, which can be further manipulated (positioned for patterning, pushed to another cell or substrate to perform SCFS), and robotic cell printing with picolitre volume dispensing. The manipulation of the physical location of cells is useful both to organise the cells in vitro for single-cell analysis and for specific cell–cell interaction analyses. Another recently developed tool is the use of the AFM cantilever as a very sensitive nanosensor that can detect the metabolic activity of living yeast cells, and even monitor protein conformational changes [305]. An optical tweezer can be used to manipulate several cells in 3D in a contactless way, and can also be applied as a micro-nanosurgery tool by using the nano-ablation option of the laser. Inside cell manipulation of structures has been demonstrated for optical and magnetic tweezers, and opens new possibilities for non-invasive cell organelle manipulation activities. Magnetic tweezers allow cell rotation, which can be important for cell surface location-dependent interactions (e.g., cell–cell interaction analysis during mating).

Electric and magnetic force can be used to trap and position cells at some physical location, to monitor cell viability and separate live from dead cells, transport cells in devices such as lab-on-a-chip to develop automated assays, and to characterise cell properties (e.g., by determining dielectrophoretic spectra of cells).

Acknowledgments: The Belgian Federal Science Policy Office (Belspo) and the European Space Agency (ESA) PRODEX program supported this work. The Research Council of the Vrije Universiteit Brussel (Belgium) and the University of Ghent (Belgium) are acknowledged to support the Alliance Research Group VUB-UGent NanoMicrobiology (NAMI), and the International Joint Research Group (IJRG) VUB-EPFL BioNanotechnology & NanoMedicine (NANO).

References

1. Roco, M.C.; Williams, R.S.; Alivisatos, P. (Eds.) *Biological, Medical and Health Applications: Nanotechnology Research Directions*; Kluwer Academic Publishers: Dordrecht, The Netherlands, 2000.
2. Roco, M.C. Nanotechnology: Convergence with modern biology and medicine. *Curr. Opin. Biotechnol.* **2003**, *14*, 337–346. [CrossRef]

3. Whitesides, G.M. The 'right' size in nanobiotechnology. *Nat. Biotechnol.* **2003**, *21*, 1161–1165. [CrossRef] [PubMed]
4. Binnig, G.; Rohrer, H. Scanning tunnelling microscopy. *Helv. Phys. Acta* **1982**, *55*, 726–735.
5. Binnig, G.; Quate, C.F.; Gerber, C. Atomic force microscope. *Phys. Rev. Lett.* **1986**, *56*, 930–933. [CrossRef] [PubMed]
6. Kada, G.; Kienberger, F.; Hinterdorfer, P. Atomic force microscopy in bionanotechnology. *Nano Today* **2008**, *3*, 12–19. [CrossRef]
7. Ando, T. High-speed atomic force microscopy. *Microscopy (Oxf.)* **2013**, *62*, 81–93. [CrossRef] [PubMed]
8. Kasas, S.; Ikai, A. A method for anchoring round shaped cells for atomic force microscope imaging. *Biophys. J.* **1995**, *68*, 1678–1680. [CrossRef]
9. Gad, M.; Ikai, A. Method for immobilizing microbial cells on gel surface for dynamic AFM studies. *Biophys. J.* **1995**, *69*, 2226–2233. [CrossRef]
10. De, T.; Chettoor, A.M.; Agarwal, P.; Salapaka, M.V.; Nettikadan, S. Immobilization method of yeast cells for intermittent contact mode imaging using the atomic force microscope. *Ultramicroscopy* **2010**, *110*, 254–258. [CrossRef] [PubMed]
11. Dague, E.; Jauvert, E.; Laplatine, L.; Viallet, B.; Thibault, C.; Ressier, L. Assembly of live micro-organisms on microstructured PDMS stamps by convective/capillary deposition for AFM bio-experiments. *Nanotechnology* **2011**, *22*, 395102. [CrossRef] [PubMed]
12. Formosa, C.; Pillet, F.; Schiavone, M.; Duval, R.E.; Ressier, L.; Dague, E. Generation of living cell arrays for atomic force microscopy studies. *Nat. Protoc.* **2015**, *10*, 199–204. [CrossRef] [PubMed]
13. Kasas, S.; Dietler, G. Probing nanomechanical properties from biomolecules to living cells. *Pflugers Arch.* **2008**, *456*, 13–27. [CrossRef] [PubMed]
14. West, M.; Zurek, N.; Hoenger, A.; Voeltz, G.K. A 3D analysis of yeast ER structure reveals how ER domains are organized by membrane curvature. *J. Cell Biol.* **2011**, *193*, 333–346. [CrossRef] [PubMed]
15. Ries, J.; Kaplan, C.; Platonova, E.; Eghlidi, H.; Ewers, H. A simple, versatile method for GFP-based super-resolution microscopy via nanobodies. *Nat. Methods* **2012**, *9*, 582–584. [CrossRef] [PubMed]
16. Formosa, C.; Schiavone, M.; Martin-Yken, H.; François, J.M.; Duval, R.E.; Dague, E. Nanoscale effects of caspofungin against two yeast species, *Saccharomyces cerevisiae* and *Candida albicans*. *Antimicrob. Agents Chemother.* **2013**, *57*, 3498–3506. [CrossRef] [PubMed]
17. Chopinet, L.; Formosa, C.; Rols, M.P.; Duval, R.E.; Dague, E. Imaging living cells surface and quantifying its properties at high resolution using AFM in QI™ mode. *Micron* **2013**, *48*, 26–33. [CrossRef] [PubMed]
18. Formosa, C.; Schiavone, M.; Boisrame, A.; Richard, M.L.; Duval, R.E.; Dague, E. Multiparametric imaging of adhesive nanodomains at the surface of *Candida albicans* by atomic force microscopy. *Nanomedicine* **2015**, *11*, 57–65. [CrossRef] [PubMed]
19. Mendez-Vilas, A.; Gallardo, A.M.; Perez-Giraldo, C.; Gonzalez-Martín, M.L.; Nuevo, M.J. Surface morphological characterization of yeast cells by scanning force microscopy. *Surf. Interface Anal.* **2001**, *31*, 1027–1030. [CrossRef]
20. Gad, M.; Itoh, A.; Ikai, A. Mapping cell wall polysaccharides of living microbial cells using atomic force microscopy. *Cell Biol. Int.* **1997**, *21*, 697–706. [CrossRef] [PubMed]
21. Touhami, A.; Nysten, B.; Dufrêne, Y.F. Nanoscale mapping of the elasticity of microbial cells by Atomic Force Microscopy. *Langmuir* **2003**, *19*, 4539–4543. [CrossRef]
22. Adya, A.K.; Canetta, E.; Walker, G.M. Atomic force microscopic study of the influence of physical stresses on *Saccharomyces cerevisiae* and *Schizosaccharomyces pombe*. *FEMS Yeast Res.* **2006**, *6*, 120–128. [CrossRef] [PubMed]
23. Pelling, A.E.; Sehati, S.; Gralla, E.B.; Gimzewski, J.K. Time dependence of the frequency and amplitude of the local nanomechanical motion of yeast. *Nanomedicine* **2005**, *1*, 178–183. [CrossRef] [PubMed]
24. Voychuk, S.I.; Gromozova, E.N.; Lytvyn, P.M.; Podgorsky, V.S. Changes of surface properties of yeast cell wall under exposure of electromagnetic field (40.68 MHz) and action of nystatin. *Environmentalist* **2005**, *25*, 139–144. [CrossRef]
25. Stephens, D.J.; Allan, V.J. Light microscopy techniques for live cell imaging. *Science* **2003**, *300*, 82–86. [CrossRef] [PubMed]
26. Roy, P.; Rajfur, Z.; Pomorski, P.; Jacobson, K. Microscope-based techniques to study cell adhesion and migration. *Nat. Cell Biol.* **2002**, *4*, E91–E96. [CrossRef] [PubMed]

27. Gitai, Z. New fluorescence microscopy methods for microbiology: Sharper, faster, and quantitative. *Curr. Opin. Microbiol.* **2009**, *12*, 341–346. [CrossRef] [PubMed]
28. Fricker, M.; Runions, J.; Moore, I. Quantitative fluorescence microscopy: From art to science. *Annu. Rev. Plant Biol.* **2006**, *57*, 79–107. [CrossRef] [PubMed]
29. Lippincott-Schwartz, J.; Snapp, E.; Kenworthy, A. Studying protein dynamics in living cells. *Nat. Rev. Mol. Cell Biol.* **2001**, *2*, 444–456. [CrossRef] [PubMed]
30. Jares-Erijman, E.A.; Jovin, T.M. FRET imaging. *Nat. Biotechnol.* **2003**, *21*, 1387–1395. [CrossRef] [PubMed]
31. Roy, R.; Hohng, S.; Ha, T. A practical guide to single-molecule FRET. *Nat. Methods* **2008**, *5*, 507–516. [CrossRef] [PubMed]
32. Strutt, J.W. On the manufacture and theory of diffraction-gratings. *Philos. Mag.* **1874**, *47*, 193–205.
33. Thompson, R.E.; Larson, D.R.; Webb, W.W. Precise nanometer localization analysis for individual fluorescent probes. *Biophys. J.* **2002**, *82*, 2775–2783. [CrossRef]
34. Sako, Y. Imaging single molecules in living cells for systems biology. *Mol. Syst. Biol.* **2006**, *2*, 56. [CrossRef] [PubMed]
35. Walter, N.G.; Huang, C.Y.; Manzo, A.J.; Sobhy, M.A. Do-it-yourself guide: How to use the modern single-molecule toolkit. *Nat. Methods* **2008**, *5*, 475–489. [CrossRef] [PubMed]
36. Dehmelt, L.; Bastiaens, P.I. Spatial organization of intracellular communication: Insights from imaging. *Nat. Rev. Mol. Cell Biol.* **2010**, *11*, 440–452. [CrossRef] [PubMed]
37. Harriss, L.M.; Wallace, M.I. Single molecule fluorescence in membrane biology. In *Single Molecule Biology*; Knight, A.E., Ed.; Academic Press: San Diego, CA, USA, 2009; pp. 253–288.
38. Müller-Taubenberger, A.; Anderson, K.I. Recent advances using green and red fluorescent protein variants. *Appl. Microbiol. Biotechnol.* **2007**, *77*, 1–12. [CrossRef] [PubMed]
39. Ando, R.; Hama, H.; Yamamoto-Hino, M.; Mizuno, H.; Miyawaki, A. An optical marker based on the UV-induced green-to-red photoconversion of a fluorescent protein. *Proc. Natl. Acad. Sci. USA* **2002**, *99*, 12651–12656. [CrossRef] [PubMed]
40. Patterson, G.H.; Lippincott-Schwartz, J. A photoactivatable GFP for selective photolabeling of proteins and cells. *Science* **2002**, *297*, 1873–1877. [CrossRef] [PubMed]
41. Chudakov, D.M.; Belousov, V.V.; Zaraisky, A.G.; Novoselov, V.V.; Staroverov, D.B.; Zorov, D.B.; Lukyanov, S.; Lukyanov, K.A. Kindling fluorescent proteins for precise in vivo photolabeling. *Nat. Biotechnol.* **2003**, *21*, 191–194. [CrossRef] [PubMed]
42. Lippincott-Schwartz, J.; Patterson, G.H. Development and use of fluorescent protein markers in living cells. *Science* **2003**, *300*, 87–91. [CrossRef] [PubMed]
43. Chudakov, D.M.; Verkhusha, V.V.; Staroverov, D.B.; Souslova, E.A.; Lukyanov, S.; Lukyanov, K.A. Photoswitchable cyan fluorescent protein for protein tracking. *Nat. Biotechnol.* **2004**, *22*, 1435–1439. [CrossRef] [PubMed]
44. Wiedenmann, J.; Ivanchenko, S.; Oswald, F.; Schmitt, F.; Röcker, C.; Salih, A.; Spindler, K.D.; Nienhaus, G.U. EosFP, a fluorescent marker protein with UV-inducible green-to-red fluorescence conversion. *Proc. Natl. Acad. Sci. USA* **2004**, *101*, 15905–15910. [CrossRef] [PubMed]
45. Gurskaya, N.G.; Verkhusha, V.V.; Shcheglov, A.S.; Staroverov, D.B.; Chepurnykh, T.V.; Fradkov, A.F.; Lukyanov, S.; Lukyanov, K.A. Engineering of a monomeric green-to-red photoactivatable fluorescent protein induced by blue light. *Nat. Biotechnol.* **2006**, *24*, 461–465. [CrossRef] [PubMed]
46. Wiedenmann, J.; Nienhaus, G.U. Live-cell imaging with EosFP and other photoactivatable marker proteins of the GFP family. *Expert Rev. Proteom.* **2006**, *3*, 361–374. [CrossRef] [PubMed]
47. Lippincott-Schwartz, J.; Patterson, G.H. Fluorescent proteins for photoactivation experiments. *Methods Cell Biol.* **2008**, *85*, 45–61. [PubMed]
48. Patterson, G.H.; Lippincott-Schwartz, J. Selective photolabeling of proteins using photoactivatable GFP. *Methods* **2004**, *32*, 445–450. [CrossRef] [PubMed]
49. Habuchi, S.; Ando, R.; Dedecker, P.; Verheijen, W.; Mizuno, H.; Miyawaki, A.; Hofkens, J. Reversible single-molecule photoswitching in the GFP-like fluorescent protein Dronpa. *Proc. Natl. Acad. Sci. USA* **2005**, *102*, 9511–9516. [CrossRef] [PubMed]
50. Verkhusha, V.V.; Sorkin, A. Conversion of the monomeric red fluorescent protein into a photoactivatable probe. *Chem. Biol.* **2005**, *12*, 279–285. [CrossRef] [PubMed]

51. Vogt, A.; D'Angelo, C.; Oswald, F.; Denzel, A.; Mazel, C.H.; Matz, M.V.; Ivanchenko, S.; Nienhaus, G.U.; Wiedenmann, J. A green fluorescent protein with photoswitchable emission from the deep sea. *PLoS ONE* **2008**, *3*, e3766. [CrossRef] [PubMed]
52. Bourgeois, D.; Adam, V. Reversible photoswitching in fluorescent proteins: A mechanistic view. *IUBMB Life* **2012**, *64*, 482–491. [CrossRef] [PubMed]
53. Zhou, X.X.; Lin, M.Z. Photoswitchable fluorescent proteins: Ten years of colorful chemistry and exciting applications. *Curr. Opin. Chem. Biol.* **2013**, *17*, 682–690. [CrossRef] [PubMed]
54. Duan, C.; Adam, V.; Byrdin, M.; Bourgeois, D. Structural basis of photoswitching in fluorescent proteins. *Methods Mol. Biol.* **2014**, *1148*, 177–202. [PubMed]
55. Nienhaus, G.U.; Nienhaus, K.; Hölzle, A.; Ivanchenko, S.; Renzi, F.; Oswald, F.; Wolff, M.; Schmitt, F.; Röcker, C.; Vallone, B.; et al. Photoconvertible fluorescent protein EosFP: Biophysical properties and cell biology applications. *Photochem. Photobiol.* **2006**, *82*, 351–358. [CrossRef] [PubMed]
56. Adam, V.; Nienhaus, K.; Bourgeois, D.; Nienhaus, G.U. Structural basis of enhanced photoconversion yield in green fluorescent protein-like protein Dendra2. *Biochemistry* **2009**, *48*, 4905–4915. [CrossRef] [PubMed]
57. Lukyanov, K.A.; Chudakov, D.M.; Lukyanov, S.; Verkhusha, V.V. Innovation: Photoactivatable fluorescent proteins. *Nat. Rev. Mol. Cell Biol.* **2005**, *6*, 885–891. [CrossRef] [PubMed]
58. Zhang, L.; Gurskaya, N.G.; Merzlyak, E.M.; Staroverov, D.B.; Mudrik, N.N.; Samarkina, O.N.; Vinokurov, L.M.; Lukyanov, S.; Lukyanov, K.A. Method for real-time monitoring of protein degradation at the single cell level. *Biotechniques* **2007**, *42*, 446, 448, 450. [CrossRef] [PubMed]
59. Patterson, G.H. Photoactivation and imaging of optical highlighter fluorescent proteins. *Curr. Protoc. Cytom.* **2011**. [CrossRef]
60. Manley, S.; Gillette, J.M.; Patterson, G.H.; Shroff, H.; Hess, H.F.; Betzig, E.; Lippincott-Schwartz, J. High-density mapping of single-molecule trajectories with photoactivated localization microscopy. *Nat. Methods* **2008**, *5*, 155–157. [CrossRef] [PubMed]
61. Gould, T.J.; Verkhusha, V.V.; Hess, S.T. Imaging biological structures with fluorescence photoactivation localization microscopy. *Nat. Protoc.* **2009**, *4*, 291–308. [CrossRef] [PubMed]
62. Greenfield, D.; McEvoy, A.L.; Shroff, H.; Crooks, G.E.; Wingreen, N.S.; Betzig, E.; Liphardt, J. Self-organization of the *Escherichia coli* chemotaxis network imaged with super-resolution light microscopy. *PLoS Biol.* **2009**, *7*, e1000137. [CrossRef] [PubMed]
63. Betzig, E.; Patterson, G.H.; Sougrat, R.; Lindwasser, O.W.; Olenych, S.; Bonifacino, J.S.; Davidson, M.W.; Lippincott-Schwartz, J.; Hess, H.F. Imaging intracellular fluorescent proteins at nanometer resolution. *Science* **2006**, *313*, 1642–1645. [CrossRef] [PubMed]
64. Rust, M.J.; Bates, M.; Zhuang, X. Sub-diffraction-limit imaging by stochastic optical reconstruction microscopy (STORM). *Nat. Methods* **2006**, *3*, 793–795. [CrossRef] [PubMed]
65. Flors, C.; Hotta, J.; Uji-I, H.; Dedecker, P.; Ando, R.; Mizuno, H.; Miyawaki, A.; Hofkens, J. A stroboscopic approach for fast photoactivation-localization microscopy with Dronpa mutants. *J. Am. Chem. Soc.* **2007**, *129*, 13970–13977. [CrossRef] [PubMed]
66. Hell, S.W. Far-field optical nanoscopy. *Science* **2007**, *316*, 1153–1158. [CrossRef] [PubMed]
67. Stiel, A.C.; Andresen, M.; Bock, H.; Hilbert, M.; Schilde, J.; Schönle, A.; Eggeling, C.; Egner, A.; Hell, S.W.; Jakobs, S. Generation of monomeric reversibly switchable red fluorescent proteins for far-field fluorescence nanoscopy. *Biophys. J.* **2008**, *95*, 2989–2997. [CrossRef] [PubMed]
68. Deschout, H.; Shivanandan, A.; Annibale, P.; Scarselli, M.; Radenovic, A. Progress in quantitative single-molecule localization microscopy. *Histochem. Cell Biol.* **2014**, *142*, 5–17. [CrossRef] [PubMed]
69. Huang, B.; Wang, W.; Bates, M.; Zhuang, X. Three-dimensional super-resolution imaging by stochastic optical reconstruction microscopy. *Science* **2008**, *319*, 810–813. [CrossRef] [PubMed]
70. Juette, M.F.; Gould, T.J.; Lessard, M.D.; Mlodzianoski, M.J.; Nagpure, B.S.; Bennett, B.T.; Hess, S.T.; Bewersdorf, J. Three-dimensional sub-100 nm resolution fluorescence microscopy of thick samples. *Nat. Methods* **2008**, *5*, 527–529. [CrossRef] [PubMed]
71. Jensen, E.; Crossman, D.J. Technical review: Types of imaging-direct STORM. *Anat. Rec. (Hoboken)* **2014**, *297*, 2227–2231. [CrossRef] [PubMed]
72. Tam, J.; Merino, D. Stochastic optical reconstruction microscopy (STORM) in comparison with stimulated emission depletion (STED) and other imaging methods. *J. Neurochem.* **2015**, *135*, 643–658. [CrossRef] [PubMed]

73. Bock, H.; Geisler, C.; Wurm, C.A.; von Middendorf, C.; Jakobs, S.; Schönle, A.; Egner, A.; Hell, S.W.; Eggeling, C. Two-color far-field fluorescence nanoscopy based on photoswitchable emitters. *Appl. Phys. B* **2007**, *88*, 161–165. [CrossRef]
74. Egner, A.; Geisler, C.; von Middendorff, C.; Bock, H.; Wenzel, D.; Medda, R.; Andresen, M.; Stiel, A.C.; Jakobs, S.; Eggeling, C.; et al. Fluorescence nanoscopy in whole cells by asynchronous localization of photoswitching emitters. *Biophys. J.* **2007**, *93*, 3285–3290. [CrossRef] [PubMed]
75. Hess, S.T.; Girirajan, T.P.; Mason, M.D. Ultra-high resolution imaging by fluorescence photoactivation localization microscopy. *Biophys. J.* **2006**, *91*, 4258–4272. [CrossRef] [PubMed]
76. Shroff, H.; Galbraith, C.G.; Galbraith, J.A.; White, H.; Gillette, J.; Olenych, S.; Davidson, M.W.; Betzig, E. Dual-color superresolution imaging of genetically expressed probes within individual adhesion complexes. *Proc. Natl. Acad. Sci. USA* **2007**, *104*, 20308–20313. [CrossRef] [PubMed]
77. Klar, T.A.; Jakobs, S.; Dyba, M.; Egner, A.; Hell, S.W. Fluorescence microscopy with diffraction resolution barrier broken by stimulated emission. *Proc. Natl. Acad. Sci. USA* **2000**, *97*, 8206–8210. [CrossRef] [PubMed]
78. Donnert, G.; Keller, J.; Medda, R.; Andrei, M.A.; Rizzoli, S.O.; Lührmann, R.; Jahn, R.; Eggeling, C.; Hell, S.W. Macromolecular-scale resolution in biological fluorescence microscopy. *Proc. Natl. Acad. Sci. USA* **2006**, *103*, 11440–11445. [CrossRef] [PubMed]
79. Willig, K.I.; Rizzoli, S.O.; Westphal, V.; Jahn, R.; Hell, S.W. STED microscopy reveals that synaptotagmin remains clustered after synaptic vesicle exocytosis. *Nature* **2006**, *440*, 935–939. [CrossRef] [PubMed]
80. Westphal, V.; Rizzoli, S.O.; Lauterbach, M.A.; Kamin, D.; Jahn, R.; Hell, S.W. Video-rate far-field optical nanoscopy dissects synaptic vesicle movement. *Science* **2008**, *320*, 246–249. [CrossRef] [PubMed]
81. De Jonge, N.; Peckys, D.B.; Kremers, G.J.; Piston, D.W. Electron microscopy of whole cells in liquid with nanometer resolution. *Proc. Natl. Acad. Sci. USA* **2009**, *106*, 2159–2164. [CrossRef] [PubMed]
82. Peckys, D.B.; de Jonge, N. Liquid scanning transmission electron microscopy: Imaging protein complexes in their native environment in whole eukaryotic cells. *Microsc. Microanal.* **2014**, *20*, 346–365. [CrossRef] [PubMed]
83. Baumeister, W.; Grimm, R.; Walz, J. Electron tomography of molecules and cells. *Trends Cell Biol.* **1999**, *9*, 81–85. [CrossRef]
84. Downing, K.H.; Sui, H.; Auer, M. Electron tomography: A 3D view of the subcellular world. *Anal. Chem.* **2007**, *79*, 7949–7957. [CrossRef] [PubMed]
85. Stahlberg, H.; Walz, T. Molecular electron microscopy: State of the art and current challenges. *ACS Chem. Biol.* **2008**, *3*, 268–281. [CrossRef] [PubMed]
86. Lucić, V.; Leis, A.; Baumeister, W. Cryo-electron tomography of cells: Connecting structure and function. *Histochem. Cell Biol.* **2008**, *130*, 185–196. [CrossRef] [PubMed]
87. Diebolder, C.A.; Koster, A.J.; Koning, R.I. Pushing the resolution limits in cryo electron tomography of biological structures. *J. Microsc.* **2012**, *248*, 1–5. [CrossRef] [PubMed]
88. Agar, H.D.; Douglas, H.C. Studies on the cytological structure of yeast: Electron microscopy of thin sections. *J. Bacteriol.* **1957**, *73*, 365–375. [PubMed]
89. Matile, P.; Moor, H.; Robinow, C.F. Yeast cytology. In *The Yeasts*; Rose, A.H., Harrison, J.S., Eds.; Avademic Press: New York, NY, USA, 1969; pp. 219–302.
90. Osumi, M. Visualization of yeast cells by electron microscopy. *J. Electron. Microsc. (Tokyo)* **2012**, *61*, 343–365. [CrossRef] [PubMed]
91. Osumi, M.; Sando, N. Division of yeast mitochondria in synchronous culture. *J. Electron. Microsc. (Tokyo)* **1969**, *18*, 47–56. [PubMed]
92. Byers, B.; Goetsch, L. Duplication of spindle plaques and integration of the yeast cell cycle. *Cold Spring Harb. Symp. Quant. Biol.* **1974**, *38*, 123–131. [CrossRef] [PubMed]
93. Byers, B.; Goetsch, L. Behavior of spindles and spindle plaques in the cell cycle and conjugation of *Saccharomyces cerevisiae*. *J. Bacteriol.* **1975**, *124*, 511–523. [PubMed]
94. Osumi, M.; Imaizumi, F.; Imai, M.; Sato, H.; Yamaguchi, H. Isolation and characterization of microbodies from *candida tropicalis* pk 233 cells grown on normal alkanes. *J. Gen. Appl. Microbiol.* **1975**, *21*, 375–387. [CrossRef]
95. Novick, P.; Field, C.; Schekman, R. Identification of 23 complementation groups required for post-translational events in the yeast secretory pathway. *Cell* **1980**, *21*, 205–215. [CrossRef]
96. Winey, M.; Goetsch, L.; Baum, P.; Byers, B. MPS1 and MPS2: Novel yeast genes defining distinct steps of spindle pole body duplication. *J. Cell Biol.* **1991**, *114*, 745–754. [CrossRef] [PubMed]

97. Baba, M.; Osumi, M. Transmission and scanning electron microscopic examination of intracellular organelles in freeze-substituted *Kloeckera* and *Saccharomyces cerevisiae* yeast cells. *J. Electron. Microsc. Tech.* **1987**, *5*, 249–261. [CrossRef]
98. Giddings, T.H., Jr.; O'Toole, E.T.; Morphew, M.; Mastronarde, D.N.; McIntosh, J.R.; Winey, M. Using rapid freeze and freeze-substitution for the preparation of yeast cells for electron microscopy and three-dimensional analysis. *Methods Cell Biol.* **2001**, *67*, 27–42. [PubMed]
99. McDonald, K. Cryopreparation methods for electron microscopy of selected model systems. *Methods Cell Biol.* **2007**, *79*, 23–56. [PubMed]
100. O'Toole, E.T.; Giddings, T.H., Jr.; Winey, M. Building cell structures in three dimensions: Electron tomography methods for budding yeast. In *Yeast Protocols*; CSHL Press: Cold Spring Harbor, NY, USA, 2016; pp. 303–312.
101. Hoenger, A.; McIntosh, J.R. Probing the macromolecular organization of cells by electron tomography. *Curr. Opin. Cell Biol.* **2009**, *21*, 89–96. [CrossRef] [PubMed]
102. Arfsten, J.; Leupold, S.; Bradtmöller, C.; Kampen, I.; Kwade, A. Atomic force microscopy studies on the nanomechanical properties of *Saccharomyces cerevisiae*. *Colloids Surf. B Biointerfaces* **2010**, *79*, 284–290. [CrossRef] [PubMed]
103. Alsteens, D.; Beaussart, A.; Derclaye, S.; El-Kirat-Chatel, S.; Park, H.R.; Lipke, P.N.; Dufrêne, Y.F. Single-cell force spectroscopy of Als-mediated fungal adhesion. *Anal. Methods* **2013**, *5*, 3657–3662. [CrossRef] [PubMed]
104. Portillo, A.M.; Krasnoslobodtsev, A.V.; Lyubchenko, Y.L. Effect of electrostatics on aggregation of prion protein Sup35 peptide. *J. Phys. Condens. Matter* **2012**, *24*, 164205. [CrossRef] [PubMed]
105. Portillo, A.; Hashemi, M.; Zhang, Y.; Breydo, L.; Uversky, V.N.; Lyubchenko, Y.L. Role of monomer arrangement in the amyloid self-assembly. *Biochim. Biophys. Acta* **2015**, *1854*, 218–228. [CrossRef] [PubMed]
106. Rangl, M.; Ebner, A.; Yamada, J.; Rankl, C.; Tampé, R.; Gruber, H.J.; Rexach, M.; Hinterdorfer, P. Single-molecule analysis of the recognition forces underlying nucleo-cytoplasmic transport. *Angew. Chem. Int. Ed. Engl.* **2013**, *52*, 10356–10359. [CrossRef] [PubMed]
107. Goossens, K.V.Y.; Ielasi, F.S.; Nookaew, I.; Stals, I.; Alonso-Sarduy, L.; Daenen, L.; van Mulders, S.E.; Stassen, C.; van Eijsden, R.G.E.; Siewers, V.; et al. Molecular mechanism of flocculation self-recognition in yeast and its role in mating and survival. *mBio* **2015**, *6*, e00427–e00415. [CrossRef] [PubMed]
108. El-Kirat-Chatel, S.; Beaussart, A.; Alsteens, D.; Sarazin, A.; Jouault, T.; Dufrêne, Y.F. Single-molecule analysis of the major glycopolymers of pathogenic and non-pathogenic yeast cells. *Nanoscale* **2013**, *5*, 4855–4863. [CrossRef] [PubMed]
109. Hwang, G.; Marsh, G.; Gao, L.; Waugh, R.; Koo, H. Binding force dynamics of *Streptococcus mutans*-glucosyltransferase B to *Candida albicans*. *J. Dent. Res.* **2015**, *94*, 1310–1317. [CrossRef] [PubMed]
110. El-Kirat-Chatel, S.; Beaussart, A.; Derclaye, S.; Alsteens, D.; Kucharíková, S.; van Dijck, P.; Dufrêne, Y.F. Force nanoscopy of hydrophobic interactions in the fungal pathogen *Candida glabrata*. *ACS Nano* **2015**, *9*, 1648–1655. [CrossRef] [PubMed]
111. Touhami, A.; Hoffmann, B.; Vasella, A.; Denis, F.A.; Dufrêne, Y.F. Aggregation of yeast cells: Direct measurement of discrete lectin-carbohydrate interactions. *Microbiology* **2003**, *149*, 2873–2878. [CrossRef] [PubMed]
112. Dupres, V.; Alsteens, D.; Wilk, S.; Hansen, B.; Heinisch, J.J.; Dufrêne, Y.F. The yeast Wsc1 cell surface sensor behaves like a nanospring in vivo. *Nat. Chem. Biol.* **2009**, *5*, 857–862. [CrossRef] [PubMed]
113. Dupres, V.; Heinisch, J.J.; Dufrêne, Y.F. Atomic force microscopy demonstrates that disulphide bridges are required for clustering of the yeast cell wall integrity sensor Wsc1. *Langmuir* **2011**, *27*, 15129–15134. [CrossRef] [PubMed]
114. Heinisch, J.J.; Dupres, V.; Wilk, S.; Jendretzki, A.; Dufrêne, Y.F. Single-molecule atomic force microscopy reveals clustering of the yeast plasma-membrane sensor Wsc1. *PLoS ONE* **2010**, *5*, e11104. [CrossRef] [PubMed]
115. Formosa, C.; Lachaize, V.; Galés, C.; Rols, M.P.; Martin-Yken, H.; François, J.M.; Duval, R.E.; Dague, E. Mapping HA-tagged protein at the surface of living cells by atomic force microscopy. *J. Mol. Recognit.* **2015**, *28*, 1–9. [CrossRef] [PubMed]
116. Takenaka, M.; Miyachi, Y.; Ishii, J.; Ogino, C.; Kondo, A. The mapping of yeast's G-protein coupled receptor with an atomic force microscope. *Nanoscale* **2015**, *7*, 4956–4963. [CrossRef] [PubMed]
117. Friedrichs, J.; Helenius, J.; Muller, D.J. Quantifying cellular adhesion to extracellular matrix components by single-cell force spectroscopy. *Nat. Protoc.* **2010**, *5*, 1353–1361. [CrossRef] [PubMed]

118. Te Riet, J.; Reinieren-Beeren, I.; Figdor, C.G.; Cambi, A. AFM force spectroscopy reveals how subtle structural differences affect the interaction strength between *Candida albicans* and DC-SIGN. *J. Mol. Recognit.* **2015**, *28*, 687–698. [CrossRef] [PubMed]
119. Benoit, M.; Gabriel, D.; Gerisch, G.; Gaub, H.E. Discrete interactions in cell adhesion measured by single-molecule force spectroscopy. *Nat. Cell Biol.* **2000**, *2*, 313–317. [CrossRef] [PubMed]
120. Helenius, J.; Heisenberg, C.P.; Gaub, H.E.; Muller, D.J. Single-cell force spectroscopy. *J. Cell Sci.* **2008**, *121*, 1785–1791. [CrossRef] [PubMed]
121. Friedrichs, J.; Legate, K.R.; Schubert, R.; Bharadwaj, M.; Werner, C.; Müller, D.J.; Benoit, M. A practical guide to quantify cell adhesion using single-cell force spectroscopy. *Methods* **2013**, *60*, 169–178. [CrossRef] [PubMed]
122. Taubenberger, A.; Cisneros, D.A.; Friedrichs, J.; Puech, P.H.; Muller, D.J.; Franz, C.M. Revealing early steps of alpha2beta1 integrin-mediated adhesion to collagen type I by using single-cell force spectroscopy. *Mol. Biol. Cell* **2007**, *18*, 1634–1644. [CrossRef] [PubMed]
123. Peters, B.M.; Ovchinnikova, E.S.; Krom, B.P.; Schlecht, L.M.; Zhou, H.; Hoyer, L.L.; Busscher, H.J.; van der Mei, H.C.; Jabra-Rizk, M.A.; Shirtliff, M.E. *Staphylococcus aureus* adherence to *Candida albicans* hyphae is mediated by the hyphal adhesin Als3p. *Microbiology* **2012**, *158*, 2975–2986. [CrossRef] [PubMed]
124. Potthoff, E.; Guillaume-Gentil, O.; Ossola, D.; Polesel-Maris, J.; LeibundGut-Landmann, S.; Zambelli, T.; Vorholt, J.A. Rapid and serial quantification of adhesion forces of yeast and Mammalian cells. *PLoS ONE* **2012**, *7*, e52712. [CrossRef] [PubMed]
125. Alsteens, D.; van Dijck, P.; Lipke, P.N.; Dufrêne, Y.F. Quantifying the forces driving cell-cell adhesion in a fungal pathogen. *Langmuir* **2013**, *29*, 13473–13480. [CrossRef] [PubMed]
126. Bowen, W.R.; Lovitt, R.W.; Wright, C.J. Atomic Force Microscopy study of the adhesion of *Saccharomyces cerevisiae*. *J. Colloid Interface Sci.* **2001**, *237*, 54–61. [CrossRef] [PubMed]
127. Götzinger, M.; Weigl, B.; Peukert, W.; Sommer, K. Effect of roughness on particle adhesion in aqueous solutions: A study of *Saccharomyces cerevisiae* and a silica particle. *Colloids Surf. B Biointerfaces* **2007**, *55*, 44–50. [CrossRef] [PubMed]
128. Hachułka, K.; Lekka, M.; Okrajni, J.; Ambroziak, W.; Wandelt, B. Polymeric sensing system molecularly imprinted towards enhanced adhesion of *Saccharomyces cerevisiae*. *Biosens. Bioelectron.* **2010**, *26*, 50–54. [CrossRef] [PubMed]
129. Martinez, V.; Behr, P.; Drechsler, U.; Polesel-Maris, J.; Potthoff, E.; Vörös, J.; Zambelli, T. SU-8 hollow cantilevers for AFM cell adhesion studies. *J. Micromech. Microeng.* **2016**, *26*, 055006. [CrossRef]
130. Hansen, K.M.; Thundat, T. Microcantilever biosensors. *Methods* **2005**, *37*, 57–64. [CrossRef] [PubMed]
131. Fritz, J. Cantilever biosensors. *Analyst* **2008**, *133*, 855–863. [CrossRef] [PubMed]
132. Braun, T.; Ghatkesar, M.K.; Backmann, N.; Grange, W.; Boulanger, P.; Letellier, L.; Lang, H.P.; Bietsch, A.; Gerber, C.; Hegner, M. Quantitative time-resolved measurement of membrane protein-ligand interactions using microcantilever array sensors. *Nat. Nanotechnol.* **2009**, *4*, 179–185. [CrossRef] [PubMed]
133. Ndieyira, J.W.; Watari, M.; Barrera, A.D.; Zhou, D.; Vögtli, M.; Batchelor, M.; Cooper, M.A.; Strunz, T.; Horton, M.A.; Abell, C.; et al. Nanomechanical detection of antibiotic-mucopeptide binding in a model for superbug drug resistance. *Nat. Nanotechnol.* **2008**, *3*, 691–696. [CrossRef] [PubMed]
134. Godin, M.; Delgado, F.F.; Son, S.; Grover, W.H.; Bryan, A.K.; Tzur, A.; Jorgensen, P.; Payer, K.; Grossman, A.D.; Kirschner, M.W.; et al. Using buoyant mass to measure the growth of single cells. *Nat. Methods* **2010**, *7*, 387–390. [CrossRef] [PubMed]
135. Lang, H.P.; Baller, M.K.; Berger, R.; Gerber, C.; Gimzewski, J.K.; Battiston, F.M.; Fornaro, P.; Ramseyer, J.P.; Meyer, E.; Guntherodt, H.J. An artificial nose based on a micromechancial cantilever array. *Anal. Chim. Acta* **1999**, *393*, 59–65. [CrossRef]
136. Braun, T.; Barwich, V.; Ghatkesar, M.K.; Bredekamp, A.H.; Gerber, C.; Hegner, M.; Lang, H.P. Micromechanical mass sensors for biomolecular detection in a physiological environment. *Phys. Rev. E Stat. Nonlinear Soft Matter Phys.* **2005**, *72*, 031907. [CrossRef] [PubMed]
137. Hosaka, S.; Chiyoma, T.; Ikeuchi, A.; Okano, H.; Sone, H.; Izumi, T. Possibility of a femtogram mass biosensor using a self-sensing cantilever. *Curr. Appl. Phys.* **2006**, *6*, 384–388. [CrossRef]
138. Liu, Y.; Schweizerb, L.M.; Wanga, W.; Reubena, R.L.; Schweizer, M.; Shu, W. Label-free and real-time monitoring of yeast cell growth by the bending of polymer microcantilever biosensors. *Sens. Actuator B Chem.* **2013**, *178*, 621–626. [CrossRef]

139. Bryan, A.K.; Goranov, A.; Amon, A.; Manalis, S.R. Measurement of mass, density, and volume during the cell cycle of yeast. *Proc. Natl. Acad. Sci. USA* **2010**, *107*, 999–1004. [CrossRef] [PubMed]
140. Godin, M.; Tabard-Cossa, V.; Miyahara, Y.; Monga, T.; Williams, P.J.; Beaulieu, L.Y.; Bruce Lennox, R.; Grutter, P. Cantilever-based sensing: The origin of surface stress and optimization strategies. *Nanotechnology* **2010**, *21*, 75501. [CrossRef] [PubMed]
141. Burg, T.P.; Godin, M.; Knudsen, S.M.; Shen, W.; Carlson, G.; Foster, J.S.; Babcock, K.; Manalis, S.R. Weighing of biomolecules, single cells and single nanoparticles in fluid. *Nature* **2007**, *446*, 1066–1069. [CrossRef] [PubMed]
142. Park, K.; Jang, J.; Irimia, D.; Sturgis, J.; Lee, J.; Robinson, J.P.; Toner, M.; Bashir, R. 'Living cantilever arrays' for characterization of mass of single live cells in fluids. *Lab Chip* **2008**, *8*, 1034–1041. [CrossRef] [PubMed]
143. Bryan, A.K.; Hecht, V.C.; Shen, W.; Payer, K.; Grover, W.H.; Manalis, S.R. Measuring single cell mass, volume, and density with dual suspended microchannel resonators. *Lab Chip* **2014**, *14*, 569–576. [CrossRef] [PubMed]
144. Nugaeva, N.; Gfeller, K.Y.; Backmann, N.; Lang, H.P.; Düggelin, M.; Hegner, M. Micromechanical cantilever array sensors for selective fungal immobilization and fast growth detection. *Biosens. Bioelectron.* **2005**, *21*, 849–856. [CrossRef] [PubMed]
145. Aghayee, S.; Benadiba, C.; Notz, J.; Kasas, S.; Dietler, G.; Longo, G. Combination of fluorescence microscopy and nanomotion detection to characterize bacteria. *J. Mol. Recognit.* **2013**, *26*, 590–595. [CrossRef] [PubMed]
146. Longo, G.; Alonso-Sarduy, L.; Rio, L.M.; Bizzini, A.; Trampuz, A.; Notz, J.; Dietler, G.; Kasas, S. Rapid detection of bacterial resistance to antibiotics using AFM cantilevers as nanomechanical sensors. *Nat. Nanotechnol.* **2013**, *8*, 522–526. [CrossRef] [PubMed]
147. Kasas, S.; Ruggeri, F.S.; Benadiba, C.; Maillard, C.; Stupar, P.; Tournu, H.; Dietler, G.; Longo, G. Detecting nanoscale vibrations as signature of life. *Proc. Natl. Acad. Sci. USA* **2015**, *112*, 378–381. [CrossRef] [PubMed]
148. Voldman, J. Engineered systems for the physical manipulation of single cells. *Curr. Opin. Biotechnol.* **2006**, *17*, 532–537. [CrossRef] [PubMed]
149. Voldman, J. Electrical forces for microscale cell manipulation. *Annu. Rev. Biomed. Eng.* **2006**, *8*, 425–454. [CrossRef] [PubMed]
150. Willaert, R.G.; Goossens, K. Microfluidic bioreactors for cellular microarrays. *Fermentation* **2015**, *1*, 38–78. [CrossRef]
151. Guo, L.J. Nanoimprint lithography: Methods and material requirements. *Adv. Mater.* **2007**, *19*, 495–513. [CrossRef]
152. Yap, F.L.; Zhang, Y. Protein and cell micropatterning and its integration with micro/nanoparticles assembly. *Biosens. Bioelectron.* **2007**, *22*, 775–788. [CrossRef] [PubMed]
153. Anselme, K.; Davidson, P.; Popa, A.M.; Giazzon, M.; Liley, M.; Ploux, L. The interaction of cells and bacteria with surfaces structured at the nanometre scale. *Acta Biomater.* **2010**, *6*, 3824–3846. [CrossRef] [PubMed]
154. Qin, D.; Xia, Y.; Whitesides, G.M. Soft lithography for micro- and nanoscale patterning. *Nat. Protoc.* **2010**, *5*, 491–502. [CrossRef] [PubMed]
155. Ekerdt, B.L.; Segalman, R.A.; Schaffer, D.V. Spatial organization of cell-adhesive ligands for advanced cell culture. *Biotechnol. J.* **2013**, *8*, 1411–1423. [CrossRef] [PubMed]
156. Singh, A.V.; Patil, R.; Thombre, D.K.; Gade, W.N. Micro-nanopatterning as tool to study the role of physicochemical properties on cell-surface interactions. *J. Biomed. Mater. Res. A* **2013**, *101*, 3019–3032. [CrossRef] [PubMed]
157. Fung, T.H.; Ball, G.I.; McQuaide, S.C.; Chao, S.; Colman-Lerner, A.; Holl, M.R.; Meldrum, D.R. Microprinting of on-chip cultures: Patterning of yeast cell microarrays using concanavalin-A adhesion. In Proceedings of the IMECE04 ASME International Mechanical Engineering Congress, Anaheim, CA, USA, 13–19 November 2004; pp. 373–374.
158. Cookson, S.; Ostroff, N.; Pang, W.L.; Volfson, D.; Hasty, J. Monitoring dynamics of single-cell gene expression over multiple cell cycles. *Mol. Syst. Biol.* **2005**, *1*, 0024. [CrossRef] [PubMed]
159. Narayanaswamy, R.; Niu, W.; Scouras, A.D.; Hart, G.T.; Davies, J.; Ellington, A.D.; Iyer, V.R.; Marcotte, E.M. Systematic profiling of cellular phenotypes with spotted cell microarrays reveals mating-pheromone response genes. *Genome Biol.* **2006**, *7*, R6. [CrossRef] [PubMed]
160. Ryley, J.; Pereira-Smith, O.M. Microfluidics device for single cell gene expression analysis in *Saccharomyces cerevisiae*. *Yeast* **2006**, *23*, 1065–1073. [CrossRef] [PubMed]

161. Narayanaswamy, R.; Moradi, E.K.; Niu, W.; Hart, G.T.; Davis, M.; McGary, K.L.; Ellington, A.D.; Marcotte, E.M. Systematic definition of protein constituents along the major polarization axis reveals an adaptive reuse of the polarization machinery in pheromone-treated budding yeast. *J. Proteom. Res.* **2009**, *8*, 6–19. [CrossRef] [PubMed]
162. Falconnet, D.; Niemistö, A.; Taylor, R.J.; Ricicova, M.; Galitski, T.; Shmulevich, I.; Hansen, C.L. High-throughput tracking of single yeast cells in a microfluidic imaging matrix. *Lab Chip* **2011**, *11*, 466–473. [CrossRef] [PubMed]
163. Kuhn, P.; Eyer, K.; Robinson, T.; Schmidt, F.I.; Mercer, J.; Dittrich, P.S. A facile protocol for the immobilisation of vesicles, virus particles, bacteria, and yeast cells. *Integr. Biol. (Camb.)* **2012**, *4*, 1550–1555. [CrossRef] [PubMed]
164. Lee, S.S.; Avalos Vizcarra, I.; Huberts, D.H.; Lee, L.P.; Heinemann, M. Whole lifespan microscopic observation of budding yeast aging through a microfluidic dissection platform. *Proc. Natl. Acad. Sci. USA* **2012**, *109*, 4916–4920. [CrossRef] [PubMed]
165. Park, M.C.; Hur, J.Y.; Cho, H.S.; Park, S.H.; Suh, K.Y. High-throughput single-cell quantification using simple microwell-based cell docking and programmable time-course live-cell imaging. *Lab Chip* **2011**, *11*, 79–86. [CrossRef] [PubMed]
166. Xie, Z.; Zhang, Y.; Zou, K.; Brandman, O.; Luo, C.; Ouyang, Q.; Li, H. Molecular phenotyping of aging in single yeast cells using a novel microfluidic device. *Aging Cell* **2012**, *11*, 599–606. [CrossRef] [PubMed]
167. Zhang, Y.; Luo, C.; Zou, K.; Xie, Z.; Brandman, O.; Ouyang, Q.; Li, H. Single cell analysis of yeast replicative aging using a new generation of microfluidic device. *PLoS ONE* **2012**, *7*, e48275. [CrossRef] [PubMed]
168. Fehrmann, S.; Paoletti, C.; Goulev, Y.; Ungureanu, A.; Aguilaniu, H.; Charvin, G. Aging yeast cells undergo a sharp entry into senescence unrelated to the loss of mitochondrial membrane potential. *Cell Rep.* **2013**, *5*, 1589–1599. [CrossRef] [PubMed]
169. Crane, M.M.; Clark, I.B.; Bakker, E.; Smith, S.; Swain, P.S. A microfluidic system for studying ageing and dynamic single-cell responses in budding yeast. *PLoS ONE* **2014**, *9*, e100042. [CrossRef] [PubMed]
170. Osada, K.; Hosokawa, M.; Yoshino, T.; Tanaka, T. Monitoring of cellular behaviors by microcavity array-based single-cell patterning. *Analyst* **2014**, *139*, 425–430. [CrossRef] [PubMed]
171. Liu, P.; Young, T.Z.; Acar, M. Yeast replicator: A high-throughput multiplexed microfluidics platform for automated measurements of single-cell aging. *Cell Rep.* **2015**, *13*, 634–644. [CrossRef] [PubMed]
172. Jo, M.C.; Liu, W.; Gu, L.; Dang, W.; Qin, L. High-throughput analysis of yeast replicative aging using a microfluidic system. *Proc. Natl. Acad. Sci. USA* **2015**, *112*, 9364–9369. [CrossRef] [PubMed]
173. Terenna, C.R.; Makushok, T.; Velve-Casquillas, G.; Baigl, D.; Chen, Y.; Bornens, M.; Paoletti, A.; Piel, M.; Tran, P.T. Physical mechanisms redirecting cell polarity and cell shape in fission yeast. *Curr. Biol.* **2008**, *18*, 1748–1753. [CrossRef] [PubMed]
174. Minc, N.; Boudaoud, A.; Chang, F. Mechanical forces of fission yeast growth. *Curr. Biol.* **2009**, *19*, 1096–1101. Erratum in: *Curr. Biol.* **2014**, *24*, 1436. [CrossRef] [PubMed]
175. Nghe, P.; Boulineau, S.; Gude, S.; Recouvreux, P.; van Zon, J.S.; Tans, S.J. Microfabricated polyacrylamide devices for the controlled culture of growing cells and developing organisms. *PLoS ONE* **2013**, *8*, e75537. [CrossRef] [PubMed]
176. Tian, Y.; Luo, C.; Ouyang, Q. A microfluidic synchronizer for fission yeast cells. *Lab Chip* **2013**, *13*, 4071–4077. [CrossRef] [PubMed]
177. Bell, L.; Seshia, A.; Lando, D.; Laue, E.; Palayret, M.; Lee, S.F.; Klenerman, D. A microfluidic device for the hydrodynamic immobilisation of living fission yeast cells for super-resolution imaging. *Sens. Actuators B Chem.* **2014**, *192*, 36–41. [CrossRef] [PubMed]
178. Nobs, J.B.; Maerkl, S.J. Long-term single cell analysis of *S. pombe* on a microfluidic microchemostat array. *PLoS ONE* **2014**, *9*, e93466. [CrossRef] [PubMed]
179. Spivey, E.C.; Xhemalce, B.; Shear, J.B.; Finkelstein, I.J. 3D-printed microfluidic microdissector for high-throughput studies of cellular aging. *Anal. Chem.* **2014**, *86*, 7406–7412. [CrossRef] [PubMed]
180. Ruiz, S.A.; Chen, C.S. Microcontact printing: A tool to pattern. *Soft Matter* **2007**, *3*, 168–177. [CrossRef]
181. Miermont, A.; Waharte, F.; Hu, S.; McClean, M.N.; Bottani, S.; Léon, S.; Hersen, P. Severe osmotic compression triggers a slowdown of intracellular signaling, which can be explained by molecular crowding. *Proc. Natl. Acad. Sci. USA* **2013**, *110*, 5725–5730. [CrossRef] [PubMed]

182. Théry, M.; Piel, M. Adhesive micropatterns for cells: A microcontact printing protocol. *Cold Spring Harb. Protoc.* **2009**, *7*. [CrossRef] [PubMed]
183. Weaver, W.M.; Tseng, P.; Kunze, A.; Masaeli, M.; Chung, A.J.; Dudani, J.S.; Kittur, H.; Kulkarni, R.P.; di Carlo, D. Advances in high-throughput single-cell microtechnologies. *Curr. Opin. Biotechnol.* **2014**, *25*, 114–123. [CrossRef] [PubMed]
184. Fritzsch, F.S.; Dusny, C.; Frick, O.; Schmid, A. Single-cell analysis in biotechnology, systems biology, and biocatalysis. *Annu. Rev. Chem. Biomol. Eng.* **2012**, *3*, 129–155. [CrossRef] [PubMed]
185. Dusny, C.; Schmid, A. Microfluidic single-cell analysis links boundary environments and individual microbial phenotypes. *Environ. Microbiol.* **2015**, *17*, 1839–1856. [CrossRef] [PubMed]
186. Groisman, A.; Lobo, C.; Cho, H.; Campbell, J.K.; Dufour, Y.S.; Stevens, A.M.; Levchenko, A. A microfluidic chemostat for experiments with bacterial and yeast cells. *Nat. Methods* **2005**, *2*, 685–689. [CrossRef] [PubMed]
187. Long, Z.; Nugent, E.; Javer, A.; Cicuta, P.; Sclavi, B.; Cosentino Lagomarsino, M.; Dorfman, K.D. Microfluidic chemostat for measuring single cell dynamics in bacteria. *Lab Chip* **2013**, *13*, 947–954. [CrossRef] [PubMed]
188. Balaban, N.Q.; Merrin, J.; Chait, R.; Kowalik, L.; Leibler, S. Bacterial persistence as a phenotypic switch. *Science* **2004**, *305*, 1622–1625. [CrossRef] [PubMed]
189. Bennett, M.R.; Pang, W.L.; Ostroff, N.A.; Baumgartner, B.L.; Nayak, S.; Tsimring, L.S.; Hasty, J. Metabolic gene regulation in a dynamically changing environment. *Nature* **2008**, *454*, 1119–1122. [CrossRef] [PubMed]
190. Yarmush, M.L.; King, K.R. Living-cell microarrays. *Annu. Rev. Biomed. Eng.* **2009**, *11*, 235–257. [CrossRef] [PubMed]
191. Nilsson, J.; Evander, M.; Hammarström, B.; Laurell, T. Review of cell and particle trapping in microfluidic systems. *Anal. Chim. Acta* **2009**, *649*, 141–157. [CrossRef] [PubMed]
192. Di Carlo, D.; Wu, L.Y.; Lee, L.P. Dynamic single cell culture array. *Lab Chip* **2006**, *6*, 1445–1449. [CrossRef] [PubMed]
193. Barbulovic-Nad, I.; Lucente, M.; Sun, Y.; Zhang, M.; Wheeler, A.R.; Bussmann, M. Bio-microarray fabrication techniques—A review. *Crit. Rev. Biotechnol.* **2006**, *26*, 237–259. [CrossRef] [PubMed]
194. Guillemot, F.; Souquet, A.; Catros, S.; Guillotin, B.; Lopez, J.; Faucon, M.; Pippenger, B.; Bareille, R.; Rémy, M.; Bellance, S.; et al. High-throughput laser printing of cells and biomaterials for tissue engineering. *Acta Biomater.* **2010**, *6*, 2494–2500. [CrossRef] [PubMed]
195. Bean, G.J.; Jaeger, P.A.; Bahr, S.; Ideker, T. Development of ultra-high-density screening tools for microbial "omic". *PLoS ONE* **2014**, *9*, e85177. [CrossRef] [PubMed]
196. Schaack, B.; Reboud, J.; Combe, S.; Fouqué, B.; Berger, F.; Boccard, S.; Filhol-Cochet, O.; Chatelain, F. A "DropChip" cell array for DNA and siRNA transfection combined with drug screening. *NanoBiotechnology* **2005**, *1*, 183–189. [CrossRef]
197. Ringeisen, B.R.; Othon, C.M.; Barron, J.A.; Young, D.; Spargo, B.J. Jet-based methods to print living cells. *Biotechnol. J.* **2006**, *1*, 930–948. [CrossRef] [PubMed]
198. Roth, E.A.; Xu, T.; Das, M.; Gregory, C.; Hickman, J.J.; Boland, T. Inkjet printing for high-throughput cell patterning. *Biomaterials* **2004**, *25*, 3707–3715. [CrossRef] [PubMed]
199. Ferris, C.J.; Gilmore, K.G.; Wallace, G.G.; In Het Panhuis, M. Biofabrication: An overview of the approaches used for printing of living cells. *Appl. Microbiol. Biotechnol.* **2013**, *97*, 4243–4258. [CrossRef] [PubMed]
200. Gonzalez-Macia, L.; Morrin, A.; Smyth, M.R.; Killard, A.J. Advanced printing and deposition methodologies for the fabrication of biosensors and biodevices. *Analyst* **2010**, *135*, 845–867. [CrossRef] [PubMed]
201. Li, J.; Rossignol, F.; Macdonald, J. Inkjet printing for biosensor fabrication: Combining chemistry and technology for advanced manufacturing. *Lab Chip* **2015**, *15*, 2538–2558. [CrossRef] [PubMed]
202. Meister, A.; Liley, M.; Brugger, J.; Pugin, R.; Heinzelmann, H. Nanodispenser for attoliter volume deposition using atomic force microscopy probes modified by focused-ion-beam milling. *Appl. Phys. Lett.* **2005**, *85*, 6260–6262. [CrossRef]
203. Deladi, S.; Tas, N.R.; Berenschot, J.W.; de Boer, J.H.; de Boer, M.J.; Peter, M.; Krijnen, G.J.M.; Elwenspoek, M.C. Micromachines fountain pen for atomic force microscope-based nanopatterning. *Appl. Phys. Lett.* **2004**, *85*, 5361. [CrossRef]
204. Kim, K.H.; Moldovan, N.; Espinosa, H.D. A nanofountain probe with sub-100 nm molecular writing resolution. *Small* **2005**, *1*, 632–635. [CrossRef] [PubMed]

205. Kato, N.; Kawashima, T.; Shibata, T.; Mineta, T.; Makino, E. Micromachining of a newly designed AFM probe integrated with hollow microneedle for cellular function analysis. *Microelectron. Eng.* **2010**, *87*, 1185–1189. [CrossRef]
206. Meister, A.; Gabi, M.; Behr, P.; Studer, P.; Vörös, J.; Niedermann, P.; Bitterli, J.; Polesel-Maris, J.; Liley, M.; Heinzelmann, H.; et al. FluidFM: Combining atomic force microscopy and nanofluidics in a universal liquid delivery system for single cell applications and beyond. *Nano Lett.* **2009**, *9*, 2501–2507. [CrossRef] [PubMed]
207. Guillaume-Gentil, O.; Potthoff, E.; Ossola, D.; Franz, C.M.; Zambelli, T.; Vorholt, J.A. Force-controlled manipulation of single cells: From AFM to FluidFM. *Trends Biotechnol.* **2014**, *32*, 381–388. [CrossRef] [PubMed]
208. Dörig, P.; Stiefel, P.; Behr, P.; Sarajlic, E.; Bijl, D.; Gabi, M.; Vörös, J.; Vorholt, J.A.; Zambelli, T. Force-controlled spatial manipulation of viable mammalian cells and micro-organisms by means of FluidFM technology. *Appl. Phys. Lett.* **2010**, *97*, 023701. [CrossRef]
209. Martinez, V.; Forró, C.; Weydert, S.; Aebersold, M.J.; Dermutz, H.; Guillaume-Gentil, O.; Zambelli, T.; Vörös, J.; Demkó, L. Controlled single-cell deposition and patterning by highly flexible hollow cantilevers. *Lab Chip* **2016**, *16*, 1663–1674. [CrossRef] [PubMed]
210. Sanford, K.K.; Earle, W.R.; Likely, G.D. The growth in vitro of single isolated tissue cells. *J. Nat. Cancer Inst.* **1948**, *9*, 229–246. [PubMed]
211. Anis, Y.H.; Holl, M.R.; Meldrum, D.R. Automated selection and placement of single cells using vision-based feedback control. *IEEE Trans. Autom. Sci. Eng.* **2010**, *7*, 598. [CrossRef]
212. Fröhlich, J.; König, H. New techniques for isolation of single prokaryotic cells. *FEMS Microbiol. Rev.* **2000**, *24*, 567–572. [CrossRef] [PubMed]
213. Gregoire, S.; Xiao, J.; Silva, B.B.; Gonzalez, I.; Agidi, P.S.; Klein, M.I.; Ambatipudi, K.S.; Rosalen, P.L.; Bauserman, R.; Waugh, R.E.; et al. Role of glucosyltransferase B in interactions of *Candida albicans* with *Streptococcus mutans* and with an experimental pellicle on hydroxyapatite surfaces. *Appl. Environ. Microbiol.* **2011**, *77*, 6357–6367. [CrossRef] [PubMed]
214. Lu, Z.; Moraes, C.; Ye, G.; Simmons, C.A.; Sun, Y. Single cell deposition and patterning with a robotic system. *PLoS ONE* **2010**, *5*, e13542. [CrossRef] [PubMed]
215. Környei, Z.; Beke, S.; Mihálffy, T.; Jelitai, M.; Kovács, K.J.; Szabó, Z.; Szabó, B. Cell sorting in a Petri dish controlled by computer vision. *Sci. Rep.* **2013**, *3*, 1088. [CrossRef] [PubMed]
216. Ungai-Salánki, R.; Gerecsei, T.; Fürjes, P.; Orgovan, N.; Sándor, N.; Holczer, E.; Horvath, R.; Szabó, B. Automated single cell isolation from suspension with computer vision. *Sci. Rep.* **2016**, *6*, 20375. [CrossRef] [PubMed]
217. Roder, P.; Hille, C. A Multifunctional frontloading approach for repeated recycling of a pressure-controlled AFM micropipette. *PLoS ONE* **2015**, *10*, e0144157. [CrossRef] [PubMed]
218. Lee, S.; Jeong, W.; Beebe, D.J. Microfluidic valve with cored glass microneedle for microinjection. *Lab Chip* **2003**, *3*, 164–167. [CrossRef] [PubMed]
219. Chung, B.G.; Lin, F.; Jeon, N.L. A microfluidic multi-injector for gradient generation. *Lab Chip* **2006**, *6*, 764–768. [CrossRef] [PubMed]
220. Stiefel, P.; Schmidt, F.I.; Dörig, P.; Behr, P.; Zambelli, T.; Vorholt, J.A.; Mercer, J. Cooperative vaccinia infection demonstrated at the single-cell level using FluidFM. *Nano Lett.* **2012**, *12*, 4219–4227. [CrossRef] [PubMed]
221. Ramser, K.; Hanstorp, D. Optical manipulation for single-cell studies. *J. Biophotonics* **2010**, *3*, 187–206. [CrossRef] [PubMed]
222. Ashkin, A. Acceleration and trapping of particles by radiation pressure. *Phys. Rev. Lett.* **1970**, *24*, 156. [CrossRef]
223. Neuman, K.C.; Block, S.M. Optical trapping. *Rev. Sci. Instrum.* **2004**, *75*, 2787–2809. [CrossRef] [PubMed]
224. Ashkin, A.; Dziedzic, J.M.; Bjorkholm, J.E.; Chu, S. Observation of a single-beam gradient force trap for dielectric particles. *Opt. Lett.* **1986**, *11*, 288–290. [CrossRef] [PubMed]
225. Neuman, K.C.; Liou, G.F.; Block, S.M.; Bergman, K. Characterization of photodamage induced by optical tweezers. In *Conference on Lasers and Electro-Optics*; Scifres, D., Weiner, A., Eds.; paper CTuR1; Optical Society of America: Washington, DC, USA, 1998.
226. Maghelli, N.; Tolić-Nørrelykke, I.M. Optical trapping and laser ablation of microtubules in fission yeast. *Methods Cell. Biol.* **2010**, *97*, 173–183. [PubMed]

227. Simmons, R.M.; Finer, J.T.; Chu, S.; Spudich, J.A. Quantitative measurements of force and displacement using an optical trap. *Biophys. J.* **1996**, *70*, 1813–1822. [CrossRef]
228. Grimbergen, J.A.; Visscher, K.; de Gomes Mesquita, D.S.; Brakenhoff, G.J. Isolation of single yeast cells by optical trapping. *Yeast* **1993**, *9*, 723–732. [CrossRef] [PubMed]
229. Arai, F.; Ng, C.; Maruyama, H.; Ichikawa, A.; El-Shimy, H.; Fukuda, T. On chip single-cell separation and immobilization using optical tweezers and thermosensitive hydrogel. *Lab Chip* **2005**, *5*, 1399–1403. [CrossRef] [PubMed]
230. Singh, G.P.; Creely, C.M.; Volpe, G.; Grötsch, H.; Petrov, D. Real-time detection of hyperosmotic stress response in optically trapped single yeast cells using Raman microspectroscopy. *Anal. Chem.* **2005**, *77*, 2564–2568. [CrossRef] [PubMed]
231. Aabo, T.; Banás, A.R.; Glückstad, J.; Siegumfeldt, H.; Arneborg, N. BioPhotonics workstation: A versatile setup for simultaneous optical manipulation, heat stress, and intracellular pH measurements of a live yeast cell. *Rev. Sci. Instrum.* **2011**, *82*, 083707. [CrossRef] [PubMed]
232. Eriksson, E.; Scrimgeour, J.; Graneli, A.; Ramser, K.; Wellander, R.; Enger, J.; Hanstorp, D.; Goksör, M. Optical manipulation and microfluidics for studies of single cell dynamics. *J. Opt. A Pure Appl. Opt.* **2007**, *9*, S113–S121. [CrossRef]
233. Dholakia, K.; Reece, P. Optical micromanipulation takes hold. *Nanotoday* **2006**, *1*, 18–27. [CrossRef]
234. Castelein, M.; Rouxhet, P.G.; Pignon, F.; Magnin, A.; Piau, J.M. Single-cell adhesion probed in-situ using optical tweezers: A case study with *Saccharomyces cerevisiae*. *J. Appl. Phys.* **2012**, *111*, 114701. [CrossRef]
235. Landenberger, B.; Höfemann, H.; Wadle, S.; Rohrbach, A. Microfluidic sorting of arbitrary cells with dynamic optical tweezers. *Lab Chip* **2012**, *12*, 3177–3183. [CrossRef] [PubMed]
236. Werner, M.; Merenda, F.; Piguet, J.; Salathé, R.P.; Vogel, H. Microfluidic array cytometer based on refractive optical tweezers for parallel trapping, imaging and sorting of individual cells. *Lab Chip* **2011**, *11*, 2432–2439. [CrossRef] [PubMed]
237. Tam, J.M.; Castro, C.E.; Heath, R.J.; Cardenas, M.L.; Xavier, R.J.; Lang, M.J.; Vyas, J.M. Control and manipulation of pathogens with an optical trap for live cell imaging of intercellular interactions. *PLoS ONE* **2010**, *5*, e15215. [CrossRef] [PubMed]
238. Tam, J.M.; Castro, C.E.; Heath, R.J.; Mansour, M.K.; Cardenas, M.L.; Xavier, R.J.; Lang, M.J.; Vyas, J.M. Use of an optical trap for study of host-pathogen interactions for dynamic live cell imaging. *J. Vis. Exp.* **2011**, *53*, 3123. [CrossRef] [PubMed]
239. Eriksson, E.; Sott, K.; Lundqvist, F.; Sveningsson, M.; Scrimgeour, J.; Hanstorp, D.; Goksör, M.; Granéli, A. A microfluidic device for reversible environmental changes around single cells using optical tweezers for cell selection and positioning. *Lab Chip* **2010**, *10*, 617–625. [CrossRef] [PubMed]
240. Arneborg, N.; Siegumfeldt, H.; Andersen, G.H.; Nissen, P.; Daria, V.R.; Rodrigo, P.J.; Glückstad, J. Interactive optical trapping shows that confinement is a determinant of growth in a mixed yeast culture. *FEMS Microbiol. Lett.* **2005**, *245*, 155–159. [CrossRef] [PubMed]
241. Charrunchon, S.; Limtrakul, J.; Chattham, N. Growth pattern of yeast cells studied under optical tweezers. In *Frontiers in Optics*; 2012/Laser Science XXVIII, OSA Technical Digest (Online); Paper FW1G.6; Optical Society of America: Rochester, NY, USA, 2012.
242. Ando, J.; Bautista, G.; Smith, N.; Fujita, K.; Daria, V.R. Optical trapping and surgery of living yeast cells using a single laser. *Rev. Sci. Instrum.* **2008**, *79*, 103705. [CrossRef] [PubMed]
243. Hu, S.; Sun, D. Automated transportation of single cells using robot-tweezer manipulation system. *J. Lab. Autom.* **2011**, *16*, 263–270. [CrossRef] [PubMed]
244. Gong, Y.; Huang, W.; Liu, Q.F.; Wu, Y.; Rao, Y.; Peng, G.D.; Lang, J.; Zhang, K. Graded-index optical fiber tweezers with long manipulation length. *Opt. Express* **2014**, *22*, 25267–25276. [CrossRef] [PubMed]
245. Gustavsson, A.K.; van Niekerk, D.D.; Adiels, C.B.; du Preez, F.B.; Goksör, M.; Snoep, J.L. Sustained glycolytic oscillations in individual isolated yeast cells. *FEBS J.* **2012**, *279*, 2837–2847. [CrossRef] [PubMed]
246. Gustavsson, A.K.; van Niekerk, D.D.; Adiels, C.B.; Kooi, B.; Goksör, M.; Snoep, J.L. Allosteric regulation of phosphofructokinase controls the emergence of glycolytic oscillations in isolated yeast cells. *FEBS J.* **2014**, *281*, 2784–2793. [CrossRef] [PubMed]
247. Habaza, M.; Gilboa, B.; Roichman, Y.; Shaked, N.T. Tomographic phase microscopy with 180° rotation of live cells in suspension by holographic optical tweezers. *Opt. Lett.* **2015**, *40*, 1881–1884. [CrossRef] [PubMed]

248. Jing, P.; Wu, J.; Liu, G.W.; Keeler, E.G.; Pun, S.H.; Lin, L.Y. Photonic crystal optical tweezers with high efficiency for live biological samples and viability characterization. *Sci. Rep.* **2016**, *6*, 19924. [CrossRef] [PubMed]
249. Tolić-Nørrelykke, I.M.; Munteanu, E.L.; Thon, G.; Oddershede, L.; Berg-Sørensen, K. Anomalous diffusion in living yeast cells. *Phys. Rev. Lett.* **2004**, *93*, 078102. [CrossRef] [PubMed]
250. Sacconi, L.; Tolić-Nørrelykke, I.M.; Stringari, C.; Antolini, R.; Pavone, F.S. Optical micromanipulations inside yeast cells. *Appl. Opt.* **2005**, *44*, 2001–2007. [CrossRef] [PubMed]
251. Tolic-Nørrelykke, I.M.; Sacconi, L.; Stringari, C.; Raabe, I.; Pavone, F.S. Nuclear and division-plane positioning revealed by optical micromanipulation. *Curr. Biol.* **2005**, *15*, 1212–1216. [CrossRef] [PubMed]
252. Jeon, J.H.; Tejedor, V.; Burov, S.; Barkai, E.; Selhuber-Unkel, C.; Berg-Sørensen, K.; Oddershede, L.; Metzler, R. In vivo anomalous diffusion and weak ergodicity breaking of lipid granules. *Phys. Rev. Lett.* **2011**, *106*, 048103. [CrossRef] [PubMed]
253. Mas, J.; Richardson, A.C.; Reihani, S.N.; Oddershede, L.B.; Berg-Sørensen, K. Quantitative determination of optical trapping strength and viscoelastic moduli inside living cells. *Phys. Biol.* **2013**, *10*, 046006. [CrossRef] [PubMed]
254. Difato, F.; Pinato, G.; Cojoc, D. Cell signalling periments driven by optical manipulation. *Int. J. Mol. Sci.* **2013**, *14*, 8963–8984. [CrossRef] [PubMed]
255. Kotsifaki, D.G.; Makropoulou, M.; Serafetinides, A. Near infrared optical tweezers and nanosecond ablation on yeast and algae cells. In Proceedings of the SPIE 17th International School on Quantum Electronics: Laser Physics and Applications, Nessebar, Bulgaria, 24–28 September 2012; Volume 8770.
256. Oddershede, L.B. Force probing of individual molecules inside the living cell is now a reality. *Nat. Chem. Biol.* **2012**, *8*, 879–886. [CrossRef] [PubMed]
257. Norregaard, K.; Jauffred, L.; Berg-Sørensen, K.; Oddershede, L.B. Optical manipulation of single molecules in the living cell. *Phys. Chem. Chem. Phys.* **2014**, *16*, 12614–12624. [CrossRef] [PubMed]
258. Toriello, N.M.; Douglas, E.S.; Mathies, R.A. Microfluidic device for electric field-driven single-cell capture and activation. *Anal. Chem.* **2005**, *77*, 6935–6941. [CrossRef] [PubMed]
259. Koyama, S.; Tsubouchi, T.; Usui, K.; Uematsu, K.; Tame, A.; Nogi, Y.; Ohta, Y.; Hatada, Y.; Kato, C.; Miwa, T.; et al. Involvement of flocculin in negative potential-applied ITO electrode adhesion of yeast cells. *FEMS Yeast Res.* **2015**, *15*. [CrossRef] [PubMed]
260. Qian, C.; Huang, H.; Chen, L.; Li, X.; Ge, Z.; Chen, T.; Yang, Z.; Sun, L. Dielectrophoresis for bioparticle manipulation. *Int. J. Mol. Sci.* **2014**, *15*, 18281–18309. [CrossRef] [PubMed]
261. Norde, W. *Colloids and Interfaces in Life Sciences*; Marcel Dekker: Monticello, NY, USA, 2003.
262. Jubery, T.Z.; Srivastava, S.K.; Dutta, P. Dielectrophoretic separation of bioparticles in microdevices: A review. *Electrophoresis* **2014**, *35*, 691–713. [CrossRef] [PubMed]
263. Kang, K.H.; Xuan, X.C.; Kang, Y.J.; Li, D.Q. Effects of dc-dielectrophoretic force on particle trajectories in microchannels. *J. Appl. Phys.* **2006**, *99*, 064702. [CrossRef]
264. Khoshmanesh, K.; Nahavandi, S.; Baratchi, S.; Mitchell, A.; Kalantar-zadeh, K. Dielectrophoretic platforms for bio-microfluidic systems. *Biosens. Bioelectron.* **2011**, *26*, 1800–1814. [CrossRef] [PubMed]
265. Koklu, M.; Park, S.; Pillai, S.D.; Beskok, A. Negative dielectrophoretic capture of bacterial spores in food matrices. *Biomicrofluidics* **2010**, *4*. [CrossRef] [PubMed]
266. Cheng, I.F.; Froude, V.E.; Zhu, Y.; Chang, H.C.; Chang, H.C. A continuous high-throughput bioparticle sorter based on 3D traveling-wave dielectrophoresis. *Lab Chip* **2009**, *9*, 3193–3201. [CrossRef] [PubMed]
267. Hunt, T.P.; Westervelt, R.M. Dielectrophoresis tweezers for single cell manipulation. *Biomed. Microdevices* **2006**, *8*, 227–230. [CrossRef] [PubMed]
268. Cetin, B.; Kang, Y.; Wu, Z.; Li, D. Continuous particle separation by size via AC-dielectrophoresis using a lab-on-a-chip device with 3-D electrodes. *Electrophoresis* **2009**, *30*, 766–772. [CrossRef] [PubMed]
269. Docoslis, A.; Kalogerakis, N.; Behie, L.A. Dielectrophoretic forces can be safely used to retain viable cells in perfusion cultures of animal cells. *Cytotechnology* **1999**, *30*, 133–142. [CrossRef] [PubMed]
270. Pethig, R.; Menachery, A.; Pells, S.; de Sousa, P. Dielectrophoresis: A review of applications for stem cell research. *J. Biomed. Biotechnol.* **2010**, *2010*, 182581. [CrossRef] [PubMed]
271. Li, S.; Li, M.; Bougot-Robin, K.; Cao, W.; Chau, I.Y.Y.; Li, W.; Wen, W. High-throughput particle manipulation by hydrodynamic, electrokinetic, and dielectrophoretic effects in an integrated microfluidic chip. *Biomicrofluidics* **2013**, *7*, 24106. [CrossRef] [PubMed]

272. Li, P.C.; Harrison, D.J. Transport, manipulation, and reaction of biological cells on-chip using electrokinetic effects. *Anal. Chem.* **1997**, *69*, 1564–1568. [CrossRef] [PubMed]
273. Kodama, T.; Osaki, T.; Kawano, R.; Kamiya, K.; Miki, N.; Takeuchi, S. Round-tip dielectrophoresis-based tweezers for single micro-object manipulation. *Biosens. Bioelectron.* **2013**, *47*, 206–212. [CrossRef] [PubMed]
274. Rosenthal, A.; Voldman, J. Dielectrophoretic traps for single-particle patterning. *Biophys. J.* **2005**, *88*, 2193–2205. [CrossRef] [PubMed]
275. Lee, H.; Hunt, T.P.; Westervelt, R.M. Magnetic and electric manipulation of a single cell in fluid. *Mater. Res. Symp. Proc.* **2004**, *820*. [CrossRef]
276. Lee, H.; Purdon, A.M.; Westervelt, R.M. Micromanipulation of biological systems with microelectromagnets. *IEEE Trans. Magn.* **2004**, *40*, 2991–2993. [CrossRef]
277. Jang, L.S.; Huang, P.H.; Lan, K.C. Single-cell trapping utilizing negative dielectrophoretic quadrupole and microwell electrodes. *Biosens. Bioelectron.* **2009**, *24*, 3637–3644. [CrossRef] [PubMed]
278. Arnold, W.M. Positioning and levitation media for the separation of biological cells. *Ind. Appl. IEEE Trans.* **2001**, *37*, 1468–1475. [CrossRef]
279. Gagnon, Z.; Mazur, J.; Chang, H.C. Glutaraldehyde enhanced dielectrophoretic yeast cell separation. *Biomicrofluidics* **2009**, *3*, 44108. [CrossRef] [PubMed]
280. Fatoyinbo, H.O.; Kamchis, D.; Whattingham, R.; Ogin, S.L.; Hughes, M.P. A high-throughput 3-D composite dielectrophoretic separator. *IEEE Trans. Biomed. Eng.* **2005**, *52*, 1347–1349. [CrossRef] [PubMed]
281. Urdaneta, M.; Smela, E. Multiple frequency dielectrophoresis. *Electrophoresis* **2007**, *28*, 3145–3155. [CrossRef] [PubMed]
282. Salomon, S.; Leichlé, T.; Nicu, L. A dielectrophoretic continuous flow sorter using integrated microelectrodes coupled to a channel constriction. *Electrophoresis* **2011**, *32*, 1508–1514. [CrossRef] [PubMed]
283. Huang, Y.; Hölzel, R.; Pethig, R.; Wang, X.B. Differences in the AC electrodynamics of viable and non-viable yeast cells determined through combined dielectrophoresis and electrorotation studies. *Phys. Med. Biol.* **1992**, *37*, 1499–1517. [CrossRef] [PubMed]
284. Hölzel, R. Electrorotation of single yeast cells at frequencies between 100 Hz and 1.6 GHz. *Biophys. J.* **1997**, *73*, 1103–1109. [CrossRef]
285. Fatoyinbo, H.O.; Hoettges, K.F.; Hughes, M.P. Rapid-on-chip determination of dielectric properties of biological cells using imaging techniques in a dielectrophoresis dot microsystem. *Electrophoresis* **2008**, *29*, 3–10. [CrossRef] [PubMed]
286. Soffe, R.; Tang, S.Y.; Baratchi, S.; Nahavandi, S.; Nasabi, M.; Cooper, J.M.; Mitchell, A.; Khoshmanesh, K. Controlled rotation and vibration of patterned cell clusters using dielectrophoresis. *Anal. Chem.* **2015**, *87*, 2389–2395. [CrossRef] [PubMed]
287. Hunt, T.P.; Issadore, D.; Westervelt, R.M. Integrated circuit/microfluidic chip toprogrammably trap and move cells and droplets with dielectrophoresis. *Lab. Chip* **2008**, *8*, 81–87. [CrossRef] [PubMed]
288. Witek, M.A.; Wei, S.; Vaidya, B.; Adams, A.A.; Zhu, L.; Stryjewski, W.; McCarley, R.L.; Soper, S.A. Cell transport via electromigration in polymer-based microfluidic devices. *Lab Chip* **2004**, *4*, 464–472. [CrossRef] [PubMed]
289. Frenea-Robin, M.; Chetouani, H.; Haddour, N.; Rostaing, H.; Laforet, J.; Reyne, G. Contactless diamagnetic trapping of living cells onto a micromagnet array. *Conf. Proc. IEEE Eng. Med. Biol. Soc.* **2008**, *2008*, 3360–3363. [PubMed]
290. Yasukawa, T.; Nagamine, K.; Horiguchi, Y.; Shiku, H.; Koide, M.; Itayama, T.; Shiraishi, F.; Matsue, T. Electrophoretic cell manipulation and electrochemical gene-function analysis based on a yeast two-hybrid system in a microfluidic device. *Anal. Chem.* **2008**, *80*, 3722–3727. [CrossRef] [PubMed]
291. Moncada-Hernandez, H.; Baylon-Cardiel, J.L.; Pérez-González, V.H.; Lapizco-Encinas, B.H. Insulator-based dielectrophoresis of microorganisms: Theoretical and experimental results. *Electrophoresis* **2011**, *32*, 2502–2511. [CrossRef] [PubMed]
292. Yue, T.; Nakajima, M.; Tajima, H.; Fukuda, T. Fabrication of microstructures embedding controllable particles inside dielectrophoretic microfluidic devices. *Int. J. Adv. Robot. Syst.* **2013**, *10*, 132. [CrossRef]
293. Shi, X.; Shi, Z.; Wang, D.; Ullah, M.W.; Yang, G. Microbial cells with a Fe_3O_4 doped hydrogel extracellular matrix: Manipulation of living cells by magnetic stimulus. *Macromol. Biosci.* **2016**. [CrossRef] [PubMed]
294. Winkleman, A.; Gudiksen, K.L.; Ryan, D.; Whitesides, G.M.; Greenfield, D.; Prentiss, M. A magnetic trap for living cells suspended in a paramagnetic buffer. *Appl. Phys. Lett.* **2004**, *85*, 2411. [CrossRef]

295. Neuman, K.C.; Nagy, A. Single-molecule force spectroscopy: Optical tweezers, magnetic tweezers and atomic force microscopy. *Nat. Methods* **2008**, *5*, 491–505. [CrossRef] [PubMed]
296. Fisher, J.K.; Cribb, J.; Desai, K.V.; Vicci, L.; Wilde, B.; Keller, K.; Taylor, R.M.; Haase, J.; Bloom, K.; O'Brien, E.T.; et al. Thin-foil magnetic force system for high-numerical-aperture microscopy. *Rev. Sci. Instrum.* **2006**, *77*. [CrossRef] [PubMed]
297. Yan, J.; Skoko, D.; Marko, J.F. Near-field-magnetic-tweezer manipulation of single DNA molecules. *Phys. Rev. E Stat. Nonlinear Soft Matter Phys.* **2004**, *70*, 011905. [CrossRef] [PubMed]
298. Chacko, J.V.; Zanacchi, F.C.; Diaspro, A. Probing cytoskeletal structures by coupling optical superresolution and AFM techniques for a correlative approach. *Cytoskeleton (Hoboken)* **2013**, *70*, 729–740. [CrossRef] [PubMed]
299. Chacko, J.V.; Harke, B.; Canale, C.; Diaspro, A. Cellular level nanomanipulation using atomic force microscope aided with superresolution imaging. *J. Biomed. Opt.* **2014**, *19*, 105003. [CrossRef] [PubMed]
300. Asakawa, H.; Hiraoka, Y.; Haraguchi, T. A method of correlative light and electron microscopy for yeast cells. *Micron* **2014**, *61*, 53–61. [CrossRef] [PubMed]
301. Kobayashi, S.; Iwamoto, M.; Haraguchi, T. Live correlative light-electron microscopy to observe molecular dynamics in high resolution. *Microscopy (Oxf.)* **2016**, *65*, 296–308. [CrossRef] [PubMed]
302. Wolff, G.; Hagen, C.; Grünewald, K.; Kaufmann, R. Towards correlative super-resolution fluorescence and electron cryo-microscopy. *Biol. Cell* **2016**, *108*, 245–258. [CrossRef] [PubMed]
303. Heinisch, J.J.; Lipke, P.N.; Beaussart, A.; El Kirat Chatel, S.; Dupres, V.; Alsteens, D.; Dufrêne, Y.F. Atomic force microscopy—Looking at mechanosensors on the cell surface. *J. Cell Sci.* **2012**, *125*, 4189–4195. [CrossRef] [PubMed]
304. Conroy, R. Force spectroscopy with optical and magnetic tweezers. In *Handbook of Molecular Force Spectroscopy*; Noy, A., Ed.; Springer: New York, NY, USA, 2008; pp. 23–96.
305. Alonso-Sarduy, L.; de Los Rios, P.; Benedetti, F.; Vobornik, D.; Dietler, G.; Kasas, S.; Longo, G. Real-time monitoring of protein conformational changes using a nano-mechanical sensor. *PLoS ONE* **2014**, *9*, e103674. [CrossRef] [PubMed]

Permissions

All chapters in this book were first published in MDPI; hereby published with permission under the Creative Commons Attribution License or equivalent. Every chapter published in this book has been scrutinized by our experts. Their significance has been extensively debated. The topics covered herein carry significant findings which will fuel the growth of the discipline. They may even be implemented as practical applications or may be referred to as a beginning point for another development.

The contributors of this book come from diverse backgrounds, making this book a truly international effort. This book will bring forth new frontiers with its revolutionizing research information and detailed analysis of the nascent developments around the world.

We would like to thank all the contributing authors for lending their expertise to make the book truly unique. They have played a crucial role in the development of this book. Without their invaluable contributions this book wouldn't have been possible. They have made vital efforts to compile up to date information on the varied aspects of this subject to make this book a valuable addition to the collection of many professionals and students.

This book was conceptualized with the vision of imparting up-to-date information and advanced data in this field. To ensure the same, a matchless editorial board was set up. Every individual on the board went through rigorous rounds of assessment to prove their worth. After which they invested a large part of their time researching and compiling the most relevant data for our readers.

The editorial board has been involved in producing this book since its inception. They have spent rigorous hours researching and exploring the diverse topics which have resulted in the successful publishing of this book. They have passed on their knowledge of decades through this book. To expedite this challenging task, the publisher supported the team at every step. A small team of assistant editors was also appointed to further simplify the editing procedure and attain best results for the readers.

Apart from the editorial board, the designing team has also invested a significant amount of their time in understanding the subject and creating the most relevant covers. They scrutinized every image to scout for the most suitable representation of the subject and create an appropriate cover for the book.

The publishing team has been an ardent support to the editorial, designing and production team. Their endless efforts to recruit the best for this project, has resulted in the accomplishment of this book. They are a veteran in the field of academics and their pool of knowledge is as vast as their experience in printing. Their expertise and guidance has proved useful at every step. Their uncompromising quality standards have made this book an exceptional effort. Their encouragement from time to time has been an inspiration for everyone.

The publisher and the editorial board hope that this book will prove to be a valuable piece of knowledge for researchers, students, practitioners and scholars across the globe.

List of Contributors

Alice Jaeger, Emanuele Zannini and Aylin W. Sahin
School of Food and Nutritional Science, University College Cork, T12K8AF Cork, Ireland

Elke K. Arendt
School of Food and Nutritional Science, University College Cork, T12K8AF Cork, Ireland
APC Microbiome Institute, University College Cork, T12YT20 Cork, Ireland

Joshua Eardley and David J. Timson
School of Pharmacy and Biomolecular Sciences, University of Brighton, Huxley Building, Lewes Road, Brighton BN2 4GJ, UK

Annereinou Dijkstra
NIZO, 6710 BA Ede, The Netherlands

Amparo Gamero
NIZO, 6710 BA Ede, The Netherlands Food Technology Area, Faculty of Pharmacy, Universitat de València, Avda. Vicent Andrés Estellés s/n, Burjassot, 46100 Valencia, Spain

Bart Smit
NIZO, 6710 BA Ede, The Netherlands FerMentor, Ede, 6718 ZB Ede, The Netherlands

Catrienus de Jong
NIZO, 6710 BA Ede, The Netherlands Wageningen Food & Biobased Research (WFBR), Wageningen University & Research, Bornse Weilanden 9,6708 WG Wageningen, The Netherlands

Alice Vilela
Department of Biology and Environment, School of Life Sciences and Environment, University of Trás-os-Montes and Alto Douro, CQ-VR, Chemistry Research Centre, 5001-801 Vila Real, Portugal

Joseph P. Barry and Mindy S. Metz
Molecular and Cellular Biochemistry Department, 212 South Hawthorne Drive, Simon Hall MSB1, Room 405B, Indiana University, Bloomington, IN 47405, USA

Justin Hughey and Adam Quirk
Cardinal Spirits, 922 S Morton St, Bloomington, IN 47403, USA

Matthew L. Bochman
Molecular and Cellular Biochemistry Department, 212 South Hawthorne Drive, Simon Hall MSB1, Room 405B, Indiana University, Bloomington, IN 47405, USA
Wild Pitch Yeast, Bloomington, IN 47405, USA

Osagie A. Osadolor, Patrik R. Lennartsson and Mohammad J. Taherzadeh
Swedish Centre for Resource Recovery, University of Borås, SE 50190 Borås, Sweden

Pasquale Russo, Giuseppe Spano and Mariagiovanna Fragasso
Department of the Sciences of Agriculture, Food and Environment, University of Foggia, Via Napoli 25, 71122 Foggia, Italy

Carmen Berbegal
Department of the Sciences of Agriculture, Food and Environment, University of Foggia, Via Napoli 25, 71122 Foggia, Italy
EnolabERI BioTecMed, Universitat de València, 46100 Valencia, Spain

Iuliia Khomenko and Franco Biasioli
Department of Food Quality and Nutrition, Research and Innovation Centre, Fondazione Edmund Mach (FEM), via E. Mach 1, 38010 San Michele all'Adige, Italy

Vittorio Capozzi
Institute of Sciences of Food Production, National Research Council (CNR), c/o CS-DAT, Via Michele Protano, 71121 Foggia, Italy

Igor G. Morgunov, Svetlana V. Kamzolova and Julia N. Lunina
G.K. Skryabin Institute of Biochemistry and Physiology of Microorganisms, Russian Academy of Sciences, pr-t Nauki, 5, 142290 Pushchino, Russia

Andreas Conjaerts and Ronnie G. Willaert
Alliance Research Group VUB-UGent NanoMicrobiology (NAMI), IJRG VUB-EPFL, BioNanotechnology & NanoMedicine (NANO), Structural Biology Brussels (SBB), Vrije Universiteit Brussel, Brussels 1050, Belgium

Khaled A. Selim and Ahmed I. El-Diwany
Pharmaceutical and Drug Industries Research Division, National Research Centre, 33-El-Bohouth St. (former El Tahrir St.), Dokki, Giza 12622, Egypt

Saadia M. Easa
Microbiology Department, Faculty of Science, Ain Shams University, Cairo 11566, Egypt

Vassileios Varelas, Evangelia Sotiropoulou, Xara Karambini and Maria Liouni
Laboratory of Industrial Chemistry, Department of Chemistry, School of Science, National and Kapodistrian University of Athens, Zografou, Athens 157 72, Greece

Elias T. Nerantzis
Laboratory of Biotechnology & Industrial Fermentations, Department of Oenology and Beverage Technology, School of Food Technology and Nutrition, Technological Educational Institute (TEI) of Athens, Egaleo 122 43, Greece

Antonio D. Moreno
IMDEA Energía, Biotechnological Processes for Energy Production Unit, Móstoles, Madrid 28935, Spain
CIEMAT, Renewable Energy Division, Biofuels Unit, Avda. Complutense 40, Madrid 28040, Spain

Mercedes Ballesteros
CIEMAT, Renewable Energy Division, Biofuels Unit, Avda. Complutense 40, Madrid 28040, Spain

David Ibarra
INIA-CIFOR, Forestry Products Department, Cellulose and Paper Laboratories, Ctra de La Coruña Km 7.5, Madrid 28040, Spain

Antoine Mialon
MetGen Oy, Rakentajantie 26, Kaarina 20780, Finland

Anne-Céline Kohler and Leonardo Venturelli
Laboratoire de Physique de la Matière Vivante, EPFL, 1015 Lausanne, Switzerland

Abhilash Kannan and Dominique Sanglard
Institute of Microbiology, Lausanne University Hospital, CH-1011 Lausanne, Switzerland

Giovanni Dietler
Laboratoire de Physique de la Matière Vivante, EPFL, 1015 Lausanne, Switzerland
International Joint Research Group VUB-EPFL NanoBiotechnology & NanoMedicine (NANO), Vrije Universiteit Brussel, 1050 Brussels, Belgium

Ronnie Willaert
International Joint Research Group VUB-EPFL NanoBiotechnology & NanoMedicine (NANO), Vrije Universiteit Brussel, 1050 Brussels, Belgium
Research Group Structural Biology Brussels, Alliance Research Group VUB-UGent NanoMicrobiology (NAMI), 1050 Ixelles, Belgium
Department Bioscience Engineering, University Antwerp, 2000 Antwerp, Belgium

Sandor Kasas and Giovanni Dietler
IJRG VUB-EPFL, NanoBiotechnology & NanoMedicine (NANO), Laboratory of the Physics of Living Matter, Ecole Polytechnique de Lausanne, Lausanne 1015, Switzerland
Laboratoire de Physique de la Matière Vivante, EPFL, 1015 Lausanne, Switzerland
Institute of Microbiology, Lausanne University Hospital, CH-1011 Lausanne, Switzerland
Plateforme de Morphologie UFAM, CUMRL, University of Lausanne, 1015 Lausanne, Switzerland

Bart Devreese
IJRG VUB-EPFL, NanoBiotechnology & NanoMedicine (NANO), Alliance Research Group VUB-UGent NanoMicrobiology (NAMI), Laboratory for Protein Biochemistry and Biomolecular Engineering (L-ProBE), Ghent University, Ghent 9000, Belgium

Index

www.ingramcontent.com/pod-product-compliance
Lightning Source LLC
LaVergne TN
LVHW060517210926
848485LV00005B/1093